Epigenetic Processes and the Evolution of Life

Anton Markoš

and

Jana Švorcová

Charles University Prague, Faculty of Sciences
Dept. of Philosophy and History of Science
Praha 2, Vinicna 7
Czechia, CZ 128 44

CRC Press
Taylor & Francis Group
Boca Raton London New York

CRC Press is an imprint of the
Taylor & Francis Group, an **informa** business
A SCIENCE PUBLISHERS BOOK

Cover illustration provided by Barbora Faiglová

CRC Press
Taylor & Francis Group
6000 Broken Sound Parkway NW, Suite 300
Boca Raton, FL 33487-2742

Printed on acid-free paper
Version Date: 20180925

International Standard Book Number-13: 978-1-1385-4192-4 (Hardback)

Visit the Taylor & Francis Web site at
http://www.taylorandfrancis.com

and the CRC Press Web site at
http://www.crcpress.com

Preface

We submitted this manuscript for publication in June 2018. During the extraordinarily hot summer that followed here in Czechia, we came across two remarkable books that can serve as a cupola above our effort. The first is *Extended Heredity* by R. Bonduriansky and T. Day (2018). We regret that it didn't appear several months earlier as it would have received due attention in this volume. Immodestly, we believe our treatise runs in parallel, or as a complement, to many points of the authors' argumentation, especially concerning epigenetics and trans-generation inheritance, i.e., the memory and experience of any living 'thing' and the context thereof, ecological and historical.

The second book is *Darwinism Evolving* by D. J. Depew and B. H. Weber (1996). Here we have no excuse for not having discovered it in the more than two decades since it appeared (Even more embarrassing is the fact that we came across the reference in an article by the Aristotelian philosopher J. G. Lennox). To explain why it fits into our present thinking, we must reveal some of our history. We have long emphasized the historical nature of life by arguing that not only lineages but also *communities* of organisms construct, keep and access vast assemblages of historical memory. These memory pools are much greater than any that can be realized within the short time period of a single lifespan of either an individual or its community, yet non the less may be searched, consulted, and activated whenever external conditions change in ways that require adopting half-forgotten modes of behavior, or even tinkering with such modes so as to invent new ways of coping with the change. External conditions often change much quicker than the appearance (phenotype) of the individual; and as the greater part of external influence comes from cohabitants in the biosphere, this influence may change with every epidemic or invasive species. Thus, the community must be able to search the fitness landscape in which it lives, and through this search co-construct the biosphere, both ways both immediate and extended. In short, living beings must be able to negotiate their entry into the future. Moreover, they can recollect and interpret past experience and select relevant strategies. Under all such circumstances, living beings play an active role in the process. Their internal dynamics and communication contribute to the settling of affairs. If so, the evolution of all forms of life is isomorphic with the history of human culture; and if so, evolutionary biologists are not *sensu stricto* scientists, but interpreters of history—historians, historiographers. Can this claim be broadened to include other forms of life, i.e., should all living beings and lineages be considered interpreters of their history? Our answer on is yes.

Today, the norm of natural science does not study history; its laws and rules are timeless, i.e., valid in all times and places ('universal truths'). It has been therefore very surprising to us that our colleagues stubbornly insist that evolutionary biology is a science of this sort, i.e., comparable to physics. Depew and Weber opened our eyes. They point to the fact that the Whig community of the first half of 19th century, to which Darwin belonged, was devotedly Newtonian, resolved to explain in Newtonian terms even phenomena that could not (then, and perhaps ever) be folded into the rubric of physics. From this perspective, a body (more precisely a mass point) will stay at rest or proceed in linear constant motion so long as a secondary cause (e.g., gravitation, collision, etc.) does not modify its position and/or movement. A mass point has two measurable quantities, mass and momentum, and (since it is a point) it is allowed no inner qualities or movements. The authors argue that devoted English Newtonians also treated living beings, including humans, as 'mass points' (i.e., 'atoms' without inner drives). Secondary qualities (natural selection in C. Darwin and A. R. Wallace; the invisible hand of the market in A. Smith; and the means of production in Marx) are commensurable (at least *per analogiam*) with Newtonian gravity. At that time, the biology of continental Europe was firmly rejected as it worked with inner drives (or even teleology) in living bodies (e.g., Lamarck, Geoffroy) and could not be reconciled with whiggish views. The authors argue that even the statistical physics of the time, which later led to quantum theory and theory of relativity, has not shattered this basic framework, neither have misinterpretations by Darwin's peers at both shores of the Atlantic. Even the Modern Synthesis, then, is an heir, or better, an interpretation of the physical structure of Darwin's original concept. In this treatise, we argue that there remains more to be discovered in Darwinian theory, and that a historical approach that allows for internal drives in organisms, lineages, and communities offers a fuller and more profound understanding of evolution.

G. Tebb, a former student of the famous molecular biologist Kim Nasmyth, recalls that while ruminating over their results, Nasmyth often asked his students: "How will this help you to discover a 'universal truth'?" (Tebb 1998) He had in mind a theory explaining life, the laws ruling its being. Let us commit a miniscule stylistic ('epigenetic') cosmetic of the letter-string: "How will this help you to discover *a* universal truth?" By italicizing the 'a' in the utterance, and removing the 'irony marks', we appropriate the claim and use it to emphasize the notion that struggles toward such an aim may well end in many, often complementary truths. This invites the reader to consider the power of interpretation along with the facts of the world and the interplay thereof: if delicate attention is paid to the relevant cues.

The central message of this book is that—in the realm of the living—allegedly universal truths or 'natural' laws represent various Norms, and that these presumed commandments or causes are rather suggestions (to living beings) or landmarks that help shape behavior, process or structure, in this or that realm of both nature and culture. Acknowledging the Norm, however, also means not piously sticking to it. This is because from the very beginning "the Norm included, not only basic metabolic and genetic processes (those that could have been established in the prebiotic phase), but also a principal novelty that could be introduced by cellular life alone—the exchange and processing of information, both within the cells and from outside" (p. 96 in this book). Living beings develop 'joyfully' around such Norms—endlessly playing, of which comes resourceful—even extravagant behavior, appearance, ornamentation, communicative tools, etc., all of which display a 'remembering' of the basic rules. Indeed, having rules is necessary to avoid a deadly downfall in a grandiose bonfire of vanities (which may happen on smaller scales). This may indeed be the universal truth of life. The historical uniqueness of an individual, of its community, lineage and ancestors, opens ever-newer state spaces for ever-newer variations of the game. Before all, this truth allows the biological sciences to consider anew the abandoned vernacular truth claim that living beings have their innerness, that they are not merely a result of the push and pull of external, deterministic, physical forces. Such ideas, of course, are not new: they may be as old as the remembered history of humankind—and yet, as with all history, they require reformulation in, and for, each generation.

We illustrate this view through recent achievements in evo-devo, epigenetics, trans-generation inheritance, and/or analogies with human language and culture. The recent avalanche of publications in these areas, together with the development of new techniques such as microscopy, brain imaging and deep sequencing (to name haphazardly a few new branches of biological technique) has amplified the available data on properties of the living; but it has proceeded without a clear idea on how to process this new data into general theories. We believe that what we call the Norm (of biology), i.e., hard biological data gathered into textbooks, presents a necessary but incomplete ground for the endless teeming of Darwin's 'tangled bank'.

We are biologists by education, and members of the Department of Philosophy and History of Science at the Faculty of Science, Charles University in Prague. The fruitful interaction of biologists, historians and philosophers has enabled our work. From this perspective, we follow recent advances in biology—especially evolutionary and developmental—and confront such achievement with ideas spawned by biologists of different historical periods. Paradigms change, often fundamentally conforming to the sociocultural air of the period, techniques become more and more sophisticated, revealing unexpected details of structures and functions of living beings. Yet, the final answer to the deepest question 'What is life?' has remained as evanescent as ever. Depending on their nature, biologists tend to present some ready-made and inveterate cliché pointing towards physical forces, or organization, or natural selection or information—but adding no value. This question cannot be answered solely within the framework of biology. In this book we try to show that answers may develop, supported by recent developments in biology and other sciences as well as from the humanities, through new interpretation of older theories, or naturally through developing new ones.

We tried to follow the most recent developments (our latest references are from Spring, 2018); undoubtedly and necessarily the selection reflects our personal bias. We draw here a map, or better a sketch, of that which we consider challenging for future research, as well as what methodologies might serve to answer those challenges.

In the preface to the previous book *Life as it own Designer* (Markoš et al. 2009, p. vii) we asked three questions concerning life and the evolution thereof: "(1) Where does the novelty come from? (2) What is the essence of evolution? and (3) What is the role of living beings in evolution? Such questions have a lot to do with the two forms of logos—narrative and rational [...]. If novelty is allowed, then history enters the stage; and if history is allowed then living beings should be taken into consideration as interpreters of the past." We hope to make evolutionary theory ever fuller and more open to new knowledge by employing the methodology of historians.

We are especially grateful to the founders of the Department in 1990—Z. Neubauer (1942–2016), Z. Kratochvíl, and S. Komárek—and to our peers and students who have passed through the Department during these last three decades. F. Cvrčková, our illustrator, and G. Ostdiek, who dampened our transgressions against the English Norm, are our learned colleagues who did more to shape this book than merely performing the jobs mentioned.

The support by the Charles University Research Centre program No. 204056 is acknowledged and appreciated.

Prague, September 2018
Anton Markoš and Jana Švorcová

Contents

In principle, the only way we could speak of life would be in terms of Humor, for being as it is the constant disruption of the expected order, life itself is comic.

(Eco 1994b, 120)

The dog trots freely in the street
and has his own dog's life to live
and to think about
and to reflect upon
touching and tasting and testing everything
investigating everything

L. Ferlinghetti, "Dog"
from *A Coney Island of the Mind: Poems*
Copyright © 1958

1

Transgressing the Norms

One Ring to rule them all, One Ring to find them, One Ring to bring them all, and in the darkness bind them...

—J.R.R. Tolkien; The Lord of the rings

There were so many dialects spoken in England, in the Later Middle Ages, that people even from neighboring shires had difficulties understanding each other. In the *Oxford history of Britain* (Morgan 1999), we read: "Geoffrey Chaucer (1340–1400) had serious misgivings as to whether his writings would be understood across England—and he wrote for a limited, charmed circle." Of course, such a situation is not exceptional for England; all modern nations arose from a pêle mêle of dialects, either by decree or by common consensus, only after a collective *norm* had been established and became *respected* by the community. Normative sets—grammar, spelling and phraseology became more or less binding—standing over all colloquial variants, and accompanied by a drastic reduction of vocabulary, grammatical forms, idioms, vocalization, etc. Similar trends towards normalization can be observed in other cultural endeavors, as law, religions, or political culture. What is obvious is the highly contingent and historical nature of such normative processes: no natural law, no strict necessity was at work here. No natural law at work, yet the spell of 'rings' of logic, tradition, scores, conventions, scripts, religious obedience, and more. The Norm constitutes the basic knowledge taught at the secondary school: not only in language courses, but also in, e.g., biochemistry, philosophy, or history.

No such historically established norm, however, is observed absolutely. Many colloquial variants from times past survive for quite long, e.g., within dialects, habits, or as pagan practices (even in areas with a long-lived Christian tradition; Ginzburg 1992a, b). By analogy, such 'palimpsests' found in the non-human life may point towards the origin of life. Soon after the establishment of the Norm, two important facts become obvious, be it in culture or life in general:

1. All members of the community recognize it and take it as a given, even if no individual is able to abide strictly by all its rules; this applies not only in cases of everyday vernacular speech and behavior. For example, Rappaport (2010) states that for religious morals of societies: "The chaos of everyday life [...] attains

some stability to the degree that it is informed by ideas representing the social facts of a shared collective existence." He introduces the concept of Ultimate Sacred Postulates that "not only stand beyond the reach of falsification by the rigorous procedures of logic of science, but are also impervious to disproof by the less formal but more compelling rigors of daily life. Their independence from ordinary experience, moreover, makes it possible for people of widely divergent experience to accept them" (Rappaport 2010). We invite the reader to treat textbook truths and norms as such *presumed* universals; it goes without saying, as we argue here, that this also applies to the whole of the biosphere.

2. If, in the course of history, different 'lineages'[1] have deviated from the norm, they never went so far that the ground would become indiscernible. Even uncommonly weird heritable modifications[2] display their grounding, either in universal postulates or in the norms of this or that life stage (as Tunicate larvae reveal their Chordate allegiance). The rings that bind the norm do not keep their grip absolutely, yet everybody feels their existence.

Once the rules are established for the game (whether ice hockey, opera, courtroom, or …), what is important is that the *performance* takes place within the framework of some such ring (chessboard, playground, or …). With this in mind, we invite the reader to compare two approaches. J. von Uexküll (2010) prods us to uncover—behind the tangled web of life—the scoring of the symphony of nature, which he sees as the very task of biology. In contrast, we argue that the ringing in the biosphere is not only the performing of symphony, but before all else, a genuine, endless jam session! Such jamming is the main topic of this present book. The statement does not suggest that the opposite does not exist—frozen communities do stick piously to the norm (sects, living fossils, etc.). They provide us a stable background for the jamming of others.

To return to our parable: The Norm becomes the acknowledged ground from which a plethora of creativity sprouts across various human activities; if we stay with the example of language, it becomes the fiction and poetry of different genres, the jargon and phraseology of the sciences, linguistics, philosophy, arts, law, diplomacy, and so on. An English laywoman may not understand a single sentence from an exposé of an English-speaking attorney, yet she recognizes that he speaks English; he may even be able to render his rumination into a form more comprehensible for her. Presumably, they will also be able to share the charm and entertainment of dinner party conversation.

For another parable pointing towards the difference between the norm and play, we take from E. Auerbach (1957) who compares classical theater with that

[1] We ask the reader to keep in mind that biological history, which is evolution, is reticular, i.e., no 'pure lineages' are available in the world of cultures, religions, holobionts, symbioses, and horizontal gene transfer. If, for simplicity, we use 'lineage' in such a context, we put it between commas.

[2] Such as, e.g., the peculiar structure of mitochondrial genes in the protist group Kinetoplastida (Feagin et al. 1988).

of 16th century: "The dramatic occurrences of human life were seen by antiquity predominantly in the form of a change of fortune breaking in upon man from without and from above. In Elizabethan tragedy on the other hand—the first specifically modern form of tragedy—the hero's individual character plays a much greater part in shaping his destiny. [...] In the introduction to an edition of Shakespeare, which I have before me, I find it expressed in the following terms: 'And here we come on the great difference between the Greek and Elizabethan drama: the tragedy in the Greek plays is an arranged one in which the characters have no decisive part. They but to do and die. But the tragedy in the Elizabethan plays come straight from the heart of the people themselves.'" Auerbach comments (1957): "In Shakespeare's work the liberated forces show themselves as fully developed yet still permeated with the entire ethical wealth of the past. Not much later the restrictive countermovements gained the upper hand. Protestantism and the Counter Reformation, absolutistic ordering of society and intellectual life, academic and puristic imitation of antiquity, rationalism and scientific empiricism, all operated together to prevent Shakespeare's freedom in the tragic from continuing to develop after him." We are the heirs of those 'countermovements' that may obstruct our feelings towards ravings of the living.

Throughout this book, our intention is to apply many such parables of human experience to an understanding of biology in the hope that we succeed in highlighting some important aspects of the processes of living. Its topic is the origins of life and its subsequent evolution, the incessant labor that makes the world a home. The story begins at some point of prebiotic evolution when zillions of organic substances, volcanism, tides, ever-changing rocks, radiation, etc. supported and drove prebiotic 'metabolic' pathways. The control of such syntheses was gradually overtaken by telluric structures: the diversity of unnecessary compounds, chiral antipodes, etc. backed away in favor of controlled metabolic pathways established on mineral and organomineral surfaces with accumulation of much smaller sets of new compounds, opening further differentiation (even Darwinian selection) of self-propagating metabolites, catalyzers, etc. Bit-by-bit, such pathways became independent on mineral structures and turned into self-propagating entities. The first cells (LUCA biosphere; middle column in Fig. 1.1) turned up, representing closures from their environment, and equipped with a handful of pathways, compounds, and structures, and this repertoire has not changed much substantially in later evolution of newly established life. What *has* changed is vital manifestations, as reliable genetic memory, amount and efficiency of catalyzers and metabolic pathways, number and diversity of cell structures, and before all, the richness and capacity of symbolic communication. The latter has enabled the establishment of ecological networks comprising all inhabitants of the biosphere.

Back to the Norm that took grip with cells: it embrances, e.g., five nucleotides, twenty amino acids, a handful of phosphorylated sugars, etc., all with the proper molecular handedness (chirality). The first cells started building biospheres with such a humble set— essentially, with what we find in the basic biochemistry and cell biology textbooks. This is the underlying 'logic of life'. As stated by the biochemist A. Lehninger (1975): "(1) There is an underlying simplicity in the molecular organization of the cell; (2) All living organisms have a common ancestor; (3) The identity of each species of organism is preserved by its possession of characteristic

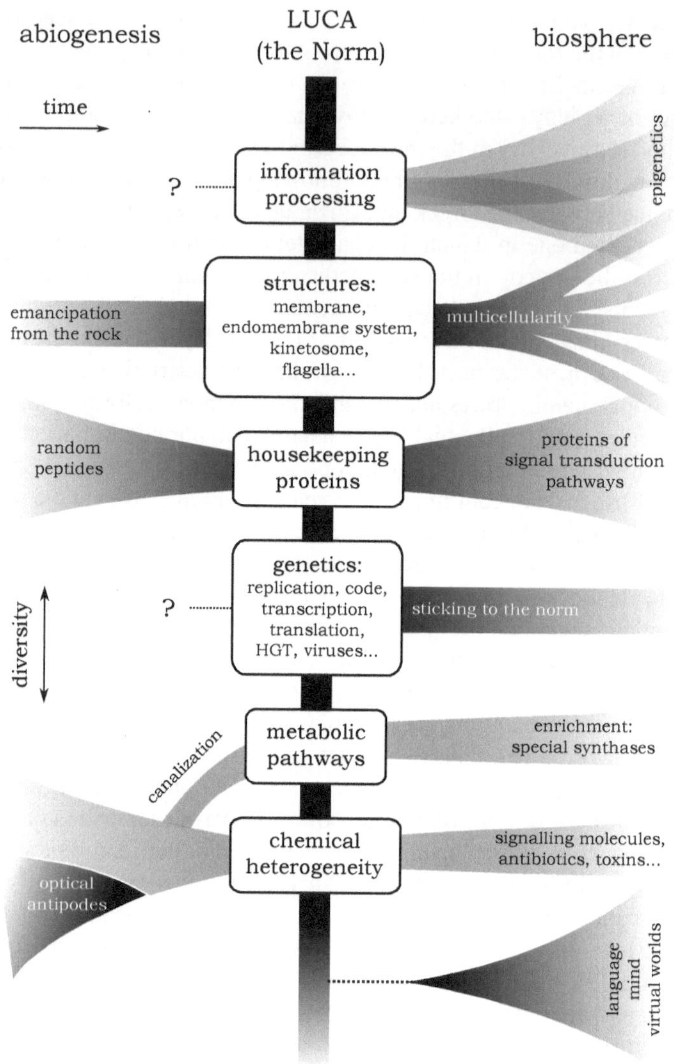

Fig. 1.1: Before and after the appearance of life. Left: abiosphere, i.e., chemical evolution driven by telluric forces and supported by mineral structures. **Middle:** The Last Universal Common Ancestor (LUCA) and its presupposed properties (the Norm). **Right:** Biotic evolution—diversification of life's functions and forms.

sets of nucleic acids and proteins; (4) There is an underlying principle of molecular economy in living organisms." Such rules represent the set of rings binding all the inhabitants of the biosphere together, and ensuring some basic level of understanding among all.

Upon the emergence of life, however, a completely new set of tools appeared, which complemented the original set to enable new forms and ways of living—derived from the Norm and rooted in it. Beside the 'big bang' of proteins and the establishment of basic functions and structures, this new set was founded in the development of sophisticated communication networks—of producing, transmitting, amplifying and decoding the many classes of signals, messages, signs. This enabled living forms (cells, multicellular organisms, communities), on one hand, to become informed of multifarious facets of their environment (consisting mainly of monitoring the presence and doings of others). On the other hand, it also allowed them to broadcast their own messages to others. Such basic 'protocols of communication'—that is a *norm of a different order* and possible only in living beings—have become, and remain to this day, shared and understood (at least to some degree) across the whole biosphere. Thanks to such bindings, living beings can rely on some historically grounded basic level of understanding that backs the plethora of ecological and symbiotic interactions.

As in our metaphor of language, evolution (in the same sense as human history) proceeded through specialization in different 'lineages' and of different ways of living, but still retaining at least parts, or remnants, of the primeval potential of mutual understanding. For reasons explained later, we call such additional storeys of life's potential *epigenetic*, to distinguish them from the genetic, or generic, norm.

Our speculations will stick to the scenario of earthly events, i.e., that organic compounds and life arose thanks to forces and structures of planetary origin. Moreover, it is obvious that we are not supporters of simple and non-explanatory statements such as: 'Life is a chemical reaction (dissipative system, thermodynamic system, etc.).' We maintain that life is a new quality that emerged upon a special, singular transition from non-living matter. Elsewhere, we have defined life as a semiotic category (Markoš et al. 2009, Markoš and Das 2016, Švorcová et al. 2018, Švorcová and Kleisner 2018); it means that *all* living beings are able to decipher meaning and react according to their memory (individual and/or inherited) and experience. *Life is not a special brand of physics, but a fundamentally new quality.* The idea is not new, and is traceable to one of Darwin's key theories on the descent with modification.

The scope of the book is broad, and of the overall image, we can provide but a short outlook on some extremely interesting facets, or even leave them out completely. We bank on the idea that life emerged on our planet, and do not bother gathering evidence for panspermia, though such evidence may exist. We did not willfully withhold any evidence that would falsify our theses, neither have we deliberately twisted the conclusions of publication cited, though we do reserve the right to re-interpret the data. We mostly avoided explaining (and citing) basic elementary concepts and procedures such as protein, mRNA, transcription, etc. Instead, we focus on recent articles that may vindicate our approach. We perceive our model reader as working with the internet, and searching therein for explanations of what he does not know while extending what he does, filling in the gaps where we may not have paid sufficient attention, and building new connections with her/his knowledge.

Chapters 2–4 deal with the establishment of life on the planet, the process from non-living 'abiosphere' to biosphere, of establishing the norm. In the second chapter, we discuss planetogenesis and events favorable for the emergence of basic biochemical pathways, and the replication of successful solutions. The third chapter reconstructs life as it may have looked shortly after the singularity of its emergence. We argue that LUCA (Last Universal Common Ancestor) contained all the elements that we call the Norm, and we give a short survey of how it may have looked. The fourth chapter then provides a survey of novelties that have evolved after the basic life processes were established: we accentuate the communicative abilities of life, which are rooted in common memory as well as the experience of a particular lineage or community. Chapter 5 summarizes the concepts of heredity and theories of evolution from Lamarck to our days. It accentuates our lines of thinking, and sets the framework for understanding chapters that follow. In Chapters 6 and 7, we further develop the 'post-Norm' evolutionary achievements, especially ways and solutions leading to information networking, heredity, signal transmission, and signal interpretation. Such properties are further illustrated on three cases: Chapter 8 is focused on morphogenesis of multicellular forms of life; in Chapter 9, we deal with symbiotic interactions; and finally, Chapter 10 discusses holobiotic ways of living. The final Chapter (11) brings a theoretical summary of our investigations throughout the book, to justify our main thesis: "Life is a semiotic category".

2
The Manifold of Prebiotic Evolution

Present study throws light on past history and vice versa. Every existing organism is, in this sense, a fossil. It carries in it, by inference, all the evidence of its predecessors; and this remains the case even if we cannot read it clearly or at all. Naturally, the study of the past [...] *can give indications as to where to look in the present to find significant things.*

—J.D. Bernal (1951)

As no direct fossils of prebiotic paths towards life have remained to our times in the geological record, and even most proxies are a matter of equivocal interpretation, we must work with conjectures—and with the 'palimpsests' retained in contemporary forms of life. This chapter brings a hypothetical scenario of events that preceded the establishment of life with its multiple closures. The enormous heterogeneity of prebiotic organic compounds, metabolic pathways and structures was, as we argue, gradually reduced to a small set of 'standard' substances that are well described in basic textbooks: five nucleotides; 20 amino acids; several phosphorylated sugars; membrane elements; metabolic pathways like fixation of CO_2, anaplerotic syntheses, etc.; and structures—capsids, proto-ribosomes, or RNA world. Effective coupling was required with both supplies of suitable forms of energy and different molecular scaffolds (minerals, rocks, organic gels, etc.), canalizing the bonfire of newly appearing substances towards manageable ends.

Preconditions

Theories that pose the origins of life directly on our planet must cope with explaining the abiotic origins of organic compounds. Their synthesis could not be haphazard (save in the initial phase), but must have proceeded in modules that—under conditions reigning on the virgin planet—either repeatedly appeared *de novo*, or persisted for long times, endowed with growth and multiplication. Such is the scenario called 'Metabolism first'. With the discovery of DNA structure and function, competitive models called 'Information first' also came to the fore. Remarkably, as we demonstrate below, both models coin an idea of 'aperiodic crystals'—of rocky chips of a kind—that 'invented' metabolism, or self-replication, or both, driven by telluric forces. Only later did the organic structures acquire the flesh that enabled their emancipation from the rock beginnings, and assumed the long path towards life. Proponents of both scenarios remain in continuous argument (de Duve 1991, Maden 1995, Pross 2004, Smith and Morowitz 2007, Morovitz and Smith 2007, Orgel 2008, Hein and Blackmond 2012). What is conspicuous, however, is the general belief that once organic compounds had appeared on the planet (in the initial random phase), the further path towards life was inevitable. This was well illustrated in the 1950s by the exultations that accompanied simple laboratory contraptions, producing organic matter from inorganic substrates (Miller 1953, Miller and Urey 1959, and a plethora of recent articles). It is also conspicuous in the recent and obstinate digging on Mars—in a search if not for life, then at least for organic compounds. Many authors thus do not take into account the long way from haphazard organic syntheses to organized structures, pathways, or information deposits that constitute the necessary precondition for the appearance of life. As an antidote, we suggest the following parable: "Suppose that in one moment the whole biosphere turns extinct—no single living cell remaining. The planet is full of organic compounds in much greater supply and much better selection than can be assumed in the primordial soup. Will anybody dare to suppose that this is a fair chance for life to reappear?" (Markoš 2016a)

Steep gradients of utilizable free energy are also indispensable conditions for prebiotic syntheses, as well as for the maintenance of evolution towards life; a mere deposit of organic compounds cannot be enough. Today, only two *primary* sources of energy can fulfill the task—oxidation-reduction reactions and visible light (400–900 nm)[3] the latter either coupled, again, to redox reactions (as in photosynthesis), or driving cell processes directly (as in photon-driven pumps). The coupling to the energy coming from our star was presumably a late innovation; if so, our first step must be to look for suitable earthbound redox batteries. Indeed, a plethora of chemolithotrophic and methanotrophic organisms meets such a scenario even today. The existence of metalloproteins among the enzymes of extant metabolic pathways (see below and Fig. 2.2) may also point us toward accepting the conjecture of the mineral origins of metabolism.

Systems situated in steep energy gradients often have a tendency to establish macroscopic and dynamic formations called *dissipative structures*, like gyres, flames, growing crystals, and some chemical reactors. They are macroscopic because they consist of a great number of molecules ordered (canalized) in a specific way (i.e., not moving randomly as in Brownian motion). They are dynamic because their very existence depends on a given gradient of energy. By building their structure

[3] It is a matter of speculation whether other forms of radiation (e.g., the radioactive decay of radioisotopes) were also at work.

at the expense of the energy potential, they dissipate the same potential as heat. The more efficient the dissipation, the more elaborated the structure, which enables even more efficient dissipation. Long-lasting dissipative structures with elaborate metabolic pathways are supposed to have existed in places similar to geothermal 'white smokers' on the oceanic bed, sites where magmatic rocks are metamorphosed upon contact with the oceanic water (Fig. 2.3). Yet, as we shall see later, the equation 'life = dissipative structure' cannot hold.

The major advantage of dissipative structures is their emergence *de novo*; their major drawback lies in the fact that upon exhausting the energy source, the structure will cease (or sometimes 'freeze' in a conventional structure, as with crystals), leaving no memory suitable for future use by some similar structure. As living beings *never* pop up *de novo*, the coupling of subsequent generations in time is a necessary precondition which must be filled before life can emerge: extant life is thus a direct descendant of living beings that appeared in Lower Archaean some 4 billions years ago. It follows that memory structures more resilient than dissipative structures have appeared before the emergence of life, encoded and replicated in various memory structures. During prebiotic evolution, however, dissipative structures may have played a decisive role in many organic syntheses.

The set of possible organic compounds that could/might appear during the initial phase of planetary evolution (imports from interstellar space, or local runaway reactions), would be mostly useless or even toxic for any form of life, both ancient and extant. Therefore, the crucial factor of prebiotic evolution must have been putting organic synthesis under control, and channeling them towards 'desirable' products. Theories of the origins of life, which emerged in the 1920s and 1930s—such as those by A.I. Oparin (2013) and J. Bernal (1951), were aware of this need. Oparin envisaged a 'pastina' in his 'primordial soup' in the form of coacervates—organic precipitates that would selectively bind organic compounds present in the soup, provide a scaffold for some reactions while avoiding others—hence catalyzing their mutual reactions. In Bernal's theory not coacervates, but clays provide a necessary scaffolding action, with similar results. Below we bring other examples of *bona fide* contraptions that could have arisen independently and subsequently found their place in the first living cells. We draw our speculations from structures in contemporary cells, which may represent possible palimpsests of those ancient events. There are two extant types of such structures. The first are those that can grow *de novo* by self-assembling elementary building blocks or by an intercalating of such parts into an existing structure, with subsequent division of such a structure. The second are structures or molecules that can serve as information deposits—databases and programs (especially linear molecules of nucleic acids). They preserve information (for the cellular 'wetware') of how to build the constituents of the abovementioned structures (RNA, proteins); moreover, they can be copied (by the same wetware) with a great precision. Today, these two kinds of structures exist mutually interwoven within cells, and cannot exist separately (Chapter 6): the more ticklish question is whether they could have arisen and replicated in the prebiotic world lacking such interwoven closures.

The whole of protometabolism may be considered as an extension of thermodynamic processes on the young planet, leading to the 'collapse to life': "The continuous generation of sources of free energy by abiotic processes may have forced life into existence as a means to alleviate the buildup of free energy stresses." Life, then, emerged "as a sort of chemical bootstrapping process from abiotic precursors in steady state." (Morowitz and Smith 2007) Below, we assume that energy, material supply, and the scaffolding of processes discussed below originated from planetary resources, and presumably took place either in tiny cavities in water-percolated rocks, or at the oceanic bottom. The principal driving force of biosphere today, the Sun, played only a little, if any, role. Indeed, analyses of biochemical pathways suggest that photosynthesis is but an extension of respiratory pathways that evolved in chemolithotrophic organisms.

Many of the problems posed may be solved in near future by the rocketing development of nanotechnologies, with their study of specific molecular patterns of surfaces or micro-spaces, and with the deciphering of physical and chemical rules, working with small and structured assemblies of molecules. This topic is, however, beyond the scope of this chapter.

Metabolism First

As previously mentioned, putative prebiotic assemblages could arise either repeatedly *de novo*, or grow and divide by adding new constituents to preexisting structures; no extra information is required apart from structures themselves. The spontaneous assembly of such jigsaw puzzles (at least in the lab) is typical for ribosomes, spliceosomes, virions, and different protein assemblies such as metabolons or signal particles; even the cisternae of endoplasmic reticulum can be reconstituted from its parts (Powers et al. 2017). Members of the second group grow by the insertion of such constituents into a preexisting structure, and multiply by division: such are membrane-bounded organelles (and, of course, cells themselves), or structures like kinetosome and derived organelles (cilia, flagella). The existing structure provides information (as a scaffold) for the correct assembly of parts. Below, we bring some hypotheses as to how such assemblies could thrive under abiospheric conditions, supported or not by mineral scaffolds that *can* be supplied *de novo*, and are driven by telluric forces.

(i) The pyrite module. In 1980s, G. Wächtershäuser and his coworkers proposed a theory of protometabolism coupled to the fixation of CO_2 (i.e., the formation of C-C bonds followed by reduction of the compound attained). The chemical assembly was bound to the surfaces of growing pyrite grains (FeS_2; Wächtershauser 1990, 1994, Drobner et al. 1990, Blöchl et al. 1992, Huber and Wächtershauser 1997); but see also comments and criticism (de Duve 1991, Martin and Russell 2007, Orgel 2008). Pyrite is a common mineral in inorganic environments that are saturated with ferrous ions (Fe^{2+}) and sulphane (H_2S). The resulting redox reaction

$\mathbf{Fe^{2+} + 2H_2S \rightarrow FeS_2 + 4H^+ + 2e^-}$, or

$\mathbf{FeS + H_2S \rightarrow FeS_2 + H_2}$ $\qquad\qquad \Delta G° = -42\ kJ/mol$

is exergonic, and in addition to energy, it also yields hydrogen, which is necessary for the reduction of CO_2. The aggregate equation for the synthesis of formic acid is as follows:

$$\mathbf{FeS + H_2S + CO_2 \rightarrow FeS_2 + H_2O + HCOOH} \qquad \Delta G° = -12\ \text{kJ/mol}$$

Formic acid (reduced CO_2) reenters the reaction with another molecule of CO_2 (C-C bonds) and by this recurrent process, even long-chain fatty acids may have appeared (but note that no functional model has been developed so far). Moreover, the surface of growing pyrite crystals is positively charged and as such shows affinity to negatively charged organic molecules floating in the environment, such as fatty acids, amino acids, or their phosphorylated esters. This would lead to protometabolic pathways such as reductive citric acid cycle, or the synthesis of long-chain fatty acids (Fig. 2.1). The precondition of this reaction is a tight coupling of redox reactions with energy flow (so it will not escape as heat) and with the generated hydrogen, which then enters the reaction in atomic form (it does not escape as hydrogen molecules). Impurities contained in the crystal, as well as the assembly of bound organic molecules and structures, may augment the efficiency of the reactions—by providing a unique surface pattern, which contributes to the catalysis. On prebiotic Earth, pyrite formation was common, so an efficient matrix for canalized organic syntheses was readily available.

In spite of concentrated efforts, however, carbon fixation by growing pyrite has not yet been demonstrated in the lab (this is also the case of other sophisticated models, as we will see). It is, however, conspicuous that the great majority of contemporary enzymes catalyzing redox reactions (oxidoreductases, such as ferredoxin, respiratory enzymes, photosystems, nitrogenase, hydrogenase, etc.) contain organometallic prosthetic groups (Fig. 2.2), many carrying elementary 'Fe-S crystal' as in pyrite (see ferredoxin in Fig. 2.2). Oxidoreductases containing organometallic prosthetic

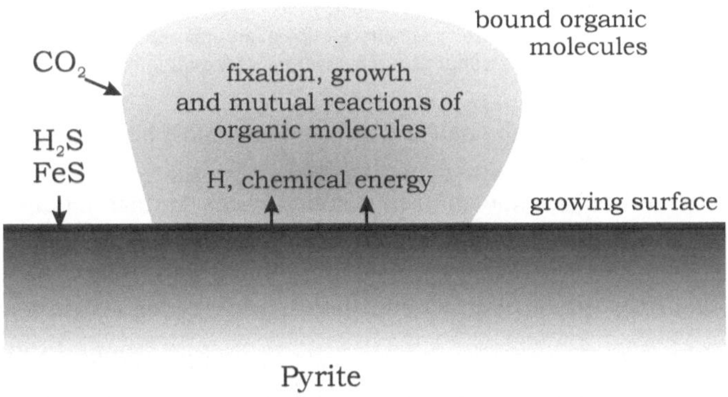

Fig. 2.1: The pyrite "pizza" as a model of carbon fixation. The exergonic synthesis of pyrite, with released energy and hydrogen, is coupled to organic syntheses. Critical in this respect (in the absence of enzymes) is the catalytic capacity of the growing surface: it bears a positive charge, and may even be patterned by the presence of various impurities that allow binding of selected organic compounds and capture CO_2. (After, e.g., Wächtershäuser (1990).)

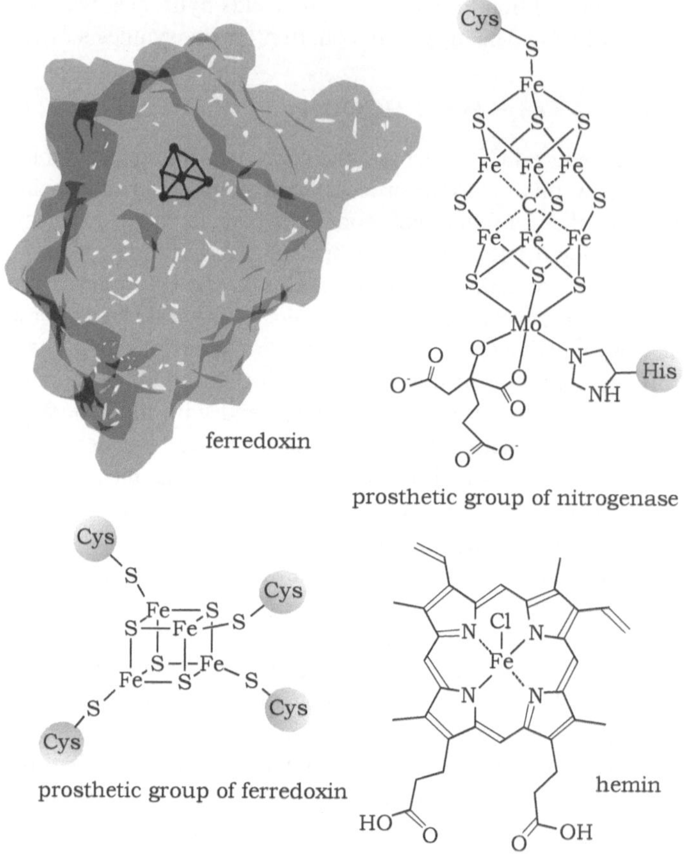

Fig. 2.2: Organometallic prosthetic groups of extant enzymes. Left: contours of protein ferredoxin with highlighted 'pyrite heart' 4Fe4S; below the scheme of its chemical binding to four cysteine residues in the protein backbone. **Right:** Fe-Mo prosthetic group of nitrogenase (nitrogen assimilation), and hemin—essential part of cytochromes (respiration).

groups apparently belong to the oldest enzymes in the biosphere, and emerged at the very beginning of life (David and Alm 2011).

The theory of surface metabolism inspired many similar models of organic syntheses towards a selected, and limited, set of compounds that gave rise to the metabolic pathways identifiable in contemporary living cells. Below we illustrate yet another model—the reaction of serpentinization as a driving force of protometabolism.

(ii) Serpentinization. The story starts with magma rising in mid-ocean ridges, thus providing material for the cycling of the oceanic crust. The rising material expands and cools, releases gases and gives rise to hard rocks, in reactions with oceanic water and its components. Below, we give a highly simplified idea of a small set of such reactions involving Fe/Mg silicates (Fig. 2.3): the process is called *serpentinization*, giving rise (besides the mineral serpentine) to two important gases, hydrogen and methane (stoichiometry in equations omitted):

Fayalite + water → magnetite + silica + **hydrogen**

$Fe_2SiO_4 + H_2O \rightarrow Fe_3O_4 + SiO_2 + \mathbf{H_2}$

Silica further reacts with silicates, giving, as a result, the mineral named serpentine, hence the name of the reaction:

Forsterite + silica → serpentine

$Mg_2SiO_4 + SiO_2 + H_2O \rightarrow Mg_3Si_2O_5(OH)_4$

In the presence of $\mathbf{CO_2}$, the reaction is also accompanied by the synthesis of methane:

Olivine + water + carbonic acid → serpentine + magnetite + **methane**

$(Fe,Mg)_2SiO_4 + H_2O + CO_2 \rightarrow Mg_3Si_2O_5(OH)_4 + Fe_3O_4 + \mathbf{CH_4}$

Sulfates present in the mixture give rise to the fourth 'biogenic' gas—sulfane $\mathbf{SH_2}$.

In addition, rock, comprising of many different minerals as well as other rocks, is porous (Zhu et al. 2011, Schiano et al. 2006), and thus contains different and interconnected channels which allow percolation by alkaline (pH 11) hot water solutions: when these find their way to the oceanic bottom, their temperature is slightly below 100°C. Such hot springs are today known as 'white smokers' (so called Lost City vents; Kelley et al. 2005).

Here begins the model of prebiotic syntheses as proposed by the working groups of M.J. Russell and W. Martin (Russell et al. 1994, Martin and Russell 2007, Russell et al. 2015, Sousa et al. 2015; for review see Sojo et al. 2016). Anoxic oceanic water, cold and acidic (about 10°C and pH 5.5), interacts with the alkaline vent rising from below and carrying the abovementioned products of serpentinization, as well as many other water-soluble compounds. Sulfides of various metals, plus silicates, plus iron hydroxides, etc., precipitate in the form of porous mineral membranes (Fig. 2.3). The membrane separates the anoxic marine waters from the water penetrating from below, with two important consequences:

(1) The pores in the precipitate (with their high surface area) allow a multitude of organic syntheses to be catalyzed by, e.g., metallic or organometallic compounds contained in their walls (essentially as in case of the pyrite model above).

(2) The membrane is a barrier between two environments of different pH, thus establishing electrochemical potential (about one volt): here, we encounter a primitive metabolic closure endowed with a powerful energy source. Such a source could provide energy for propelling essential metabolic reactions: the most important being the synthesis of pyrophosphate (PO_3-O-PO_3^{4-}), which is an equivalent of ATP in extant cells. In addition to an energy resource, the reaction also requires the exclusion of water from the body of the membrane. Another putative energy resource for protometabolism is acetyl phosphate, synthesized as explained below and in Fig. 2.3. The process depicted by this model could have started even without phosphate esters, with thioesters instead (see also Huber and Wächtershäuser 1997, Goldford et al. 2017). Warning again: neither this model nor those that follow have been realized in laboratory.

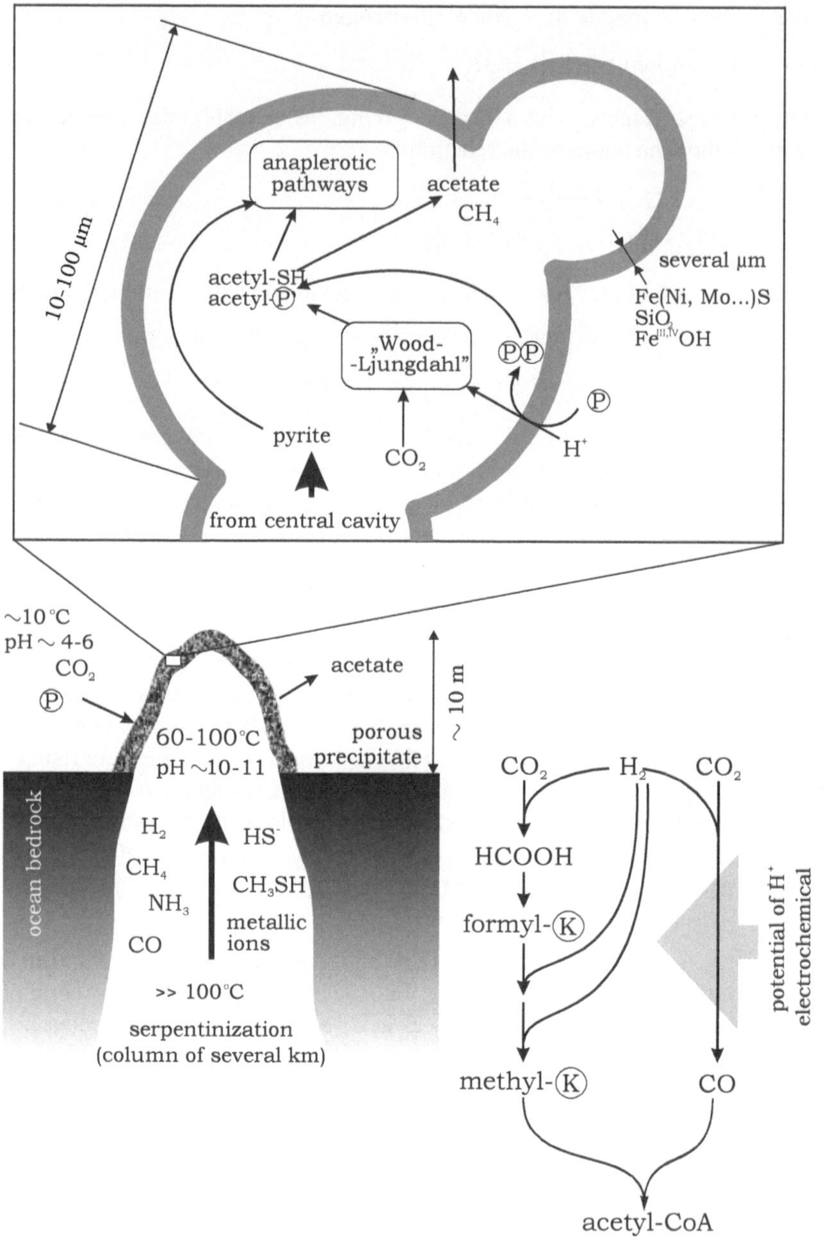

Fig. 2.3: Hypothetical protometabolism on oceanic 'white smokers' coupled to serpentinization. Uprising rock undergoes metamorphosis to serpentine, resulting in warm alkaline solution containing hydrogen, sulfane, ammonium, methane, and a mixture of ions of transition metals (Fe, Ni, Mo, W, etc.). Injected into cold acidic oceanic water enriched in CO_2 and perhaps soluble phosphates, its components precipitate at the interface of both solutions, into porous material allowing establishment of electrochemical potential across such a boundary, theoretically enabling "respiration" and phosphoester syntheses that drive subsequent reactions of CO_2 fixation. **Below right**: Extant Wood-Ljungdahl pathway of CO_2 fixation: its variants might have worked in white smokers. (After Nitschke and Russell (2013), Lane and Martin (2012).)

Reactions in the lumen of the pore set up a pathway similar to Wood-Ljungdahl reaction of carbon fixation—present in some extant bacterial metanotrophs and Archaean methanogens, with an output of acetyl-CoA from CO_2 (Fig. 2.3). The pathway, albeit exergonic, requires also the production of ATP by other mechanisms (see also Barge et al. 2014). Under protobiochemic conditions, the authors suppose the production of acetyl sulphide or acetyl phosphate; the extra energy would be provided by pyrophosphate—the synthesis of which is driven by the transmembrane potential, as mentioned above. The waters ascending through the vent would provide most of the material for building the whole contraption and establishing its pathways. Of course, extant cells do the same job without being dependent on telluric forces (Schuchmann and Muller 2014). Martin and Russell (2007) point to the fact that oceanic water percolates through the newly emerging crust up to depths of 5–7 km—thus protometabolism may not have been bound only to vents, but may have proceeded in billions of microcells in porous 'serpentinization factories'. Thus, rising waters could have appeared on the surface loaded already with sets of organic compounds, different from those made in the surface precipitates.

Protometabolism in percolated rocks ('cells') is also in the focus of other authors, e.g., Branciamore et al. (2009), and Gold (1999); the latter emphasizes the fact that even today, Earth's percolated crust is populated by chemolithotrophic organisms, the limits of which seem bounded only by the high temperatures in the depths. This suggests that conditions favorable for organic synthesis reign in darkness and may have reigned even in the far past (without cells, of course). For experimental testing of particular reaction regimes, see McCollom and Seewald (2007). Likewise, Dodd et al. (2017) report that proxies of life exist exactly where expected—places of ancient (3.8 or even 4.3 billions of years) hydrothermal vents, and Nutman (2016) reports patterns interpreted as 3,7 billions of years old stromatolites.

(iii) Chirality. For the researchers in the field, one of the most annoying enigmas is how preferences for one out of two (or more) possible 'mirror images', or enantiomeric compounds appeared (e.g., preference for D-sugars or L-amino acids). In extant life with sophisticated enzymes, the synthesis of only one enantiomer is no mystery. Known procedures of non-enzymatic syntheses, however, always lead to a mix of both forms (Cronin and Reise 2005).

A miniscule preference (later amplified) for one enantiomer was sought, either in inherent asymmetry of elementary particles (Davankov 2009, McGuire et al. 2016), or in the synthesis and a subsequent biased propagation of one form (Cronin and Reise 2005, Breslow and Cheng 2009). Hein and Blackmond (2012) see the solution in mutual potentiation of L-amino acids and sugar precursors that could lead to the awaited bias. Root-Bernstein (2007) sees the divide only in later phases of evolution—in a mutual interconnection of genetic code, L-amino acids, and D-sugars in protoribosomes (see Hein and Blackmond 2012). However, to our knowledge *no* theory or speculation is able to explain the selection and its maintenance, so a 'frozen accident' may be as good an explanation as any. Moreover, the improper mirror enantiomers interfere with contemporary metabolic pathways and may even be toxic. Thus, the selection of one of both forms must have taken place at the very dawn of prebiotic evolution and in the absence of efficient catalysts (see Fig. 1.1, p. 4). How such an accident became preserved and even imposed through billions of parallel events must lie in as of yet unknown properties of rocks present in the virgin Earth. In templates built of even achiral compounds yet to be crystallized in two chiral forms, e.g., silica and other silicates, a miniscule shift in the occurrence of one of optoisomers may become amplified and finally prevail in the Earth core. Research in nanotechnologies may bring solutions to the enigma in the future, but at this point, we are groping blindly without an answer.

(iv) Organic forms. In pre-Darwinian biology, the Platonic view of pre-given rules determining the pattern of organisms was highly popular—hence the opposition against Darwin's claims that such patterns are not lawful and result but from historical contingencies maintained by natural selection. An analogy drawn with the periodic table of chemical elements served as paragon and challenge for the endeavors of Darwin's peers (mostly morphologists or embryologists). A belief in laws of forms has resurfaced many times even since Darwin's time. These include, e.g., the mathematical morphology by Thompson (1961 [1917]), the movement of biological structuralism (Goodwin et al. 1989, Webster and Goodwin 1996), views of the paleontologist S. Conway Morris (2003), or biophysical theories (Newman et al. 2006, Newman and Bhat 2009). Surprisingly, such views have reemerged recently not in the domain of the living (most biologists stick today to historical explanations of pattern evolution), but in the realm of biological macromolecules that repeatedly tend to attain very similar forms; this holds especially for 'aperiodic crystals' such as proteins (Denton et al. 2002, Hou et al. 2005, Ahnert et al. 2015). A practically endless number of possible polypeptide chains collapses into a small set of simple secondary structures, motifs like α-helix, β-sheet, β-barrel, helix-turn-helix, sandwich, etc.; see some examples in Fig. 2.4.

The number of such building blocks of proteins seems to be very low indeed: in 2015, the databases SCOP and CATH kept the files of 1282 and 1393 such structures, respectively, and none have been added since 2008. Such data, of course, tempt speculations, as the database set may represent an example of 'rational generative morphology'—obviously at a molecular level unknown to 19th-century morphologists. Denton et al. (2002) summarize: "[T]his might mean that wherever random polypeptides are synthesized anywhere in the cosmos—in the laboratory, in a pre-biotic soup or in a primeval cell—all the basic folds utilized by life on earth would be bound to be generated after only a few thousand trials. In addition, this leads us to surmise: if at some stage in cellular evolution, 'random polypeptides' were synthesized in great numbers, then all the folds might have been discovered quite easily by chance. This would mean that in the right environment, just as atoms are assembled in the stars and crystals form when rocks cool slowly, so the protein folds would also be formed automatically, wherever conditions permitted the synthesis of any quantity of polypeptides to occur." Similarly, Keefe and Szostak (2001) report on selection of functioning ATP-binding polypeptides out of billions of random peptides 80 amino acids long.

From the point of prebiotic evolution, we are poised to conclude:

* *If* the environment produces random polypeptide strings packed into limited varieties of shapes; and

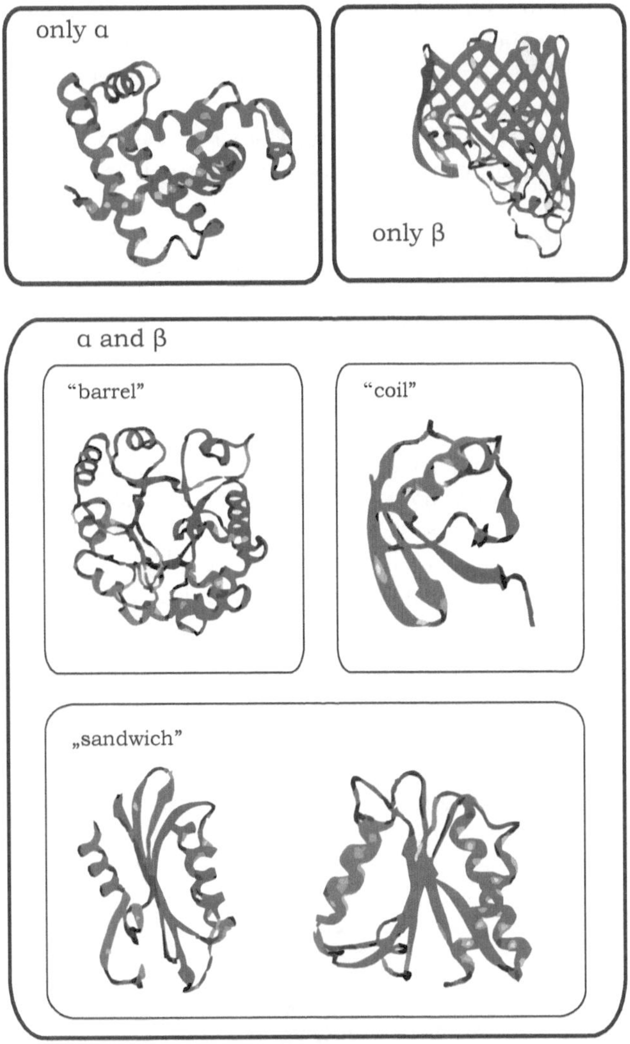

Fig. 2.4: Protein motifs (folds): α-helix and β-sheet and some of their combinations. (After Denton et al. (2002), Hou et al. (2005).)

* *If* there exists an environmental scaffold that recognizes not sequences but shapes of such polymers;

* *Then* higher-order complexes should assemble repeatedly, according to the properties of the specific scaffold (be it, e.g., a mineral, membrane, prosthetic group, or RNA). What holds for polypeptides can be true for other polymers such as nucleic acids, lipids, or polysaccharides. As the environment can supply an enormous heterogeneity of such matrices (see the examples of clays, minerals, and rocks in this chapter), a *parallel* testing of different assemblages might have proceeded—and selected all this organic manifold—without a need to store information in a medium different than structures themselves. In other words,

the genetic background as we know it in extant living forms could be lacking, the replicators having been realized by the structures themselves, arising many times (with modifications) in parallel *de novo*. This theory will undoubtedly require experimental testing of the assembly of existing or even *de novo* designed proteins (Huang et al. 2016, Garcia-Seisdedos et al. 2017, Rocklin et al. 2017). As Kauffman (2000) points out, a single test tube can contain as many as 10^8 different short polypeptides, in principle serving whatever catalytic task. As polypeptides could theoretically spring from any evaporating pothole containing amino acids and such polypeptides could subsequently survive when they happen to enter a solution, abiotic syntheses could have provided whatever shapes would be required for catalyzers, even in the absence of genetic machinery and code.

An indirect support for redundant sequence space comes from Starr et al. (2017), who undertook a reconstruction of an ancient precursor of an extant protein domain. They show that there could have existed many alternative histories leading to sequences with the same function, as had the ancient sequence, without ever passing through non-functional proteins. The extant sequence is thus one of many possible and may not even be optimal in all possible contexts. Here, of course, the issue is not one of abiotic evolution, and yet the argument does provide a parallel to our discussion of protein shapes.

(v) Virions can serve as an example of organic forms self-assembled *in vitro* from isolated protein components, e.g., membranes from lipids, or ribosomes from proteins and RNA; hence, virions may be a candidate of prebiotic structure. Moreover, among protein motifs there exists a group common to all viruses—the capsid protein encapsulin. Again, their sequences and those of their genes show no recognizable common ancestry (Nasir and Caetano-Anollés 2015; Fig. 2.5). "The defining feature of a virus is its protein capsid, whose structure is strictly constrained

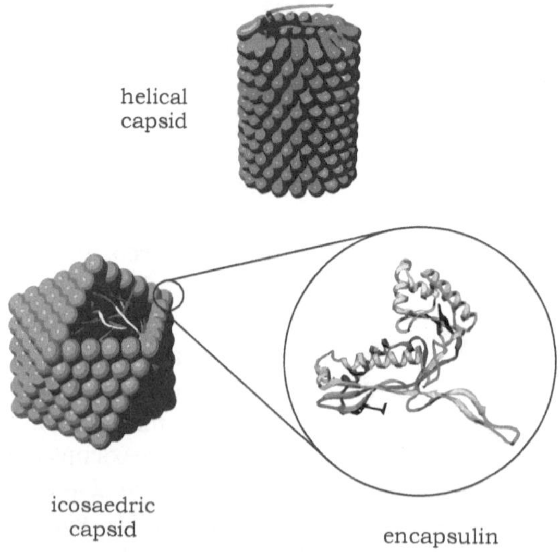

Fig. 2.5: Virions. Two motifs of the capsid setup, and the basic structure of the encapsulin tripod.

by rules and genetic parsimony. Capsid structure is therefore far more persistent than viral genome or protein sequences." (Abrescia et al. 2012) Thus, virions may represent palimpsests of *structures* present in the ancient abiosphere, assembled from certain types of proteins on the nucleic acid scaffolding, and containing some cargo inside (not necessarily of nucleic acids). What is striking is that structures similar to viral capsids are present in all the three cell domains, built from proteins with structures analogous to virions; their cargo, however, is not nucleic acids, but storage compounds, or else they contain some metabolic pathways (Sutter et al. 2017, Docampo and Moreno 2011).

'Replicators'—Code First

The key event in prebiotic evolution came when organic assemblages became emancipated from mineral scaffolds: this requires some kind of memory, because construction rules could no longer be imposed from outside. Organic forms came to be complemented by organic codes (Barbieri 2003, 2015) that did not follow from mere chemical affinities, but must be 'negotiated'; in Barbieri's terms, they are results of natural conventions. Aperiodic crystals appearing repeatedly in rocks were replaced by aperiodic linear crystals of nucleic acids that require processes such as transcription coupled with the synthesis of a small set of building blocks. Later, translation followed putting an end to the spontaneous generation of random polypeptides. The scaffold not only selected appropriate shapes, it also encoded optimal steps towards such shapes. With grammatical rules of this kind, prebiotic evolution entered a new stage—examples will be given by the 'RNA world' and 'protoribosome' below. However, we first head back to the rock surfaces.

(i) Clays as information deposit. Inspired by J. Bernal's (1951) theory of clay metabolism, A.G. Cairns-Smith (1982, 1985) adapted the theory for the age of molecular biology. The scales of clays crystallize layer after layer—from the mother liquor, interspersed by aquatic solution; the resulting structure resembles a sandwich, or a column of coins (Fig. 2.6). An 'ideal' scale (e.g., of china clay) has a summary formula $Al_2(Si_2O_5)(OH)_4$. The aluminum layer, however, is almost never pure because the mother liquor contains many other ions that tend to enter into the crystal lattice. Clays differ in the amount and ratios of such impurities; for example, the clay sauconite has an overall formula $(Si_{6,94}Al_{1,06})(Al_{0,44}Fe_{0,34}Mg_{0,36}Zn_{4,80})O_{20}(OH)_4$. The argument is that the impurities form a pattern in the crystal that manifests itself as a specifically rugged surface of the scale. The newly growing scale has a tendency to 'feel' such surface patterns, and this will facilitate ordered insertions of impurities to identical coordinates on the new one. The old pattern thus serves as a matrix for patterning the new one; the daughter clay crystal preserves in its pattern the information content of the mother flake. Again, attempts to prove the theory in the laboratory have so far been unsuccessful; what follows is pure (yet reasonable) speculation.

According to the theory, what enters the stage with clay replicators is a 'Darwinian evolution of non-living organisms'. Some scales may catalyze self-replication more efficiently, thus their progeny will successfully outcompete other (differently

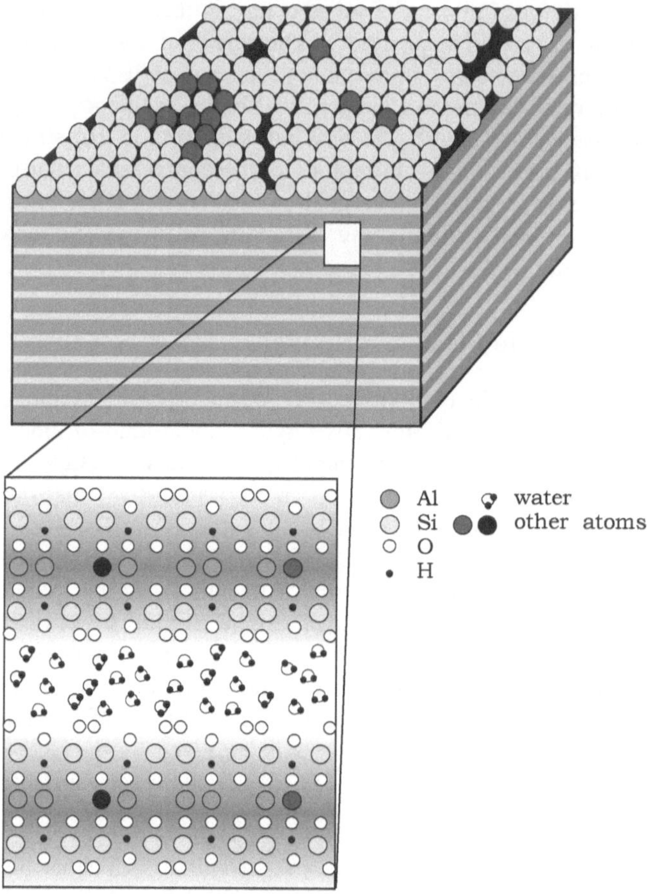

Fig. 2.6: Clays in a role of 'non-living organisms'. Above: a stack of clay crystals displaying, on the top, the Al-layer with impurities; the pattern of impurities has a tendency to be copied into the new layer. **Below:** section through two crystals (scales) with water solution in between. The watery layer and its content (ions, organic compounds) may increase the rate and accuracy of replication of both the crystal and its pattern. (After Cairns-Smith (1982, 1985).)

patterned) scales for the precursors contained in the liquor so as to produce more new copies. As a result, such variant patterns will prevail in the final product. If the clays dry up, the columns will break up into single scales, which will float with the wind only to start growing again somewhere else.

The aqueous layer between the scales also plays a role: it may contain various ions and even organic compounds. They bind non-randomly to the non-random crystalline pattern, and may further influence the efficiency of the replication process, i.e., the final yield. As in case of pyrite model discussed above, they can polymerize and further improve the catalytic efficiency. (The major difference is that the growth of pyrite also provides energy and reducing equivalents driving the reaction, whereas growing clay scales do not.) Evolution may proceed towards more and more complex organic aggregates, up to the point that such aggregates will no longer need the inorganic matrix, and start replicating autonomously. Good candidates for such a process are linear molecules—e.g., RNA or similar aperiodic polymers that are able to catalyze their own copy (or complementary string). The model of such an 'RNA world' is presented below.

The simple laboratory model of Kreysing et al. (2015) gives an idea of how such emancipated abiotic polymerization and replication may have proceeded: the authors were able to develop polymers as long as 75 nucleotides in rock capillaries were situated perpendicular to the thermal gradient (Fig. 2.7). Of course, the reaction is dependent on the supply of activated nucleotides—there is currently no idea about how the precursors and a suitable energy source may have looked under prebiotic conditions.

(ii) RNA world. In the 1960s after the discovery of genetic code, the construction of an artificial cell seemed to be a question of some 10 years. Solving the mystery of DNA-protein dualism was to be solved with one single kind of compounds—RNAs. On one hand, they can be copied as DNA, on the other, they can attain 3D structures as proteins, and as such catalyze their own replication. Thus, in theory, RNA can work as an RNA replicase, catalyzing a synthesis of its own copies, or of some other RNA molecules. The discovery of ribozymes in contemporary cells and investigation in the field of artificial ribozymes enhanced the optimism even more.

In the 1970s, the 'RNA world' theory was formulated, justified by physical chemistry, Darwinian theory, and a theoretical model of hypercycles (Eigen and Winkler-Oswatitsch 1992, Vaidya et al. 2012, Eigen 2013, Fig. 2.8). Briefly, a team

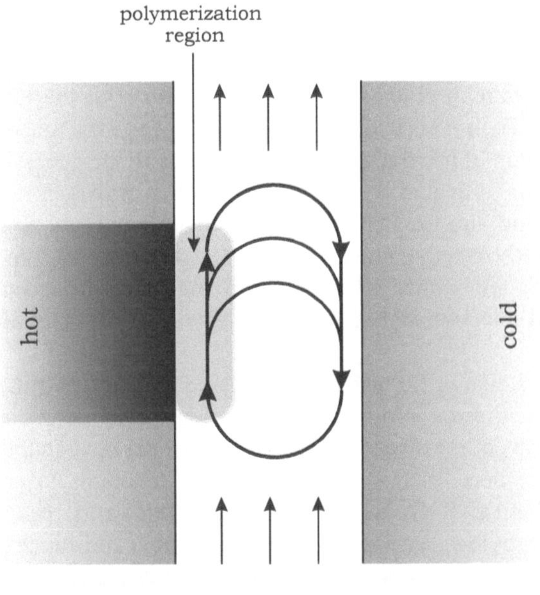

Fig. 2.7: Model of nucleic acid replication in a pore in the rock. There is a thermal gradient across the rock. The solution with precursors, together with short strings of nucleic acids, flows upwards along the hot surface, where prolongation and replication takes place, followed by escape of products, or by cooling, decreasing, and starting a new round of prolongation/replication. (After Kreysing et al. (2015).)

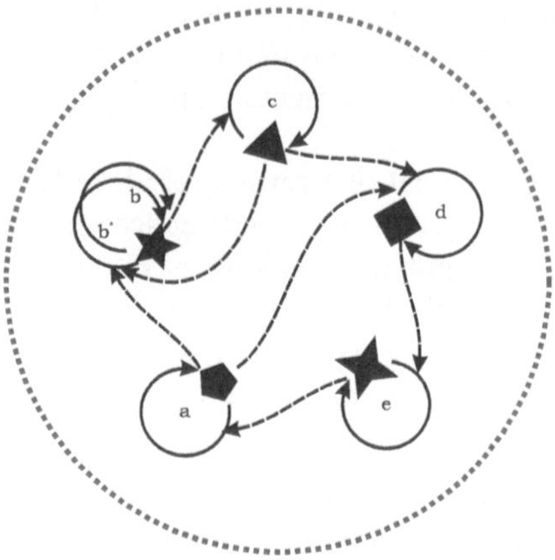

Fig. 2.8: The hypercycle. The dashed line represents the closure—membrane with controlled permeability for different compounds (flows not shown). Ribozymes a–e catalyze synthesis of intermediates utilizable also by other members of the team. At least one member of the team has a property of RNA-dependable RNA polymerase. The ribozyme **b** shows evolution in time – mutant **b', b"** ... **b**^{n} will appear and become subject to natural selection and may replace the original copy. (After Eigen (2013).)

of different RNAs endowed with different catalytic functions (ribozymes) ought to be able to fulfill the informational, metabolic, and self-propagating tasks of modern cells. There appeared the concept of first closures, followed by a controlled exchange of both matter and information, capable of producing descent with modification, and hence undergo natural selection. Moreover, Vaidya et al. (2012) demonstrated experimentally that such assemblies of heterogeneous RNAs could cooperate, evolve, and emerge spontaneously *in vitro*.

Yet, there remains problems in connection with both the RNA world and the hypercycle model. Before all, it is the stability of RNA and the subsequent dialectics of precise conformation of the ribozyme on one hand, and the necessity of denaturation of the RNA when replicated on the other. Another drawback is presented in the accuracy of copying: the frequency of errors in contemporary enzyme, i.e., *protein* RNA-dependent RNA polymerases is 10^{-4} (compare to 10^{-9} in DNA-dependent DNA polymerase); apparently, RNA polymerizing ribozymes accumulated so many errors that natural selection had no chance to test the variants (evolution by Muller's ratchet).

The closure is also important for hypercycle assemblies, but in the first place, it is not clear how these could be attained, unless the hypercycle assumed a state comparable in terms of closure to modern cells (see the next chapter).

The last three decades have been devoted to attempts to create a laboratory model of hypothetical RNA world(s): roughly, this work proceeds along three parallel paths (for review, see Szostak 2011, Robertson and Joyce 2012, 2014, Blain and Szostak 2014). First, attempts are made to synthesize and replicate RNA under laboratory condition, wherein *any* empirical procedure is allowed to achieve the goal (except, of course, that of adding protein RNA polymerase). Efforts to synthesize artificial ribozymes, especially of RNA-dependent RNA polymerases are under way

(Wochner 2011, Robertson and Joyce 2014). A ligase connecting two strings of RNA into one was found, with a catalytic activity of about 800 per minute (Shechner et al. 2009—for comparison: extant protein enzymes have turnover rates of up to the order of 10^6 per *second*).

Second, similar attempts are made, which concentrate on achieving the same goal, but under *bona fide* prebiotic (i.e., not laboratory) scenarios. Third, attempts are made to produce a 'protocell' (a closure, an autonomous agent) matching theoretical assumptions of the hypercycle theory.

Questions also appeared concerning whether or not activated nucleosides (such as ATP, GTP, etc.) could be at hand without the presence of any catalyzers. The formation of phosphate-activated nucleotides and their proper ligation seems to be improbable, yet some authors suggest such syntheses from low-molecular precursors might be plausible (Powner et al. 2009, Saladino et al. 2012, Patel et al. 2015).

Such activated monomers could then ligate into a random polymer, or be connected to an existing polymer, thus prolonging its chain. In a laboratory with purified reactants and controlled conditions, such polymerization does take place but many different strings appear, differing in the character of the bond (3'-5', 2'-5', branched, etc.). A 'normal' linear string with 3'-5' bonds will appear only with 3'-5' phosphodiester precursors, and with the help of a double-string RNA (Szostak 2011, Adamala et al. 2015). Ligation of two single strands can also be accomplished in the presence of a complementary template. The pairing of nucleotides is quite imprecise, and the efficiency of the synthesis is dependent on the sequence of the template—with G and U preferred.

At the beginning, a mineral matrix could perhaps serve for RNA synthesis, but it is necessary to keep in mind that many more nucleotides would be expected at the beginning instead of the canonical four today, established much later. Such atypical nucleotides, on the other hand, may have increased the efficiency of the ribozyme catalysis.

We conclude with a quote: "The discussion has focused on a straw man: The myth of a small RNA molecule that arises *de novo* and can replicate efficiently and with high fidelity under plausible prebiotic conditions. Not only is such a notion unrealistic in light of current understanding of prebiotic chemistry, but it should strain the credulity of even an optimist's view of RNA's catalytic potential." (Robertson and Joyce 2012) The hypothesis of the RNA word is very popular today; however, a plethora of problems connected with its reconstructions may one day lead to its rejection, even if its theoretical prerequisites look sound.

(iii) Ribosomes in the abiosphere? Root-Bernstein and Root-Bernstein (2015) performed a sequence analysis of genes coding for three bacterial ribosomal RNAs (23S, 16S and 5S). Ribosomal RNAs are not translated, but make up a scaffold for the proper attachment and orientation of tens of proteins comprising the ribosome; a small part of rRNA is also contained at the center of peptide prolongation. The authors discovered a surprising fact that *genes* for rRNAs (more precisely, both strings of DNA, not only that which is transcribed) contain sequence motifs homologous to genes coding for several tRNAs, tRNA synthases, ribosomal proteins as well as other proteins involved in ribosomal function. Moreover, such coding sequences can be found at many places overlapping over all six reading frames of rRNA genes (Fig. 2.9).

The discovery is highly surprising. As rRNA is not translated, there is no reason for it to contain any reading frame, let alone tens of them. Second, such reading frames also occur in the complementary string of DNA coding for the rRNA gene, i.e., that part of the gene that is not even supposed to be transcribed. Third, there is the problem of overlaps of reading frames: in such cases, a single string of DNA can code for 2–3 proteins of markedly different sequences and functions (if any). Such overlapping genes have been found in some viruses (Keese and Gibbs 1992) and interpreted as necessary to save space in the virion capsid, and yet, any explanation of how such an arrangement came into existence is extremely difficult because, in principle, it does not allow for evolution by mutation-selection; thus an overlap in all six reading frames is extremely rare.

Do we witness here a palimpsest of a structure that was capable of self-assembly under prebiotic conditions—a proto-ribosome that would be a replicator while simultaneously representing an extremely compressed genome? Or is it not a genome *sensu stricto*: could 'rRNA' merely provide highly specific docking sites for (randomly emerging) polypeptides from elsewhere? The last conjecture may be true: contemporary proteins whose parts are coded by the rRNA gene show affinity to the very sequence coding for them—also a very unusual finding (Root-Bernstein and Root Bernstein 2016). As discussed above, the space of protein appearances (conformations) is rather limited, so prebiotic processes could have involved the co-evolution of replicators with specific peptides. What is surprising, however, is

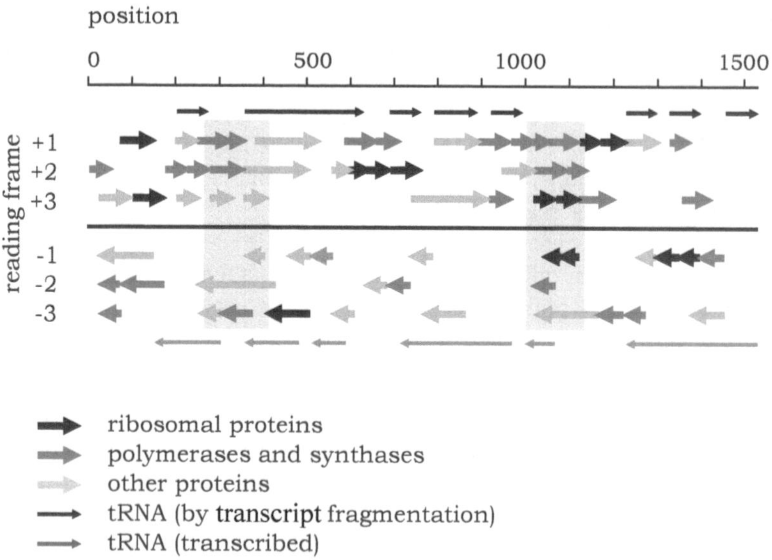

Fig. 2.9: Reading frames on the extant 16S rRNA. The sequence contains motifs homologous to genes coding for many tRNAs and sets of proteins connected with translation (arrows). The horizontal line divides complementary DNA strands of the rRNA gene. Gray columns highlight areas of very dense overlaps of coding sequences. (After Root-Bernstein and Root-Bernstein (2015).)

the fact that the above-cited works by Root-Bernstein and Root Bernstein seem to be almost neglected (7 and 13 citations, respectively, on PubMed in February 2018), despite the fact that we are experiencing a boom of RNA research in cell signaling and regulation (Schimmel 2018).

The replication of RNA certainly could also involve 'reverse transcription' into DNA, with a subsequent emancipation of DNA as information storage, with 'itemizing' strings and 'zipped', or condensed information present as rRNA, into autonomous genes. Of course, the whole story contains many questions, the most important being the delivery of anaplerotic components necessary for synthesizing the polymers involved. Last but not the least: what is so specific in the ribosomal structure that it has remained highly conservative up to our days?

3

Life from Nonlife

Establishing Rules and Codes

That the ring shall be made on turf, and shall be four-and-twenty feet square, formed of eight stakes and ropes, the latter extending in double lines, the uppermost line being four feet from the ground, the lower two feet from the ground. [...]. *That a blow struck below the waistband shall be deemed foul, and that, in a close, seizing an antagonist below the waist, by the thigh or otherwise, shall be deemed foul....*

—London Prize Ring Rules of 1838

This chapter attempts to develop a basic outlook on extant life. Its evolution can be illustrated by the reciprocal dialectics of *autonomous agents* and their *biospheres*, as presented by S. Kauffman (2000). An autonomous agent is any entity that is able both to perform work and to reproduce (it is a *replicator*). To accomplish both tasks, however, it must master communication with its environment as well as command its internal memory and experience; in other words, we argue that it must be able to build an internal representation of its world. Autonomous agents live in biospheres, communities both of their kin and of agents belonging to different lineages, each with their own representation. Evolution is accomplished by *negotiating* (adjusting) transitions into the *adjacent possible* (of both the biosphere and its agents). Such negotiation will lead to a variety of consequences: optimal and/or novel solutions, but also to the emergence of dead alleys, paradoxes, conflicts and catastrophes. Those who negotiate must, of course, understand their partners, at least to some degree. Kauffman offers examples of human individuals and their econosphere as well as cells and their nested communities[4]—ranging up to the whole biosphere.

[4] Nested in the sense that a community of cells may become an agent of its own, i.e., a multicellular body, a lichen, or a holobiont.

What is essential for the functioning of the arrangement is the *closure* separating the interior of the agents from the external world. A sharp boundary delimits the qualitative difference of both spaces (as to their compositions and workings); at the same time, it allows communication between both.

It is important to discern the Kauffmanian model from two seemingly similar systems. The first is composed of thermodynamic systems with microstates and macrostates. Atoms and molecules are not autonomous agents, they are inert and do not change in time, but are pushed and pulled by external forces. The macrostate of the system (temperature, pressure, etc.), then, is a physical contraption based on the statistical properties of atoms involved, and on the energetics of the environment—it is not negotiated. This also holds for self-organizing dissipative systems that—in some aspects—may resemble biospheres.

The second model is a neo-Darwinian mechanistic view of cells as machines 'made' by evolution: "We are survival machines—robot vehicles blindly programmed to preserve the selfish molecules known as genes," proclaims R. Dawkins (1976) in the preface to *The selfish gene.* Cells in such a model have no interior and no say in their fate; i.e., they do not negotiate anything: things simply happen to them—as well as to their communities. We do not agree with these conceptions: in our view, the origin of life was a singularity marked by a parallel appearance of cells and their communities, not robots. At this place, however, we discuss mainly 'synchronous' properties of cells, leaving evolution for the next chapter. We begin with a short characterization of proteins.

Proteins

Even if, as stated in the previous chapter, short random polymers of amino acids may have satisfied the demands of abiospheric structures, life is possible only with a sophisticated genetic apparatus which saves—with utmost precision—information about the sequence of amino acids in proteins, and which enables translation of such information into the protein molecule. This also allows the attaining of much greater sizes of protein molecules—often up to several hundreds of amino acid residues in the sequence (the average is about 500). Normalization and reliability of the whole process guarantees its repeatability in terms of the identity of copies and of their functioning in a given task. Briefly, thanks to its specific binding center, a protein is able to recognize, out of the background of thousands of different molecules, its *ligand* (a small molecule, a specific part on other protein(s), a photon of a given wavelength, mechanical inputs, etc.). Such an interaction leads to a change of the overall conformation (shape) of the protein or protein complex, and thanks to such change, it induces a specific *action* exerted to a ligand (e.g., catalyzing its chemical transformation; transporting it from the site A to B; amplifying the message carried; establishing protein complexes, etc.). Moreover, the rest of the huge molecule is prone to fine adjustments, in order to change the affinity of ligand binding, or the flexibility or speed of protein action (see Chapters 6 and 7). Such adjustments can be attained by changing the physical environment of the protein (as by temperature, or acidity), or by the presence or absence of some specific signaling molecule(s) that bind to a specific (allosteric) site(s) on the protein molecule (different from its active center). The third option is chemical modification of amino acyl residue(s) in the protein, by some specific action of other proteins.

'Universal Cell'

We suppose that the reader is acquainted with the basics of cell biology—what we call here the Norm. What follows is only a short summary of crucial, defining traits of *any* cell, traits that had to have existed from the very singularity of life's emergence. Even if some items from the list below may have appeared in the prebiotic world (e.g., membranes, metabolic pathways, ribosomes: see Fig. 1.1, p. 4), the origins of life mark the appearance of new structures and functions, such as proteosynthesis and genetic regulation; protein assemblages such as nucleosome, metabolon, signal particles, cytoskeleton; and structures such as spliceosome, kinetosome, chromatin, cell wall, organelles, etc. The most important invention, however, was systems of symbolic communication that coordinated internal and external events. All this constitutes an entirely new quality—autonomous agent or, for purpose of this text, simply cell—incommensurable with physical systems in its characteristic qualities.

First, cells never come to existence *de novo*; they are descendants of existing cells—of their division or fusion (as in, e.g., fertilization). Thus, the cell represents a basic type of *replicator*, producing entities similar (not identical) to itself: it is apt to pass to its progeny its constituents and/or instructions by which it might produce such constituents in case of need. Here we use the term 'replicator' in a sense different from that of molecular biology, as inspired by R. Dawkins. For us, a replicator is an entity that can *organize* its own replication: it is not a macromolecule, a virus particle, a sheet of xeroxed text, and the like. Our usage differs in yet another sense: as nothing in this world is digital, not even DNA, our view is that descendants are never identical (Markoš et al. 2009, Markoš and Švorcová 2009, Chapter 6 in this book).

Cellular constituents are either replicated, or built *de novo*, but never exist as individual replicators, independent from cellular whole. Other domains of life can also exist as replicators of some 'higher order', but these always contain cells; in these situations, cells replicate as constituents, in accordance with the rules of a given 'lineage' or community (culture). Thanks to this uninterrupted chain of being, cells harbor (and to some extent modify) the memory of the experience of the 'lineage'.

Second, as mentioned above, a necessary condition for the existence of cells is their closure against the external world (mostly, but not always, including other cells). The primary role is played by physical barriers, especially membranes that surround the internal milieu of the cell to confine—though not absolutely—the integrity of its overall material composition: metabolic regime, genetic thesaurus, structures and their dynamics, and signal transmission; practically, no parameter of the interior is present outside, and vice versa. The membrane is also the boundary maintaining the difference of (electro)chemical potentials of many chemical entities (ions, metabolites, metabolic intermediates, etc.). The ability to form membrane vesicles, which enable endo- or exocytosis of various materials, is also important. The membrane also protects the interior against the majority of unwanted parasitic elements from the surroundings.[5]

The third power of cellular closure is in information processing, semantics and semiosis, conspicuous features that could have appeared *only* with the first cells. The establishment of different codes and the interpretation thereof and therein is definitely one of the achievements of life—and life alone. The genetic code comes immediately in mind, but creating, receiving, and interpreting various messages is a common property of life in any domain of organization (for reviews on various code systems, see the special issue of *BioSystems* 2/2018, e.g., Maraldi 2018 for lamins, Gabius 2018 for sugars, Buckeridge 2018 for extracellular matrix, Prakash and Fournier 2018 for histones, or Marijuán et al. 2018 for prokaryote communication). Most proteins (up to 90% in eukaryotic cells) are engaged in information games organized into complicated networks. Coding systems represent the software (or better: wetware), which enables and supports automatic processes (e.g., translation). In contrast, semiotic processes, i.e., the extracting of meaning from a situation, also require memory, experience, and knowledge of the overall situation (e.g., timing of the ontogenetic stage) of a given organism. A fruitful analogy is provided by the difference between programming (machine), and spoken languages.

Information processing triggers cell's responses to a great manifold of cues: messages must be recognized, amplified, and confronted with the overall context (i.e., the presence of other signals and physical cues) with the past memory and experience (embodied in cellular structures and their setup), to steer cellular behavior towards particular ends. As in the case of genetic machinery and metabolism, some of these signaling pathways are very old—hence used in many, or even all, lineages of life. Where they differ is in the input and output of such pathways, and the possibility to set sensitivity thresholds to various stimuli. This never-ending whirl of activity ensures crosstalk among the signaling pathways, as well as mutual enhancement, attenuation, etc.

This brings us to a very important point: the conservative character of regulatory/ signaling pathways suggests that members of different 'lineages' of life may be able to exert influence upon each other because they, at least to a certain extent, *feel the presence* of other forms of life, and are able to establish, at least rudimentary, or even weird, ways of communication. Their regulatory pathways have the capacity to become mutually coupled, able of cooperating, exploiting, cheating, etc., when confronted with their partners in different communities (examples in Chapters 7–10). Symbioses may well depend on such interactions rather than on metabolic influence; wherever we look, we find mosaics of communities of cells, multiplex in structure. Bacterial colonies and mats, multicellular organisms, cancer cells in such organisms, intracellular symbionts, holobiont: all represent examples of what Darwin (1876) tellingly calls a tangled bank.

[5] Horizontal gene transfer (HGT) is an important way of spreading information between different cellular lineages. As in the case of the internet, however, such information flow may contain parasitic spam or malware and must be controlled. See the next chapter.

Life Is...

As biologists, we often feel offended by what many philosophers and scientists call the 'machine (or program) metaphor of life'. Very early, one comes to understand that for many this is even not a metaphor, but a strong ontological statement. Take a quote randomly chosen from a single issue of *Science*: "Cells are complex machines constructed from genetic blueprints generated by genetic mutations and evolutionary forces, whose information is expressed under the influence of cellular environment" (Horwitz and Johnson 2017). On the opposite end of the scale, we commonly see a continuum of cosmic evolution; for example: "There's a continuum between life and nonlife, and the black and white distinction between the two has to be minimized" (Linewaver 2006).

Such views are deeply ingrained into the community of biologists, according to which, species, individuals, cells, etc., remain locked in strictly defined (or definable) and unique umwelten, and evolutionary change is driven, or supported, only by random *physical* events: mutation and selection on one hand, and chemistry, energy and thermodynamics on the other. Elements of 'Platonism', i.e., reliance on a single, basic level of an immutable element, are present in the thinking of founders of both molecular biology and ethology. J. Monod explicitly claims: "In science there is and will remain a Platonic element which could not be taken away without ruining it. Among the infinite diversity of singular phenomena science can only look for invariants." (Monod 1972) For K. Lorenz, mutation and selection represent "the great 'constructors' which make genealogical trees grow upwards." (Lorenz 1966)

From such vistas, living beings offer a view of an embodiment of virtual information (a Platonic ideal), of a set of instructions of how and in what order to perform chemical reactions. Ontogenesis and flexibility are merely questions of subroutines, which are likewise set in advance. Also, Darwin's Descent with modification (Chapter 5) is gnawed clean to bone-white, user-friendly and simple to handle, standing before your eyes in its totality.

What is left unnoticed is the fact that not only such 'bones', but also living flesh must be present for life to thrive, that it necessarily must *understand and interpret* such instructions. As is obvious from what was said earlier, we take the living being as an uninterrupted interplay (i.e., evolution) among different domains of organization, each with its own 'logic', memory, experience, quest for novelty, etc.

Pillars and Itinerary of the Biosphere

The most important pillar is undoubtedly, as discussed above, the fact of closure: the sharp divide of inner events from external milieu, with a plasma membrane at the interface. Disrupting the membrane, hence destroying the closure, inevitably leads to the death of the cell. A cell is usually also closed towards other cells: cells communicate via metabolites, signaling molecules, and/or a mutually constructed scaffold of an extracellular matrix. Yet cell-to-cell closure is often surmountable; the anatomy of vascular plants may give a good illustration (Fig. 3.1). Intracellular dwellers (parasites or other symbionts) may also serve as an example of cross-species

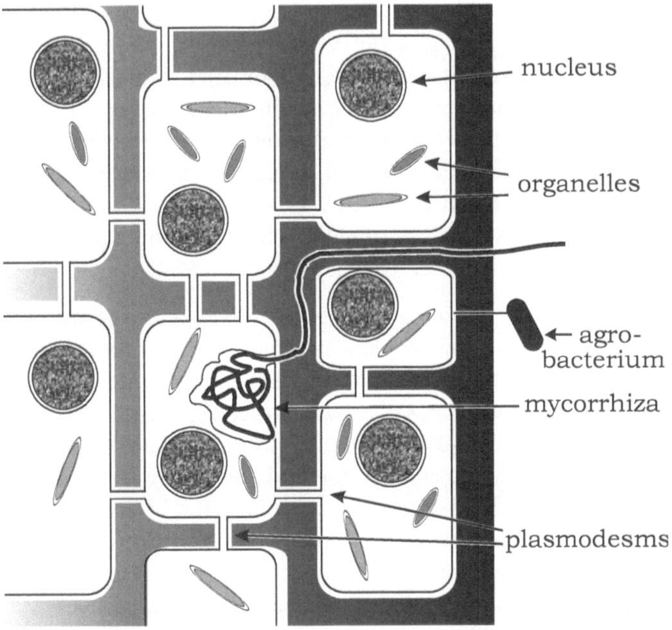

Fig. 3.1: A schematic section through a plant root. Plant cells as rectangles delineated by plasma membrane (thin line) against the dark background of extracellular matrix (that includes cell walls). The closure of individual plant cells is partly—and in a controlled way—upset by plasmodesms—cytoplasmic bridges connecting them. Membrane-bounded organelles in the cytoplasm lost their closures almost completely and became part of the cell. A thread of symbiotic mycorrhizal fungus enters deeply into the cytoplasm of plant cell, yet both organisms strictly observe their closures. The parasitic bacterium (*Agrobacterium*) builds a special bridge interconnecting cytoplasm of both organisms. Intracellular symbiotic bacteria are not shown.

breaking of the closure: they thrive in the cytoplasm of their host. Such symbioses are frequent, and often very intimate.

The plant model as sketched in Fig. 3.1 gives a more general illustration of cell interactions based on intimate contacts.

(i) Plasmodesms—channels interconnecting the cytoplasm of most plant cells. Topologically, a plant is thus a single multinuclear cell, only locally interspersed by 'normal' cells (e.g., stomata). The break of closures is, however, not absolute: plasmodesmal transfer is strictly controlled. Even so, some elements (e.g., viruses) can freely penetrate from cell to cell: any such possibility may become dangerous. As gardeners know, the network can also be easily coupled to foreign grafts. So too, a parasitic plant, dodder (*Cuscuta*), grafts itself to its host in just such a way. Through such interspecies plasmodesms, both partners exchange thousands of different transcripts (Kim et al. 2014, Shahid et al. 2018). Densely grown non-parasitic plants of different species (e.g., trees in a forest) can also be interconnected through their roots.

Of course, cytoplasmic bridges (called also gap junctions) are not restricted only to plants. Animal oocytes are coupled to their nurse cells. The fact has been known in insects for several decades, but recently was also described in mice, where even organelles can be supplied in such a way (Lei and Spradling 2016). Bacteria are not an exception: metabolic or electric interconnections via 'nanowires' have been described in sediment or soil bacteria (Ntarlagiannis et al. 2007, Pfeffer et al. 2012, McGlynn et al. 2015). Coordination in multicellular bacterial bodies such as colonies or mats may also involve such coupling (Gorby et al. 2006, Reguera 2009). In animal cells, there exists yet another form of cell-to-cell communication: cell threads (or better tubes—cytonemes; Inaba et al. 2015, Osswald et al. 2015, Baker 2017). Such conduits between cells can make coordination easier, without the need of developing extracellular signaling devices (see Chapter 7).

(ii) Mycorrhiza. Another example of cell-to-cell contact, drawn in Fig. 3.1, is arbuscular mycorrhiza—a form of plant symbiosis with a fungus. Fungal hyphae (which are much thinner than the diameter of plant cells) penetrate the root cells where they jointly produce a mycorrhizal structure, which assists the plant in extracting nutrients from the soil, while also providing nutrient for the fungi. Yet, in spite of such a very intimate contact, the cytoplasms are not interconnected as in the previous example: their interactions take place solely via plasma membranes, across cell closures. Fungal mycelia have the capacity to interconnect all plants at a given locality (a forest, a grassland, etc.; Klein et al. 2016). Behie et al. (2012) even report a curious symbiosis with the fungus decomposing a cadaver and conducting its nitrogen compounds to the plant. The capacity of fungi (and also bacteria, as shown in Chapter 9) to establish symbiotic assemblages is easily malleable and opportunistic: for example, Hom and Murray (2014) established a tight symbiosis of an alga (*Chlamydomonas*) with yeast cells (*Saccharomyces*) and even with a filamentous fungus—simply by co-cultivating the organisms. We will return to mycorrhiza and other symbioses in Chapter 9.

The examples we give in this section are drawn from the frequent cases of symbiosis where cellular closure is not breached, and intimate cooperation is established by message transmission and interpretation (via hormones, pheromones, cell-to-cell contacts, historically established coordinates, etc.).

(iii) Parasite. Our list will continue with symbioses where cell closure is transgressed. Such an example of cell-to-cell interaction is represented in Fig. 3.1 by the infection of roots by the parasitic bacterium (*Agrobacterium*). The bacterium breaks both metabolic and genetic closures by building a special channel to interconnect both cytoplasms. Special bacterial genes then enter the plant cells, and instruct it to synthesize a special nutrient for the bacterium. Here we have an example of horizontal gene transfer between unrelated 'lineages'. Such genetic transformation (i.e., transgressing the genetic closure) is common (Gladyshev et al. 2008, Schönknecht et al. 2013b, Crisp et al. 2015), especially in prokaryotes: it has been estimated that in pelagic oceanic bacteria, about 5–7% of cells are genetically transformed during a single life cycle (Proctor and Fuhrman 1990).

(iv) Intracellular symbionts. An extreme case of intracellular symbiosis is the turning of intracellular symbionts (bacterial or eukaryotic) into an organelle, and the concurrent loss of the symbionts' ability to live or disseminate outside its host. The organelle maintains only the remnants of the original closure of its ancestors. Such symbiogenetic events seem to be extremely rare in evolution—only mitochondria and a handful of plastid types are known (see the next chapter).

(v) Virions do not have their own closure (i.e., they are not alive), so their entrance within the cell enclosure does not establish symbiotic (parasitic) interaction. Viruses may, however, reprogram the cell and turn it into a 'virus factory' (Forterre 2010a, 2013) and eventually even destroy it, as is the case of phages in lytic phase of their dissemination.

Biospheric Games

In this part of our narrative, we deliberately leave aside energy metabolism and metabolic pathways. Whereas these may have played a dominant role in the abiosphere, in the biosphere they remain necessary, but a condition of life that is far from satisfactory. Here we concentrate on the added value that could and did appear *only* with life, as it requires a sophisticated and dependable protein ecosystem.

As mentioned, the common origins and universality of shared tools (protocols) allow the inhabitants of the biosphere some degree of *mutual understanding*. On one hand, the shared, universal protocols define the cell itself; on the other, it allows every cell (individual) the ability to reach out and establish an overlap with other living beings thus enabling communication. Only in this way is the biosphere able to *negotiate* its 'adjacent possible', i.e., its immediate future state (see above; Kauffman 2000) and thereby avoid resemblance with sand dunes dragged by chaotic blasts of wind. Below, we present some of the communicative games available within the extant biosphere, and with this palimpsest at hand, we shall, in the next chapter, attempt to reconstruct the first biosphere and chart its further development.

Hairballs. Recent years have been characterized by variations on the slogan: "From bean bag genetics to hairball biology". The bean bag represents a caricature of genes (and their proteins) as autonomous units working as if on their own, with 'soluble' proteins interacting only through their metabolites or other products. The hairball points at multiprotein complexes working with inputs and outputs of whole pathways, as ribosomes, transcription initiation complexes, metabolons as respiratory chain, fatty acid oxidation, signal particles, cytoskeletons, etc. (for just a few examples of such structures and approaches, see Mahamid et al. 2016, Nixon-Abell et al. 2016, Thul et al. 2017, Valm et al. 2017, Banani et al. 2017). The same holds for chunks of organelles, of cells in tissues, etc. What is new is the recognition of such hairballs as very conservative sets common to big groups of lineages (Boyle et al. 2014, Wan et al. 2015), interconnected within 'small world' networks through cell-, species-, lineage-specific connections (Costanzo et al. 2016, Valm et al. 2017, Turgay et al. 2017, Sutter et al. 2017, Thul et al. 2017). Changing such networking may result in conspicuous shifts in cell differentiation, disease, ontogeny, or even evolution: the same basic modules interconnected differently thus become involved in different tasks (Huttlin et al. 2017). In Chapter 6, we suggest extending the rather disgusting hairball model to a more palatable hairdo metaphor, as the structure and function allow for quick epigenetic transformations of such modules (Fig. 6.3, p. 103).

Membrane metamorphoses. The dialectics of membranes as impermeable barriers between compartments, yet a seat of specific gates allowing communication between such compartments, plays a decisive role in maintaining cellular and/or organelle closures. Here we shall focus on another aspect of membrane dynamics: fusion and detachments of membrane vesicles which enable yet another type of communication.

The endomembrane system (interconnected with the cytoskeleton) secures various forms of budding, movements, cell division and/or fusion (as in fertilization), exocytosis and endocytosis of vesicles with a plethora of contents: food vacuoles, intracellular symbionts and/or viruses, etc. A bizarre example of such traffic is endocytosis of bacteria and even yeast cells by plant roots (Paungfoo-Lohnienne et al. 2010).

Exomembrane derivatives are represented by enucleated cells, such as mammalian erythrocytes or blood platelets. What is even more important is cisterns emptying their contents into surroundings (as synaptic vesicles in neural connections), and export vesicles pitched off from cells, observed in all three domains and abundant in some environments (Biller et al. 2014, Melentijevic et al. 2017). Virions also leave their host cells in a vesicular emballage. It turned out that the primary tumor sends such vesicles around the body: they lower an anchor in some tissues, manipulate its cells, and thus prepare a niche for the metastatic cells that come in the wake (Kaiser 2016).

Cytokinesis and cell fusion (as in fertilization, or plasmodia) also belong to this category of membrane activity. Of course, all such movements are enabled through coordination with the cytoskeleton and membrane skeleton, with membrane motors and special protein complexes enabling docking or releasing of the vesicles.

Genetic games. The view on the function of closure can also be applied to the genetic apparatus of cells: we find strictly controlled replication and distribution of genetic information to divvy out cell progeny (the segregation of chromosomes, mitosis, meiosis and related processes). Yet, we also observe openness towards an internal reorganization of the genome (e.g., transposons), or a transfer of genetic material between cells, belonging even to different domains of life, i.e., horizontal gene transfer (HGT), or even horizontal genome transfer. From the point of evolution, the endosymbiotic transfer of genes (EGT) is likewise very important, i.e., the transfer between endosymbionts or semiautonomous organelles, and a nucleus.

As in the case of membranes, well-elaborated protocols are at work here (genetic code and translation, replication and transcription, mitosis, meiosis, recombination, conjugation, plasmid transfer, etc.). Such protocols are shared across the biosphere, or at least through a substantial part thereof. We stated above that processes involving membranes are the masterpiece of eukaryotes; gene flow across the biosphere reached its heights in Bacteria and Archaea. EGT, in turn, provides a way towards new brands of cellular life.

First of all, let us consider the gene flow; i.e., the recombination of genetic material in frames of cellular genome or genomes: these processes embrace sexual processes of crossing over, plus various non-standard somatic recombinations (insertions, deletions) caused by inherited malfunctions, 'jumping genes' and exon shuffling, viral insertions or deletions, etc., as well as random mutations.

To offer an example, the LINE-1 transposons, occupying about 17% of human genome, are the only units in the genome capable of autonomous mobility between loci; the rest are, at present, dormant, controlled by the cell. But even LINE elements are to a great extent regulated and are not allowed to jump at random (Liu et al. 2018). Here, and elsewhere, epigenetic silencing is often at work (see the example with palm oil, where just such silencing failed; Ong-Abdullah et al. 2015 and Chapter 6). Even more elaborated is the decision between lytic and lysogenic phage regimes in bacteria. It turns out that the bistable element regulating the entry of the virus to this or that phase (Ptashne 1987) is itself regulated by phage-specific short peptide signals released from cells bearing a lysogenic phage. The concentration of the signal in the environment may bias the regulators in infected cells to switch into the lytic phase of the cycle. We are confronted with a peculiar quorum sensing system (arbitrium, as called by the authors; Erez et al. 2017); for quorum sensing, see Chapter 7.

As to mutations, cellular repair systems usually keep such random disturbances at a very low level. Only within strictly defined conditions and at defined parts of the genome is mutagenesis allowed to rise to several orders of magnitude. For example, such experiments are allowed in vertebrate locuses where new genes for antibodies and T-receptors are constructed. In bacteria, the SOS reaction is triggered by conditions incompatible for life: it consists of extensive mutagenesis launched in the hope of finding a solution to overwhelmingly threatening circumstances. Most of such experiments are lethal, yet even mutated DNA released from dead cells may serve the survival of other bacteria—see the discussion of transfection below.

The redistribution of genetic material between genomes dwelling in a shared cytoplasm is even more interesting. Of particular importance is recombination between plasmids, especially in prokaryotes: plasmids can be made 'to order' so as, e.g., to confer multiple resistance to antibiotics, create pathogenicity islands, or enrich metabolic pathways—thus enabling the bearer to overcome some adverse environmental difficulty or expand into a new niche. A plasmid or its part can be also (reversibly) inserted into chromosome(s) and become part of the main genome; on the other way round, it can gain some genes contained in the chromosome. A viral genome can also enter a cellular genome(s) and become latent, in the sense that in such a phase it presents no danger to the cell; it can even render the cell immune to other viruses (as in case of lysogeny or retroviruses).

Shapiro (2016) points out the high plasticity of genomes, not by chance mutations, but through sophisticated tools available to the cells. Whereas the classical view treats genomes as if analogous to a computer's ROM (read-only memory), with similar deleterious effects when mutations occur, his model of genome is rather a 'read-write' (RW) analogy of computer functioning. Shapiro argues that the most genome evolution is of a RW mode, i.e., life is not waiting on the mutational jackpot, but steers its evolution actively. While writing this text, we came across a felicitous editorial in *Radiology* (Gatenby 2017): the author comments on the surprising findings that not only cancer cells, but also normal somatic cells accumulate hundreds or thousands somatic mutations during the lifetime of the body. The difference, the author argues, is in the fact that somatic cells do not compete for proliferation, being sensitive to, and disciplined by, the local niche in which they live: "cancer cells can display a seemingly paradoxical state in which their mutational burden is similar to and perhaps even lower than that of adjacent normal cells." A cancer cell may differ by "simply accessing the vast information stored in the human genome that, for example, is normally deployed during wound healing or fetal development." Research, however, is focused on sequencing only tumor cells and their clones, while ignoring 'normal' surrounding tissues, the author says. In the light of the fact that somatic cells (and probably cancer cells) may sometimes even find their way into gonads and become gametes, they may provide a powerful, however infrequent, tool for the evolution of the lineage.

In eukaryotes, the vertical transfer of genes was marked by the invention of sex. In the frame of a species, sexual intercourse leads to a unique genotype (caused by crossing-over and/or random distribution of homologous chromosomes)—within a set of identical or very similar genomes—with unique individuals resulting from the process. Hence, the process maintains the species-specific gene pool, yet allows it endless variations. Sexual processes, however, may also play 'Chinese whispers' (in the U.S., 'telephone') in interspecies hybridization processes, which are typical in plants.[6]

Horizontal gene transfer, i.e., the transfer of DNA not in terms of a single cell, nor in progeny, but by trading among contemporary dwellers in the biosphere (as in case of *Agrobacterium* above), is an extremely important biospheric process for all its cohabitants. The transfer is mediated, in prokaryotes, essentially, in three ways. In the process of conjugation, plasmids or parts of a chromosome go from one cell to another (transduction) through a special channel built for the purpose. In the process of transfection, viruses can hike fragments of DNA from their previous host. Finally, through the process of grazing (or scavenging), DNA released from dead cells (transformation) joins in the function of living cells. In the light of horizontal gene transfer, some authors speak of a pan-domain genome of the biosphere (Schönknecht et al. 2013a, b, Crisp et al. 2015). It turns out that animal genomes contain tens to hundreds of thus acquired genes (not to mistake them with EGT mentioned above); the group Bdelloidea, with about half of its genes accumulated in this way, provides an extreme example (Gladyshev et al. 2008). In other animal 'lineages', such transfers are much slower, yet through longer periods, they are frequent enough. Crisp et al. (2015) concluded: "Although observed rates of acquisition of horizontally transferred genes in eukaryotes are generally lower than in prokaryotes (but horizontal *genome* transfers may occur, at least in plants, in occasional grafting; Desai and Walczak 2015), it appears that, far from being a rare occurrence, HGT has contributed to the evolution of many, perhaps all, animals and that the process is ongoing in most 'lineages'. Between tens and hundreds of foreign genes are expressed in all the animals we surveyed, including humans. The majority of these genes are concerned with metabolism, suggesting that HGT contributes to biochemical diversification during animal evolution." In other eukaryotes, the situation may be similar.

[6] Take a set of species A, B, C…X, where each can give fertile hybrids with its neighbors in the row (i.e., B with A and C, but not with, Q); through such hybridization chain, genes from A can reach the species Q (and vice versa).

The acceptor of new genes may get a sudden advantage in a given environment: many eukaryotes thus received genes for antibacterial proteins (Chou et al. 2015). In this way, ticks keep at bay bacteria of genus *Borrelia*. Similarly, the extremophile red alga *Galdieria* thrives in an acidic environment containing heavy metals, thanks to genes acquired from extremophile bacteria: about 5 per cent of its genome code for the proteins that enable its metabolic detoxification, all acquired from prokaryotes (Schönknecht et al. 2013a, b).

As prokaryotic genomes are rather small and cannot bear too many genes, HGT enables them to use such universal protocols to 'google' required information out of the biospheric web. New techniques that enable the sequencing of genomes of single cells reveal that bacterial populations, previously taken for clonal, in fact represent a very heterogeneous genomic mosaic. The frequency of allele transfer and recombination in prokaryotes (and to a lesser extent, even asexual protists) is comparable to vertical transfer in sexual eukaryotes (Shapiro et al. 2012, Kashtan et al. 2014, Rosen et al. 2015). In a commentary on a paper by Shapiro et al. (2012), Papke and Gogarten (2012) state: "When 99% of genes from a population of very closely related strains do not have the same common ancestor, the only reasonable conclusion is that prokaryotic speciation does not have much to do with divergence from common ancestors—a startling anti-Darwinian outcome." What is surprising is the frequency of mosaicism; what is anti-Darwinian on it remains elusive to us.

The 'genetic internet' of HGT is one of the key processes that unify all cellular closures in the biosphere. A second set of such unifying processes is presented by symbolic communication.

Symbolic games. The transfer and amplification of signals and/or interpretation of signs belongs to the properties that could have appeared and been differentiated only with cells. It is estimated that the basic operations of a cell (i.e., genetic function plus metabolism) can proceed with about 600–1000 proteins. With the (generally not valid) assumption that 'one gene = one protein', we see approximately the same number of protein-coding regions in the genome. Only a handful of cells can do with such modest equipment—those that thrive only as intracellular parasites, with their host securing the delivery of most metabolites and biospheric interactions. C. Venter with co-workers (Hutchinson et al. 2016) constructed (by systematic knockouts of 'superfluous' genes not necessary for survival under lab conditions) a 'synthetic' bacterium based on such an intracellular parasite—*Mycoplasma mycoides*: the genome of this Frankenstein contains 473 protein-coding areas. It is much more than the originally estimated 250 genes (Yus et al. 2009). Why the reduction to this level did not work is not clear; moreover, in one third of 'superfluous genes' the function remained unknown—even if it is of vital importance and cannot be knocked out.

The genomes of most Bacteria and Archaea bear about 3,000–5,000 protein-coding genes, eukaryotic genomes contain usually 15–30 thousand of such regions—i.e., more than one order of magnitude than necessary for the housekeeping functions. Moreover, if we take into account possible alternatives of transcriptions, alternative splicing of transcripts (especially in eukaryotes), and post-translational modifications of proteins, a single gene can code for a great variety (even tens) of different proteins working differently and in different contexts (Pelechano et al. 2013;

more in Chapter 6). The human genome contains about 20 thousand protein-coding genes; about 10 thousand of the set is expressed in particular cells. It follows, from what was said above, that about 90% of gene products, and a much greater fraction of protein variability, serves functions other than housekeeping (e.g., Kuzmin et al. 2018); in prokaryotes, the fraction is smaller (50–80%); when in need, however, a prokaryotic genome is able to adjust quickly (via HGT) from biospheric depots.

What is the function of all that protein richness, if mere several hundred is enough to support basic survival and propagation of the cell? It serves to broadcast, receive, amplify and coordinate information flow (crosstalk), and thus enable the cell (organism, community…) to adjust its functioning in accord with the overall context (Langeberg and Scott 2015). The network is sprawled inside the cell and across the whole biosphere—with an abrupt qualitative change at the boundary of the closures. The web thus embraces all its extant dwellers (who bear the memory and experience of their predecessors) and their manifestations, which are prone to be intercepted and interpreted. It contains contributions to the web from cells, organisms, and communities, as signals (hormones, pheromones, flashes of light, etc.), and also extracellular matrix and mobile genetic elements, etc. (Chapter 7). Finally, yet importantly, physical factors (temperature, length of daylight, humidity, etc.) also play a role as such signals and signs. The whole of the biosphere, as well as cells, either reacts passively, or changes state according to context. In cells, such adaptation very often leads to activation/inhibition/modification of the composition and activity of protein cohorts, or their interconnections. For this to work, the partners in a biosphere must know each other to *some* extent (see the overlap of *umwelten*, Chapter 4) and employ such understandings in various setups—e.g., altruism, parasitism, apoptosis, or anabiosis. Below, we mention some biospheric signals shared in this way; the cell must *somehow* perceive such cues, and distinguish them from the background surroundings.

The first group of cues consists of physical impacts (light and temperature, compounds, electric pulses, etc.) that by their mere presence or absence, by gradient of concentration, frequency of impulse or oscillation, mutual proportion to other such effectors, etc., elicit some specific reaction in the receiving cells/organisms. The response is not passive (as when a body gets warmer when exposed to sunlight), but actively calls up some specific behavior. A second group, often, but not necessarily coupled with the previous one, is signals—hormones, pheromones, kairomones, scents, displays, mimicry, and bioluminiscence. These are often produced with great effort and through special metabolic pathways.[7] Special compounds that are not metabolites are much more suitable for the task. They can also occur in concentrations much lower (often several orders of magnitude) than metabolites. Their often

[7] Some role is played also by common metabolites such as glucose or particular amino acids. These are, however, quite unreliable, as their concentrations may fluctuate in response to the nutritional state of the organism, i.e., not in connection with some specified, more sophisticated, messages. Moreover, metabolites generally occur in relatively high concentrations, which are usually not amenable to signaling specification.

bizarre molecular structures, they can easily be recognized on the background of thousands other molecules, signals or not. When they exert their influence as signals, they necessarily elicit an unequivocal response. In contrast, messages requiring interpretation (signs) will also be confronted with additional factors present in the memory and experience of the receiver. Hence, the challenge a sign offers can be obeyed, disobeyed, circumvented, etc., in endless games of cheating, coercing, imposing, etc. (Cohen et al. 2017).

The carriers of such information are not as universal as the genetic code, yet they are shared within big groups of organisms, often unrelated; let us mention pheromones of quorum sensing in bacteria, or the universal usage of many eukaryotic hormones (Chapter 7).

Symbiotic games. The principle of symbiosis is a necessity—for partners to understand each other—to participate, at least marginally, in each other's information web; from negotiations of inhabitants of, say, a meadow, through different tight liaisons as in mycorrhiza or rhizosphere, up to sophisticated host-parasite interactions and intracellular symbioses. Below, we list just a couple of examples, more details in Chapters 9 and 10.

The symbioses of Bacteria or Archaea with a eukaryotic host are pervasive and often obligatory (for review, see Sachs et al. 2011). Textbook examples include the symbiosis of *Rhizobium* cohabiting with leguminous plants (Denison and Kiers 2011, Markmann et al. 2012); the light organ of squids inhabited by luminescent bacterium *Vibrio fischeri* (McFall-Ngai et al. 2012); the ecosystems of alimentary tracts (Chapter 10); intracellular symbioses in insect mycetomes (Douglas 2010); and sophisticated examples of intracellular symbioses in protists (Gast et al. 2009, Thompson et al. 2012, Zehr 2015). Kiers and West (2015) discuss the evolution of new species via long-term obligate symbioses, especially when symbionts are directly (vertically) transmitted to the progeny (as in some insects).

First, however, let us focus on multicellular organisms, where cooperation of cells is highly elaborated. In most cases, multicellular organisms constitute themselves as bodies that grew from a single cell, zygote, or spore. Such is the case of animals, plants, fruiting bodies of fungi, and some microbial colonies. Here, we pay no attention to vegetative propagation by budding, body fragments, or special propagulae as in lichens. Neither have we discussed reproduction or bodies arisen by aggregation of free-living cells—as in slime molds.

The early ontogenesis of germs is, to a great extent automatic, according to established programs, and mostly in isolation from the surrounding biosphere. Indeed, early embryogenesis is perhaps the single case that requires the strict isolation of an individual from its surroundings (Pátková et al. 2012; Chapter 6). Only later does the growing germ become busy with establishing contacts with its living neighborhood. The body is a single clone of cells; however, ontogenesis is accompanied with epigenetic process of differentiation. Differentiation is probably a proficiency of all dwellers of the biosphere, but in multicellular forms of life, we witness its elaboration to a high degree. The progeny of a single cell (zygote or spore) knows each other intimately; hence, subtle, elaborated negotiations among them can take place.

Let us focus on the phases of ontogenesis when the germ emerges from its early insulation. From this moment, what enters the world is a holobiont, a biological individual: multicellular forms of life enter, or grow into, incessant interactions with other inhabitants of the biosphere. Its resulting appearance (or phenotype) is derived from such relations (not haphazard but well elaborated): the holobiont is a new type of ecosystem (Gilbert et al. 2012, Chiu and Gilbert 2015, Gilbert and Epel 2015, Coyte et al. 2015, Rakoff-Nahoum et al. 2016, Stappenbeck and Virgin 2016; Chapter 10). We have repeated evidence that the composition and activity of holobiotic companions substantially influence the internal processes and ontogenesis of the macroorganism (Eisenstein 2016, Perry et al. 2016). The classical explanations for such coexistence suppose totally naive players who, having come into mutual contact, laboriously learn to live together through maneuvering from an incessant 'position of conflict' towards one of negotiation with the others. In contrast, we highlight the fact that such holobiotic partners recognize each other from the very beginning, thanks to common origin and long interaction throughout ages of evolution. Communication and a common understanding of experienced partners is the precondition of such a partnership.

The composition of symbiotic populations largely influences the well-being of the host (metabolism, ontogenesis, immunity, inclinations to diseases, etc.), be that host an animal or a plant. For example, the differentiation of the alimentary tract into microhabitats is reviewed in Donaldson et al. (2016), and complicated holobiotic interactions are demonstrated on germ-free antibiotic-treated animals (Kernbauer et al. 2014). For more about holobiotic interactions in humans, see Chapter 10.

Compared to animals, plants present an inside-out example of holobionts: they lack internal cavities and their symbionts are generally recruited and organized at their periphery, as rhizosphere, phytosphere, etc. The rhizosphere, which is comprised of roots and their cohabitants in soil, structures itself only through substantial effort in terms of energy expenditure, as well as special compounds such as extracellular matrix and signaling molecules, on the part of both plants and their symbionts (Berendsen et al. 2012, Zhalnina et al. 2018). Microbial mats, consortia, or stromatolites, represent in fact another version of holobionts, even if the macroorganism is not present in these cases (Chapter 9).

Semiautonomous organelles (mitochondria, plastids, and their derivatives) result from the transformation of 'normal' intracellular symbionts into a special type of organelle. This process involves (i) a massive transfer of genes into the nucleus of the host; (ii) a massive transfer of proteins into the organelle, with the establishment of a sophisticated protein apparatus enabling such import; and (iii) practically, a complete abolishment of the closure of the organelle, as to metabolism and energy production. Usually the organelle retains a reduced genome and proteosynthesis, but some derivatives of mitochondria may lack even those, as in case of hydrogenosomes living anaerobically in protists (Mentel and Martin 2008, Mentel et al. 2014). It would be a mistake to consider mitochondria and plastids as mere and miserable remnants of once full-fledged cells. Many metabolic pathways were delegated to them; moreover, a recent body of information suggests they play a role in many inherited and inborn diseases in humans.

Mitochondria and primary chloroplasts obviously represent direct descendants of symbiotic bacteria; discussions, however, persist about the characteristics of such free-living bacteria, of precursors, or even whether there existed some 'forerunners' facilitating the symbiogenetic process. In case of mitochondria, there is a consensus that they are descendants of α-proteobacteria, from the vicinity of Rickettsiae, a group containing obligatory intracellular symbionts of many eukaryotes, so forebears of mitochondria may have been pre-disposed for extending the process even further, towards an absolute inability of leaving the cell and becoming an organelle. Recently, however, Martijn et al. (2018) challenged the rickettsial origin of mitochondria, and argue that the root is to be sought much deeper, in the common ancestor of α-proteobacteria and mitochondria.

In case of primary plastids, the cyanobacterial origin is not questioned, but debates go on whether another obligatory intracellular symbionts, Chlamydia, had not preceeded them, and 'smoothened' the way for mutual tuning of presumptive symbionts. The 'pro' arguments point to the fact that plants contain proteins homologous to chlamydial ones, and speculate that in early phases both chlamydia and cyanobacteria coexisted in presumptive plant cells; other authors, however, question such conjectures (Brinkman et al. 2002, Deschamps 2014, Ball et al. 2015).

The capture and transformation of both of these endosymbionts took place at the very beginning of eukaryotic evolution.[8] No extant eukaryotes are directly descendent of cells from the pre-mitochondrial era: extant amitochondrial lineages are 'tarnished' by residues of their presence. In the lineage Archaeplastida, primary plastids were acquired somewhat later, only after mitochondria. All other eukaryotic lineages equipped with chloroplasts bear secondary chloroplasts, descendants of green or red algae (e.g., photosynthetic eukaryotes) turned into an organelle; in some lineages, the process recurrently leads even to tertiary or quaternary chloroplasts (for a review, see Bhattacharya et al. 2004, Reyes-Prieto et al. 2007). These recurrent events are ancient; they probably took place some 1.2 billions of years ago. In different lineages of such secondary (and tertiary) photosynthesizers, the original alga is reduced to various degrees: it always lacks mitochondria, but often keeps remnants of the nucleus (so called nucleomorph), and, of course, the original chloroplast. The number and size of interacting genomes thus varies between the lineages: Curtis et al. (2012) who studied secondary chloroplasts (with a nucleomorph) in Cryptophyta and Chlorachniophyta, found a complicated genetic and biochemical mosaic involving multiple and multi-site transfers of genes and metabolic pathways between cell compartments. One lineage of dinoflagellates, which champions in acquiring secondary and tertiary plastids, transformed even their semiautonomous organelles into a light sensor (ocelloid): the role of the cornea is played by the transformed mitochondrion, whereas that of the retina is formed from a chloroplast (Gavelis et al. 2015). On the other end of the scale, the genome of many protists that today lack plastids (Ciliates, Apicomplexa), witnesses a photosynthetic period in their lineage (for ciliates, see Reyes-Prieto et al. 2008).

---

[8] With the exception of some species of the genus *Paulinella*, whose lineage acquired cyanobacterial chloroplast some 60 millions of years ago (Nowack 2014).

Games played by viruses. We insert this part rather hesitantly, but the role of—albeit non-living—viruses is much greater than that of any other abiotic factor in the biosphere, and deserves mentioning.

We understand the word 'virus' to mean virions in the environment, that is, non-living particles which are carriers of genetic information and proteins required for their propagation in living cells. Such is the usual understanding; Forterre (2010a, 2013), however, points towards a 'living' form of viruses. He specifies an uninfected cell as a 'ribocell', i.e., a cell equipped with ribosomes, and an infected cell, producing capsid-enveloped virions (and often nothing else) as a 'virocell'. The virocell is, then, according to the author, the living, vegetative phase of a viral 'life' cycle. As cell populations tend to be massively infected, it comes out that about 40% of what we call bacteria have been, in fact, reprogrammed, transformed into 'vampire' virocells. Transitive forms also exist: in 'ribovirocells', the virus is latent (as in the lysogenic phase of phages), or the cell produces and propagates virions without doing so absolutely: in such cases, a special part of the cell—a 'viral factory' —often produces virions, with the rest of the ribovirocell doing its usual job.

Krupovič and Bamford (2010) argue that plasmids can be considered derivatives of viruses that lack the capsid, and only propagate via cell conjugation. From such a perspective, viruses and plasmids represent a family of genetic carriers that play a role in HGT and in regulating both size and composition of particular populations or ensembles thereof. This role is not negligible—we bring but two examples:

(i) In oceans, virioplankton (occurring in densities up to 10^{10} particles per ml) causes a decrease in prokaryotic heterotrophic production of up to 80% (Danovaro et al. 2008): viruses thus play a decisive ecological role in food webs, as well as in regulation of oceanic biodiversity (Fuhrman 1999, Wommack and Colwell 2000, Rohwer and Thurber 2009).

(ii) Endogenous retroviruses (or better, retroviral genetic elements in the genome) make up about 8 per cent of human genome; these were long considered to be remnant of ancient infections, ballast that cells could not eliminate. However, it recently turned up that they are often recruited into many cellular functions. For example, retroviral proteins assist in building the syncytial tissue of placenta (Stoye 2009, Chuong 2013). Grow et al. (2015) report that one of the 'younger' retroviruses, shared with both humans and chimps, becomes activated in the 8-cell stage of the germ, and its genes remain exprimed up to the stage of constituting the embryo proper from the blastocyst. Viral proteins confer on the embryo immunity against exogenous viral infections, and interact with other RNAs inside the cell. The role of retroviruses in antiviral immunity (the activation of interferon production) has also been observed (Chuong et al. 2016). In the light of the long-known fact that lysogenic phages protect 'their' bacterium against infection with other phages, the discovery is not particularly surprising (Chapter 10).

Strategic games. Cell replication consists in handing over information (i) as inscribed (encoded) into an easily replicable medium different from the components of the cell; and (ii) the cell itself propagates its structure in such a way as to hand off new copies to the daughter cell(s). The very simplified scenarios below illustrate some models of the whole procedure.

1. The most accepted model today states that all of the information important to building the body is inscribed in the genetic memory, with the cell executing the program. The execution itself is based on a system of organic codes, with the genetic code as the best-known paragon (Barbieri 2015). It can lead to new properties of the cell, new configuration of the wetware that has access to new subroutines of the program, etc. If the environment remains constant, and if *all* properties of the cells are stored in the inscription of the code, the only source of variability rests in random irreparable mutations of the script, which leads to novel variations in program execution, and thus, new properties of the body. Through natural selection, the environment will evaluate the new quality in terms of fitness. If, on the other hand, properties of the cell do not *all* spring up from the script, the wetware may introduce an epigenetic source of variation; for simplicity and elegance of the model, such bodily excesses are taken as resettable in the new entity arising from the replication process. From this view, a new cell or creature starts from its 'bios' but then goes through a bootstrapping process. Sophisticated models based on the assumptions above allow us to treat the cell as a code-controlled machine; the neo-Darwinian model of new synthesis is the paradigm of such models: "Never were so many facts explained by so few assumptions. Not only does the Darwinian theory command superabundant power to explain, but its economy in doing so has a sinewing elegance, a poetic beauty that outclasses even the most haunting of the world's origin myths" (Dawkins 1995, see also Dawkins 1982). Most biology lives—and puts together its interpretation of life—with an air of deep fascination with this paradigm.
2. The environment is not constant, and it changes more quickly than any possible delivery of new—and suitable—mutations. Recall that the major factors of an environment are primarily not physical (temperature, climate, etc.); on the contrary, they are constituted by other dwellers in the biosphere. The dynamics of such factors are neither easy to understand nor to program all possible answers in advance (as an analogy, consider human political and/or economical systems). The selection of subroutines may be large, yet not endless, and the cell itself decides—according to cues from its environment—which will be executed and when. The famous operon model of genetic regulation belongs to this category of models (Monod 1972, Ptashne 1987). The cell actively enters into the process through its set of receptors, regulatory proteins, and overall state—such tools are usually the products of many genes, other than those subject to regulation. For examples of such subroutines launched by environmental cues, see Gibson (2008). Note that all the subroutines must be contained in the genetic system in advance for the encounter with the environmental cue: haphazard responses do exist as well, but are not easily palatable in frames of the model; see examples below.

   Another model in this category is called the insurance effect (Yachi and Loreau 1999, Boles et al. 2004): some individuals in the population become prepared for tasks that *may* be foreseen—even if those individuals may not revert to becoming members of the majority again. As an example, we can draw an analogy with social animals with soldiers and other casts. Soldier casts in termites, ants, or mole rats come into mind, but examples abound; after all, the authors who introduced the term, and are cited above, worked with bacterial populations. Two examples:

Certain ciliates, frequently fed upon by mosquito larvae, take countermeasure for the case that larvae would appear: part of their population differentiates into forms that are able to parasitize and kill the larvae. If the larvae do not show up, such soldiers die within 24 hours. If the larvae appear in plenty, the fraction of the ciliates that take parasitic form increases (Washburn et al. 1988).

Finally, the immune systems of vertebrates may well be described as a highly complex and successful insurance policy of an organism.

3. The third category of models is the protein ecosystem in the cell and, recursively, the ecosystem of cells in bodies, and then that of individuals in communities or humans in various cultural settings. Massive parallel information processing takes place here, which leads to a great variety of behavior as well as epigenetic forms of memory—propagated through many cellular generations. The cell, and those cellular structures that cannot appear *de novo*, serve as good examples. Every cell is the progeny of an incessant chain of similar structures, from the very beginning of life.

   One of the very efficient mechanisms of epigenetic memory is the regulation of nucleic acids and proteins by covalent, mostly reversible, modifications of nucleotide and/or amino acid residues in the molecule (performed by other proteins from the ecosystem). In some cases, the protein may contain tens of different amino acid residues, instead of the original 20, as they are incorporated into native proteins. Such epimutations comprise phospho serine, trimethyl lysine, acyl cysteine, glycosylated proline, etc. (for more details, see Chapter 6). In this way, one set of identical products of a single mRNA may give rise to a plethora of different proteins involved in different tasks. What is important is that such changes are stable, yet reversible: all is in the 'hands' of a great number of proteins and structures that situate epigenetic labels on the protein molecule, or remove them according to need. Moreover, proteins usually move in teams, and the conformation of such a multiprotein particle may be subtly regulated again, often by modification of a mere single amino acid residue. Such covalent modifications, those realized by protein ensembles, could also be introduced to DNA (e.g., methylation code), RNA, or polysaccharides—see Chapter 6. The nucleosome can serve as a paradigmatic example of such dynamic protein rebuilding (Danchin 2013, Danchin and Pocheville 2014, Voss and Hager 2014, Pombo and Dillon 2015, Dann et al. 2017; Fig. 3.2).

DNA in chromatin is parsed to sections of about 200 nucleotides, wrapped around a protein core of histone octamer (four histone doublets); at this moment, we shall ignore tens of other proteins and many kinds of RNAs bound to the structure, stabilizing it or functioning in its dynamic metamorphoses. There exists around 20 histone genes, with the protein products of H2A, H2B, H3 and H4 in a routine

```
MARTKQTARKSTGGKAPRKQLATKAARKSAPATGGVKKPHRYRPGTVALR
**************************************************

EIRRYQKSTELLIRKLPFQRLVREIAQDFKTDLRFQSSAVMALQEACEAY
****************************************A*****A***

LVGLFEDTNLCAIHAKRVTIMPKDIQLARRIRGERA
************************************
```

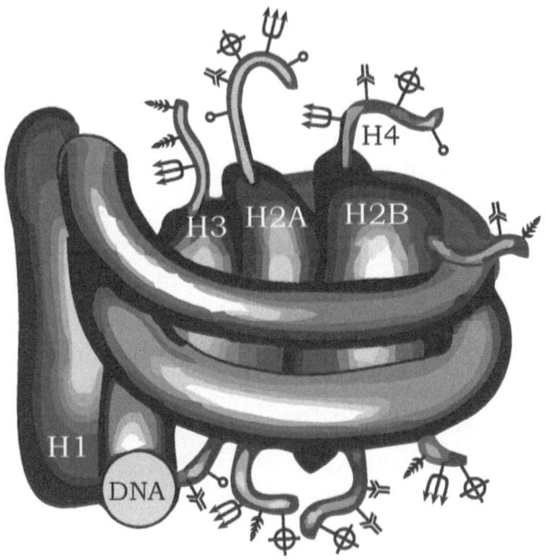

Fig. 3.2: Scheme of a nucleosome. Above: Histones are very conservative as to native proteins: comparison of histone H3 from humans (upper line) and from corn (*Zea*; lower line) reveals differences in mere two positions. **Below:** In contrast, posttranslational modifications of amino acyl residues in histone molecules are frequent and diverse. Symbols on protein ends stretching of the histone core mimic such protein modifications. Each such label contributes to the conformation of the whole structure. A stretch of DNA (about 180 nucleotides long) wraps itself around the octamere of proteins—histones.

use. Depending on the context, histones will be multiply modified, as described above (Chen and Dent 2014, Tessarz and Kouzarides 2014). In principle, each one of the millions of nucleosomes present in the nucleus may be specifically marked; moreover, this 'bar code' may, in some areas, be promptly rewritten but in other areas function as 'read only'—its changes being blocked, each according to the cellular context. Moreover, such epigenetic modifications can be preserved even to the next generation, and influence, among other things, early embryogenesis (Burton and Torres-Padilla 2014). The distribution of histone modifications is studied today even at the level of a single cell (Sekelja et al. 2015, Veluchamy et al. 2015, Bintu et al. 2016, Prakash and Fournier 2018).

We observe here the principle that constitutes the leading braid of this book: the comparatively monotonous basis (native histone, i.e., the Norm) allows a manifold of subtle baroque modifications, 'differences that make the difference'. A constant background with a creative extension is, we argue, both typical and definitional for the varying appearances of life, but also to a great extent, thanks to the common grounding that these make possible, a mutual understanding among the most varied of forms. Even bacteria are able to manipulate the chromatin code (Alvarez-Venegas 2014)—and thus we return to the umwelten and the information web in biosphere.

Throughout this book, we come upon even more examples of the consequences of shared information—based on the reciprocal shaping of proteins in the protein hairballs; here we briefly mention the games that are played at the level of nucleic acids.

Genome tinkering. DNA is often presented as a database of extraordinary stability, thanks to the powerful mechanisms of faithful replication and repair. Yet, this is not true for all parts of genome, and/or longer strings of it; on the contrary, the genome may be seen as a very dynamic entity. We leave aside the 'standard' processes such as genome shuffling in meiosis (crossing over), jumping genes of various kind, and horizontal gene transfer. Here, however, we turn to examples of building new protein-coding sequences from fragments of DNA.

We inherit about 10^4 protein-coding sequences, yet our immune system creates some 10^8 or even more new genes (for antibodies or T-receptors). This is achieved by combinatorial joining of shorter elements, each taken from plentiful sets of such sequences; in addition, extensive mutagenesis takes place at their junction. Of course, the process takes place only in cells that are to become lymphocytes, and only two such genes are constructed in each. It is the sheer number of such cells that attains the overall number of new genes.

The second example is the genome of ciliates. These huge cells cannot be served by a single diploid nucleus; many more gene copies are therefore needed. The dilemma is solved by the existence of two types of nuclei within such cells. The micronucleus contains the standard diploid genome that serves purposes of heredity; however, this nucleus mostly does not contain standard genes for protein coding. These are constructed from fragments, and short chromosomes containing several such genes are transferred into the macronucleus and massively multiply therein. The macronucleus serves somatic genetic functions, and is replaced after each sexual process (conjugation).

Whereas the examples above—however haphazard at some level—are under the thorough control of cellular processes in eukaryotes, our last example concerns the SOS reaction in bacteria. In conditions incompatible with life, bacteria activate an extensive mutagenesis in their genomes, in a 'hope' to avoid extinction. Obviously, such roulette will mostly end badly anyway, yet some individuals may win the jackpot.

As to RNA, we again recognize some broadly spread modifications, such as the processing of primary transcripts and splicing exons into mRNA with standard caps from modified nucleotides. mRNA can be manipulated by external powers connected with translation, so as to adjust the reading frame and for the beginning of translation: by 'stuttering'—changing the reading frame in the midstream of translation, or even redefining the code by recognizing some triplets as homonymous (for insertion of rare amino acids such as selenocysteine, formylmethionine, or pyrolysine). A massive overwriting of the primary transcripts, however, is very rare; yet, it can be found, as in the following two examples:

In the protistan order Kinetoplastida (trypanosomes), the mitochondrial protein-coding genes are highly incomplete, and the primary transcript must be reconstructed by a complicated set of mechanisms (Feagin et al. 1988). As an example, take the sequence:

...AG~~TT~~UUUGCAuuGuuAuuuAuuACAuuAAGuuGGG~~TTTT~~UGuuuuuGGu...

The upper case letters represent a short string from a protein-coding *gene*; the 'A's resulting from outcrossed 'T's will be deleted from the transcripts as kind of 'miniexons', whereas the lowercase 'u's will be inserted into the primary transcript—all in a strictly defined order of procedures. Only after such manipulation will the string code for a protein that is homologous to proteins synthesized from 'normal', full-fetched genes—normal, in this case, meaning common to all other living beings bearing mitochondria. By introducing such extravagant rules, the whole lineage found itself on a dead-end trajectory—yet it has persisted for hundreds of millions of years.

Our second example refers to transforming adenosine moiety in RNA to inulin (A→I). This occurs universally, and requires that about 100 nucleotides upstream and downstream of the edited site be packed into a double helix. In most organisms, the process occurs mostly in non-coding transcripts (hundreds of thousands were described in a single genome), and the reason for such modifications is largely unknown. Only extremely rarely, it occurs in the coding sequence of mRNA, where it causes a mutation of the codon: in translation, 'I' is recognized as 'G'. It was surprising then when Liscovitch-Brauer et al. (2017) described such widely distributed editing in cephalopods (in the groups including octopuses, cuttlefish and squids). About 65% of such mutations are non-synonymous, i.e., leading to a different amino acyl residue in the amino acid chain of the protein. Such alternative epimutations are especially frequent in the brain, thus increasing the repertoire of functions of such gene products, i.e., proteins coded by the same gene. The authors speculate that the process may correspond somehow to the cognitive abilities of animals. All this comes at a price of 'freezing' the evolution: the necessity of the double helix flanking the edited site hampers mutations, which is to say, does not allow the sequence evolution in such regions. As the group has been around for some 300 millions of years, it seems that no harm is caused from the 'drawback'; yet speculations may arise as to why related groups of ammonites and belemnites went extinct.

Around 150–200 epigenetic modifications on different types of RNA were described; the most abundant are methylations in adenine and cytosine or isomerization of uridine to pseudo-uridine. A great part of these modifications is reversible, and play a regulatory role in differentiation, development, or cancer (Hsu et al. 2017), all in a manner similar to the described processes in proteins.

4

Towards a New Manifold
The Biosphere

Play is always a matter of context. [...] *In play, we manifest fresh, interactive ways of relating with people, animals, things, ideas, images, ourselves* [...] *We toss together elements that were formerly separate. Our actions take on novel sequences. To play is to free ourselves from arbitrary restrictions and expand our field of action. Our play fosters richness of response and adaptive flexibility. This is the evolutionary value of play—play makes us flexible. By reinterpreting reality and begetting novelty, we keep from becoming rigid. Play enables us to rearrange our capacities and our very identity so we can be used in unforeseen ways.*

—S. Nachmanovitch 1990, 43

As demonstrated in previous chapters, life brought with it a clean emancipation from the happenstance of mineral processes: through cells *and* the symbiotic interconnections of cells, *via* structural, metabolic, *and*, before all, symbolic communication tools. Once again, we remind the reader: LUCA (Last Universal Common Ancestor) was a biosphere, not a single Cell-the-Urmother.

In this chapter, we shall follow the journey of the biosphere and notable landmarks thereof, from its very beginning some 4 billions of years ago to the present days. Equipped with a basic knowledge of extant life, we seek their possible 'antediluvian' origins. The assumption that the branches, or domains, of extant life (Bacteria, Archaea, and Eukarya) all had a last universal common ancestor, or LUCA, is solid science; it is unlikely to be overturned. However, the characteristics of this hypothetical ancestor *will* be submitted to incessant revisions and particularizations, as will the order of the branching of particular domains. Below, we provide our vision of putative early forms of life.

Dating Life's Appearance

Paleontological findings and geological proxies are very scarce, subject to many bold conjectures, and raising many controversies. Thus, the dating of the first biosphere may vary by orders of hundreds of millions of years, and so do opinions concerning its properties. The age of the oldest proxies (based on biogenic organic carbon contained in rocks) indicates that life appeared around 4 billion years ago or even earlier (Sleep et al. 2012, Bell et al. 2015), perhaps even 4.28 billion years ago (Dodd et al. 2017, Hassenkam et al. 2017). What can be interpreted as genuine fossils of (cyano)bacterial filaments is dated 3.5–3.8 billion years. Heated discussion proceeds, however, if the specimens represent fossils at all (Schopf 1983, 1993, Schopf and Klein 1992, Rasmussen et al. 2008, Wacey 2009, Noffke 2013, Hormann et al. 2016), and if so, whether they represent cyanobacteria. This uncertainty is supported by the fact that phylogenic reconstruction based on genome comparisons places a much later date on the origin of oxygenic photosynthesis—about 2.5 billion years ago (Soo et al. 2017).

Prokaryotes first? In any case, all putative fossils from the Archaean are interpreted as remnants or products of prokaryotic life—of Bacteria or Archaea. What may be remnants of eukaryotes is dated much later, some 1.5–3.2 billion years ago (note the range of scatter), as so called Acritarcha (Javaux et al. 2010). The reason Acritarcha are classified as eukaryotes, however, is merely on account that their size (tens of micrometers) falls into the range of typical eukaryotic cells. However, as referenced below, even extant eukaryotic picoplankton may be of bacterial size. On the other hand, giant bacteria can also flourish in the contemporary biosphere. For example, sulfobacterium *Thiomargarita* (300–750 μm in diameter; Schulz et al. 1999), which lives in subsurface marine sediments full of storage vacuoles, is also in the 'eukaryotic' range and may even give an impression of such. It is easily recognized as belonging among bacteria, but fossil specimens of such size would not be so easy to recognize.

Whatever fossil specimens are remnants of, they became preserved because of their cell walls. Organisms without such envelopes had no chances to leave fossil traces, even if they existed from the very beginnings of life. Zhu et al. (2016) claimed discovery of multicellular macroscopic eukaryotes (30 × 8 cm) dated 2.56 billion years; Bengtson et al. (2017) interpreted 1.6 billion years old fossils (from a different deposit) as red algae. A truly reliable and uninterrupted paleontological track of eukaryotes, then, commences with only first protists enveloped with shells, as diatoms, heliozoans, etc. (some 700 Ma); they were followed by multicellular specimens similar to embryos or rather algae (similar to *Volvox*; Chen et al. 2014). Yet later came enigmatic 2D fossils known as fauna Ediacara (McMennamin and McMennamin 1990, Yuan et al. 2011). It is not surprising that some authors claim that all three branches of life may be of similar age (Glanssdorf et al. 2008); later we will press this point as ours, too.

To illustrate the difficulties our quest inheres, take the contemporary biosphere. For example, the rich world of picoplankton, tiny eukaryotes (picoplankton under 1 μm—smaller than most bacteria) flourishing in oceans, was never expected and till recently never sought (Massana 2011, Seenivasan et al. 2013). Moreover, these creatures often enter symbiosis with cyanobacteria (Thompson et al. 2012); in this same environment, we also find tiny bacteria (under 0.2 μm; Luef et al. 2015), difficult to discover even in *contemporary* biosphere, and there is no chance that life this small could have left fossils to provide us keys to the early evolution of life.

Origins: Biosphere LUCA

The term Last universal common ancestor (LUCA) commonly denotes the hypothetical ancestral population of cells that gave rise to *all* extant form of life. This much is agreed, if nothing else; as to its basic characteristics, the views differ. Many authors take the view that LUCA belonged to the RNA world. The division of labor between DNA–RNA–protein, the argument goes, took place later (Koonin et al. 2006, Glansdorff et al. 2008), perhaps even after LUCA split into the three extant domains of cellular life (Forterre 2010b). Such scenario supposes that viruses played the role of DNA 'donors'. We prefer the alternative view, which argues that the biosphere's LUCA already showed the basic properties typical of contemporary life, as listed in previous chapter: genetic processes and code, metabolism and information processing based in proteins, symbiotic interactions, etc. We also envisage LUCA as a common ancestor of *all* three branches of cellular life (i.e., not prokaryotes first, eukaryotes later).

Gene sequences are comparable, yet the only ones shared by all extant life are those of ribosomal RNAs and proteins of the translation apparatus. A well-supported empirical rule suggests that genes for ribosomal components do not travel horizontally, i.e., a comparative analysis of gene sequences of ribosomal RNA and proteins may reveal the truest—and deepest—topology of the tree. In the case of all other genes this clarity is lost because of (i) homologs absent in some branches; (ii) multiple and frequent horizontal gene transfer (David and Alm 2011); and (iii) multiple symbiogenetic events leading to a fusion even of whole genomes or their parts (Martin and Embley 2004, Rivera and Lake 2004, Lake 2009, Ku et al. 2015). For example, Lake (2009) explains the origin of gram-negative proteobacteria by the fusion of cells belonging to lineages of Clostridia and Actinobacteria. Multiple theories also exist as to the symbiotic origins of Eukarya (see below); moreover, the concept of eukaryotes as 'cells with organelles' is imprecise as some bacterial lineages also harbor membrane-bounded organelles (e.g., Planctomycetes). As stated above and in the absence of paleontological findings, dating the appearance of Eukarya thus remains uncertain.

The status of viruses, entities lacking ribosomes, cannot be decided at all: no sequence of virions (except those apparently acquired by HGT) properly fits any hypothesized evolutionary tree. Are they latecomer 'inventors' of DNA, or do they represent remnants of a fourth, now extinct cell lineage that belonged to LUCA?

Despite all this, progress as to the reconstruction of deep branching has recently been achieved. Metagenomic[9] studies from various microbial communities revealed the existence of a plethora of groups previously unknown to science. Such groups have as of yet only a virtual status because they were not isolated and cultivated, i.e., described as living beings; the only witness of their existence are the signatures of their reconstructed genomes (Fig. 4.1). One group of Archaea—the newly defined superphylum Asgard—has attracted serious attention (Spang et al. 2015, Hug et al. 2016, Zaremba-Niedwiedzka et al. 2017). Genomes from this group bear bona fide gene sequences (not those coding for ribosomal proteins) homologous to many eukaryotic genes previously taken as specific to Eukarya; in this, gene coding for proteins engaged in membrane vesicle transport and function are important signifiers. By this analysis, the whole domain of Eukarya is merely a branch of the Asgard superphylum (Dey et al. 2016, Zaremba-Niedwiedzka et al. 2017).

Evolutionary trees, of course, cannot reveal properties of LUCA as a biosphere. However, as no life can thrive over longer periods without a community, the ecological facet of LUCA should always be taken into account. As is stated in Markoš (2014, 2016a) and throughout this book, both individuals *and* communities bear multiple facets of the memory and experience of the biosphere—thanks to this, dwellers therein can understand each other, at least to some extent. Below, we suppose that in the time when LUCA thrived, such understanding was universal and optimal. We communicate three scenarios of evolution—not necessarily mutually exclusive—from LUCA to present.

1. From simple to complicated. This scenario supposes, not completely wrongfully, that earlier stages of evolution should be 'objectively' simpler than later ones. The structure of a bacterial or archaeal cell, the argument goes, is undoubtedly 'simpler' than that of a eukaryotic cell (for example, prokaryotes have no nucleus 'yet'). Moreover, if we ignore the bacterial biosphere, the prokaryotic way of life may also appear more primitive compared to, say, animals with morphogenesis. In addition, not long ago most knowledge of prokaryotes resulted from the study of laboratory monocultures, and their characteristics could be summarized as primitive unicellular organisms'. All of the scenarios mentioned in this section therefore work with the hypothesis 'Prokaryote first'. It considers prokaryotes as having mastered metabolism at all imaginable levels, and then remaining frozen at this level up to our days. Only a small fraction of bacteria lost their cell walls, increased their volume and discovered endocytosis, i.e., the ability to engulf other bacteria. Thanks to such ability, they began to receive nutrient in packages, without the need to scavenge for them from the environment, molecule after molecule, as most prokaryotes do. They

[9] (1) Isolation of DNA from a raw sample, e.g., marine sediment; (2) fragmentation of DNA, its amplification and sequencing; and (3) reconstitution of sequenced fragments into bona fide genomes. A great majority of genomes thus obtained belong to known species or groups—some, however, must be assigned to living beings not known to science—except by their genomes.

were also able to capture bacteria and archaea, and to enter intracellular symbiotic and symbiogenetic bonds—thus gaining internal organelles (see below).[10] Such predators gave rise to eukaryotes: thanks to predation they could reduce their metabolic skill, and instead (i) master membrane processes, such as vesicle or even cell fission and fusion; (ii) develop a sophisticated cytoskeleton; (iii) develop a nucleus with karyogamy (mitosis, meiosis); and (iv) with the invention of sex, get on the path that eventurally lead to multicellular organisms. More recent adherents of the scenario see viruses as a crucial factor in eukaryogenesis, as they might be the first to bear the know-how for membrane fusion and fission (Bell 2009).

When mitochondria and plastids were confirmed as descendants of bacteria, the scheme became legitimized even more firmly: not every bacterium was digested upon engulfing; some entered into intracellular symbioses (as many do today), and in at least two cases, the bacterial cell closure was subsequently breached and the engulfed bacterium set on a trajectory towards semiautonomous organelle. Again from this argument, four groups of extant eukaryotes (Diplomonadina, Microsporidia, Parabasalia, and Archamoebae) were grouped under 'Archaeozoa' and taken as remnants of pre-mitochondrial eukaryotic life; the first sequence comparisons seemed to confirm the conjecture (Müller 1992, Sogin et al. 1996). Many models appeared, seeking for more bacterial precursors of eukaryotic traits, e.g., for kinetosomes and flagellae, hydrogenosomes, and even of nucleus and cytoplasm. While this was happening, many other speculations died out. In the meantime, Archaea were discovered as a separate cell type, and it was shown that all groups of 'Archeozoa' are descendants of mitochondriate protists. The common attractor of many competing theories became the model of *serial endosymbiosis*, which explains the emergence of eukaryotes from multiple symbiotic events. First came the merger of Archaean and Bacterial cells, one giving rise to nucleus, chromatin, and ways of dealing with genetic material, the second to metabolism and membrane (Margulis and Fester 1991, Margulis 1996, Martin and Müller 1998). Bandea (2009) argues that viruses represent palimpsest of parasitic cells that fused with their hosts and cannot be recognized today.

One contemporary version of the theory treats Eukarya as a sister group of Archaea, or even members of the Archaeal domain (Fig. 3.1, p. 31). In many ways, Archaean genetic apparatus appears similar to that of Eukarya—for example, the presence of chromatin with proteins homologous to eukaryotic histones (Mattiroli et al. 2017): and so we see that serial endosymbiosis would then only appear later, with the capture of mitochondria (Williams et al. 2013, Spang et al. 2015). On the other hand, Pittis and Gabaldón (2016) keep the original concept of a Bacterial-Archaean chimera, no longer recognizable in any extant living cell. Bacterial origin can be demonstrated only in mitochondria and plastids—as latecomers to the scheme of life.

The discovery of giant viruses brought back the hypothesis that viruses play a role in the process, with a viral factory as the source point of the cellular nucleus (Forterre 2010b, Forterre and Gaïa 2016, Chaikeeratisak et al. 2017). As

[10] Archaea were unknown, or too little known to be taken into the scheme at that time.

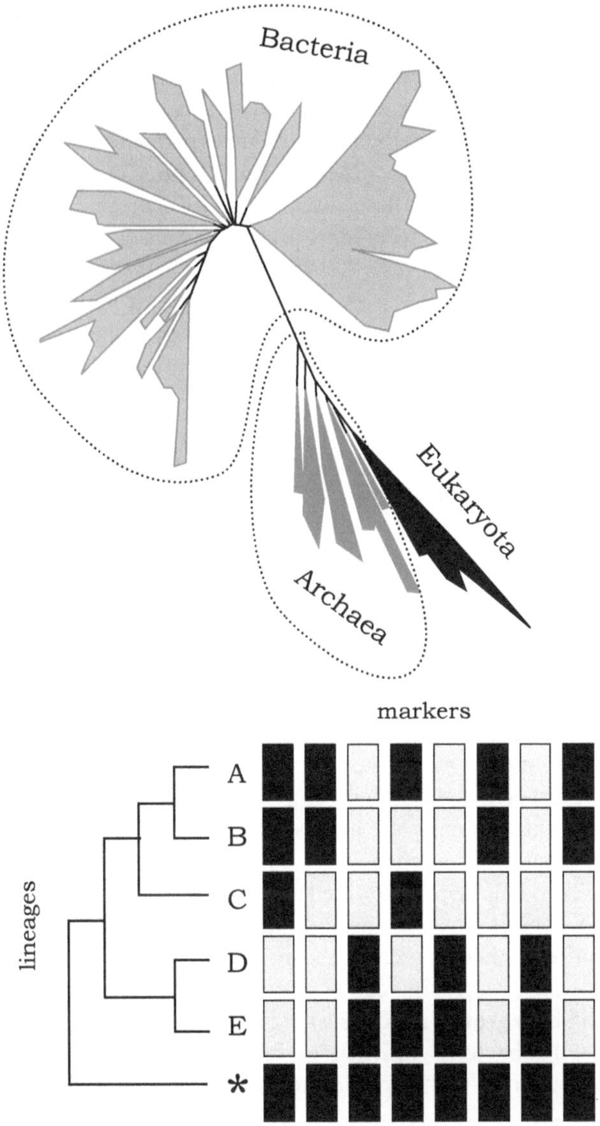

Fig. 4.1: Tree of life as based on sequences of genes for rRNA and ribosomal proteins. Above. Essentially, two branches can be recognized—Bacteria and Archaea, with Eukarya as but a scion of Archaea (92 named bacterial, and 26 archaeal phyla, but merely five eukaryotic supergroups). **Below.** Justification of taking eukaryotes as a mere branch of Archaea: principle of constructing archaeal tree based on homology of genes for ribosomal proteins and proteins for vesicle transport: asterisk is eukaryotes, A–E archaeal groups. (After Hug et al. (2016), Zaremba-Nedwiecka et al. (2017).)

mentioned in Chapter 1, virions may represent palimpsests of *structures* present in the ancient abiosphere, assembled from certain types of proteins on the nucleic acid scaffolding, and transporting within them some sort of cargo (DNA or a special metabolic pathway). After the origin of cells, they became intercellular messengers and/or parasites (Koonin et al. 2006, Brüssow 2009, Holmes 2011). Other authors, however, prefer scenarios with cells coming first. Thus, Nasir and Caetano-Anolles (2015) hypothesize a fourth, now extinct domain of cellular life with idiosyncratic propagation wherein virions act as a kind of 'spores': later, propagation was delegated to other cellular domains and the lineage of 'viriocells' faded away. Yet other authors take viruses for derivatives of cellular life, reduced parasites (Bandea 2009), or even rapidly evolving cellular derivatives (Moreira and López-Garcia 2009). Interestingly, in all three domains of extant cellular life there exist nano-compartments analogical to virions, bounded by capsids formed by proteins with a conformation similar to the encapsuline tripod (Fig. 2.5, p. 18). Acidocalcisome is just such an 'organelle', and the only one shared by all three domains of extant life. It harbors calcium ions and also—what is even more important for our quest—pyrophosphate (Docampo and Moreno 2011). It is not clear whether the structures originated virally or vice versa, whether virions are derivatives of such structures.

With a single exception, all organisms with primary chloroplasts (or more generally, plastids) belong to the group Archiplastida (Green plants, red plants, and glaucophytes)—obviously the capture of cyanobacterium and its transformation to chloroplasts occurred only once in early evolution (after gaining mitochondria in the root of eukaryotes), and characterizes the whole lineage. The exception from the rule is *Paulinella chromatophora* (and perhaps some other species of the genus), a protist from the domain Rhizaria that bears in its cytoplasm two chloroplasts of cyanobacterial origin, not related to plant chloroplasts. In this case, the capture of bacterium occurred much later, some 60–200 million years ago (Nowack 2014).

2. Proto-eukaryote. The models discussed in this section assume that Bacteria and Archaea are not primitive, but to the contrary, are derived or reduced forms of life; thus, LUCA represented a biosphere of proto-eukaryotic cells. Glansdorff et al. (2008) argue that extant eukaryotic cells synthesize a great variety of RNA classes, with many functions in the cell: LUCA could have descended directly from the RNA world, or even belong to it, with eukaryotes as its direct descendants. Bacteria and Archaea, in contrast, passed through the bottleneck of the hyperthermic way of life: this required reducing thermo-sensitive RNAs to an imperative minimum,[11] developing cell walls, decreasing cell size, etc. The authors also support their theory by the fact that in both groups the deepest branches of the evolutionary tree belong to extremophiles. Other bacterial and archaeal branches, after having broken loose from the hyperthermic environment, steered their further evolution with newly acquired cellular patterns.

In this scenario, LUCA appears as a morphologically and metabolically heterogeneous assemblage of cellular life, readily exchanging both structures and/or genetic material. As C. Woese stated: "Organismal lineages, and so organisms as we know them, did not exist at these early stages. The universal phylogenetic tree, therefore, is not an organismal tree at its base but gradually becomes one as its peripheral branchings emerge. The universal ancestor is not a discrete entity. Rather, a diverse community of cells survives and evolves as a biological unit. This

[11] Even today, the half-life of bacterial mRNA is in order of minutes.

communal ancestor has a physical history but not a genealogical one. Over time, this ancestor refined into a smaller number of increasingly complex cell types with the ancestors of the three primary groupings of organisms arising as a result." (Woese 1998, see also Woese 2002)

Going-along with others.[12] We suggest the reader to consider a perspective well known from the evolution of (Indo-European) languages. Linguists have reconstructed the 'tree', with its roots in a hypothetical 'ur-languages', and have described the branching, fusions, etymologies, generation of novelty (new meanings), phonetic and grammatical shifts, horizontal transfer of words between branches, etc. The level of understanding among members of particular groups depends on the degree (time) of divergence, as well as on the history of mutual contact. If understanding is not at all possible, body language—gestures, mimics, anger, etc.—take their place: *to some level*, all human beingS are able to understand each other. What is important is that they are able to raise such emergency tools back to the level of language rather quickly—e.g., different pidgin languages, or the communication of children from different language groups. "Such contact languages begin to emerge when people first develop their own individual ways of communicating, often by using words and phrases they have learned from other languages [...] they think others might be familiar with." (Siegel 2008) In 2–3 generations the pidgin may develop into a creole, with its own grammatical and lexical rules, and vocabulary. Lacey and Danziger (1999) give a fascinating and persuasive account of how English emerged as a pidgin—to ease communication between local Anglo-Saxons and Norsemen foraying from the North-East: "Before the Viking invasions, both *Englisc* and Norse were strongly inflected languages [...] The solution was the rubbing away through day-to-day usage of complicated word endings. [...] By the year 1000, a hybrid language had been stirred together by the integration of the two great waves of invaders, and a common tongue existed that was at least roughly understood in every corner of the country" (Lacey and Danziger 1999). Keep the parable in mind during the discussion below. We consider LUCA as a biosphere of highly promiscuous cells that were capable of horizontal exchange of modules (genes or structures), which gradually differentiated into three (or more?) great domains.

Bacteria and Archaea have retained, to a high degree, the ability of horizontal gene transfer: this enables them to share or communicate metabolic dexterities, defenses (e.g., the ability to synthesize antibiotics as well as acquire resistance to them), pathogenicity, etc. This allows great plasticity, with *ad hoc* teams capable of populating most different niches. In the domain Eukarya, we witness the rather early appearance of distinguishable lineages with a predominantly vertical transfer of genetic information (safeguarded by mitosis and meiosis), with elaborated membrane processes and transformations connected not only with the fission and fusion of vesicles, but even of whole cells (as in fertilization). This also enabled a plethora of symbioses—extracellular as well as intracellular.

[12] An allusion to Heidegger's concept of *Mitgehen* (1995). He, however, meant it as an exclusively human property, whereas we think it as a property of all life. More 'scientifically-sounding' is 'co-construction of the biosphere' (Kauffman 2000). For more details, see Markoš et al. 2009.

As the three major cellular domains branched off (Fig. 4.2), their mutual understanding became more and more blurred, fragmented, misspelled or misunderstood: conflict and numerous bizarre forms of common life followed. The last 'deep' understanding was transformation of endosymbiotic bacteria into organelles: first to mitochondria in the very root of Eukaryotic tree, and later as primary chloroplasts in one of its branches (Archaeplastida). Within the framework of single domains, mutual understanding lasted for longer: the best example is the emergence within different eukaryotic branches of secondary chloroplasts, not from bacteria any more but from eukaryotic algae of Archaeplastid origin. Later symbioses with bacteria (*Paulinella* as the exception) or with eukaryotes often also ended in tight cohabitation (lichens, mycorrhiza, holobionts, ciliate assemblages, etc.), but never the transformation of one partner into an organelle.

With time, mutual understandings came to be more and more focused (i) on symbolic or semiotic 'games' and (ii) on more and more elaborated cellular closures (e.g., neurons); in both cases, these understandings have been brought to high degree of reliability (see also Fig. 1.1, p. 4). Mutual understanding across the biosphere is retained only in the deepest levels (such as genetic and perhaps, also other codes or domains that control animal ontogenesis); often, however, the shared and overlapping areas of their mutual umwelt tend to be very small, limited, e.g., to a single signaling protein (a 'command' of a kind) in some pathogens.

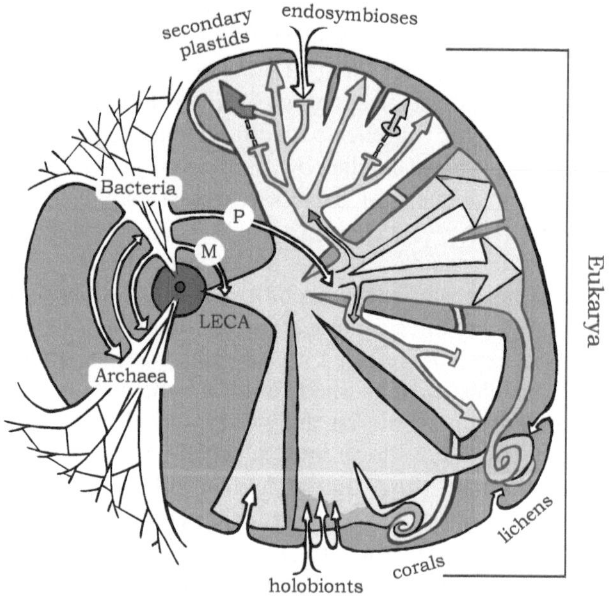

Fig. 4.2: Diversification of the LUCA biosphere: branchings, reticulations, and symbioses. LUCA is the black point in the center, from which radiates time axes. On the left side is the reticulate evolution of Bacteria and Archaea; the grey background of the scheme represents the prokaryotic biosphere pervading all life. LECA—last eukaryotic common ancestor; arrows M and P mark the capture of mitochondria and plastids early in eukaryotic evolution; other arrows (connecting eukaryotic branches with each other, or with the prokaryotic background) symbolize examples of endosymbiotic (secondary chloroplasts, corals) and exosymbiotic assemblages (lichens, mycorrhiza, holobionts). Compare this reticulate scheme with the tree in Fig. 4.1.

Reconstructing the Past, Making Sense of the Present

Now we move to the area of biosemiotics, and it requires, again, a parable from our experience as humans, semiotic beings. U. Eco states: "Our perceptual relationship with the world works because we trust prior stories. [...] We accept a story that our ancestors have handed down to us as being true, even though today we call these ancestors scientists. No one lives in the immediate present; we link things and events thanks to the adhesive function of memory, both personal and collective (history and myth). [...] Living with two memories (our individual memory, which enables us to relate what we did yesterday, and the collective memory, which tells us when and where our mother was born), we often tend to confuse them, as if we had witnessed the birth of our mother (and also Julius Caesar's) in the same way we 'witnessed' the scenes of our own past experiences." (Eco 1994a)

We believe that *all* life necessarily comprises organic memory and experience—of individuals, of 'lineages', and of communities which enables them to interpret their surrounding world and create a context allowing them to behave accordingly (Markoš and Švorcová 2009, Markoš et al. 2013). All dwellers of the biosphere share common roots and an identical evolutionary time: they diverged into many 'lineages', yet share many basic structures, functions, and ways (habits?) of how to behave with others, i.e., how to understand others. Such ways of mutual understanding (but also misunderstanding, wrong expectations, etc.) allows them to enter into more or less sophisticated manifolds of symbiotic interactions. This book is based on the assumption that no living being can exist alone: it bears the memory and experience of its ancestors, its kin, and all its cohabitants in the biosphere. Luck, chance, genetic and extragenetic factors, family, community, etc., all these influence the appearance of both the individual, as well as its community.

Moreover, whether cells or individuals, 'units' of complex communities cannot be compared to undistinguishable 'atoms' of non-living systems: they are born and will die; they are unique as to their memory and experience. Yet, their bodies continue the uninterrupted existence of their lineage from the beginnings of life, and their umwelten are much richer than can be manifested through single individuals, or through communities living here and now.

Umwelt. The concept was coined by the theoretical biologist J. von Uexküll (1864–1944; von Uexküll 2010) and defines the inner world of a living being or community, well tuned, optimized for the life. We will not go into Uexküll's views as to how umwelten came into existence, but we suspect he was much looking for the harmony lost with the teaching of Darwin. The world beyond the umwelt is unknown for a given creature or ensemble, i.e., does not exist for it. "[T]he countless Umwelts represent the keyboard upon which nature plays its symphony of meaning, which is not constrained by space and time. In our lifetime and in our Umwelt we are given the task of constructing a key in nature's keyboard, over which an invisible hand glides." (von Uexküll 2010)

Here we try to broaden the concept, to explain our view on the semiotic space of the individual, lineage, or a community (Švorcová et al. 2018; Fig. 4.3). The basic presupposition is that (1) the umwelt is a repository of memory and experience of a given individual or ensemble; (2) no such living entity has a capacity to draw

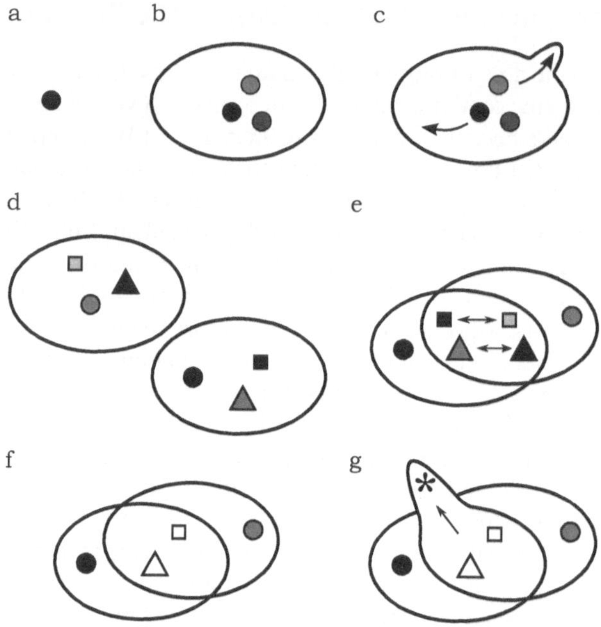

Fig. 4.3: Views of the *umwelt.* a. The circle represents the closed Uexküllian model of *umwelt*, a stable, well-tuned pattern of behavior according to a *written* score. **b.** *Umwelt* as represented by the outlines of the ring that delineates the endowment (experience) of a given lineage, with but its small part being accessible to, or chosen by, a given cell or individual or population. **c.** Evolutionary trajectories within the state space of a given *umwelt* marked out by the experience of the lineage (left arrow), or claiming new space from the surroundings by inventing novelties (right arrow). **d–g.** Symbiotic overlaps of *umwelten.* Patterns inside ovals correspond to different life strategies or features in a given state space. **d–e.** Two lineages come into contact, each with 3 features; features depicted as triangle and square are homologous, represent 'dialects' of a common theme. There is a partial 'understanding' in the overlapping area; **f.** 'Symbiotic negotiating' leads to a state where homologous features assume common coordinates in the overlapping state space. **g.** The symbiont claims a new realm on the surrounding world, by reinterpreting some existing features, or *via* a new feature invented (asterisk).

from the totality of the umwelt, but it may change its coordinates within it—during ontogeny, pushed by external conditions, or simply contingently; (3) as to the common ancestry, all umwelten find a partial overlap, allowing communication and 'going along' with others.

The view of evolution as changes—in time—of wholly 'autistic' lineages (perhaps best articulated in the ideal trees of cladistics) has recently witnessed a shift towards a reticulate scheme of mutually interacting entities—via symbiotic interactions, hybridization events, or horizontal gene (genome) transfer; moreover, vesicle traffic across lineages may also contribute to such exchanges, by carrying structures, viruses, plasmids, etc. Even 'objectively existing' entities such as species have been called into question, and replaced with notions of vague shapes of a spatiotemporal 'chunks' of traits that are being realized here and now (Flegr 2013). Proponents of these views, however, see causes of 'speciation' either in hybridization events, or in the establishment of a new holobiotic assembly. We propose a third way which consists in moving in the space of a species' umwelt. New interpretation drawn from the umwelt pool, the overlapping of different umwelten, inventions allowing the enrichment of the umwelt—all this creates a new 'worldview', 'narrative' for a given individual or community—which can serve to isolate a lineage, and generate a new 'species'.

Historiography. For reconstruction of evolution (here taken as genuine history), historians, linguists, and biologists must rely, on one hand on the state of contemporary cultures, languages, or life forms, and discover in them 'palimpsests' of the past. On the other hand, they can analyze rare, desperately incomplete and scattered relicts of the past—artifacts from archives, museums, or excavation deposits of archaeology, geology, or paleontology (Fig. 4.4).

Comparative studies forage for accumulated deviations from a presumed 'logic' or initial state (the Norm in the sense of Chapter 1)—of languages, religions, metabolism, ontogeny, etymology, gene sequences, pronunciation, mating rituals, etc. Old strata discovered in traditions, languages, or geological and archeological

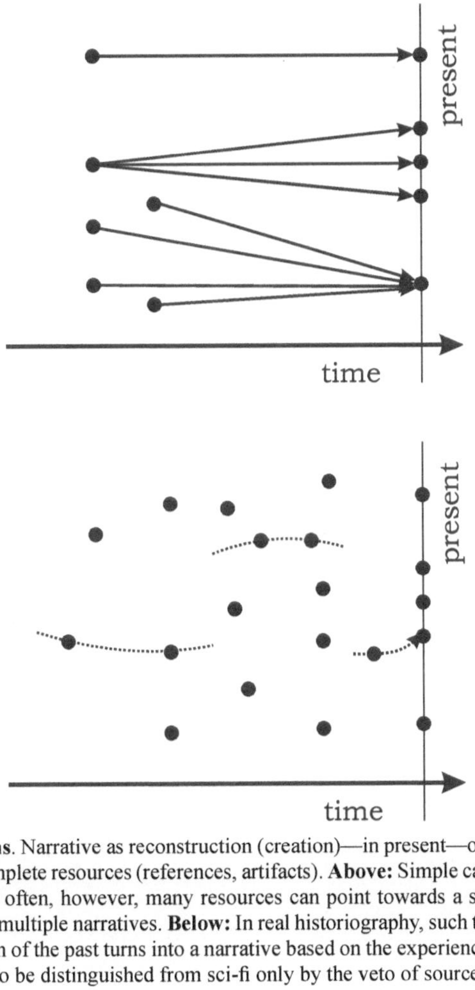

Fig. 4.4: Interpretations. Narrative as reconstruction (creation)—in present—of a historical event based on scattered and/or incomplete resources (references, artifacts). **Above:** Simple causal chain are seemingly readily interpreted; very often, however, many resources can point towards a single interpretation, or a single resource suggests multiple narratives. **Below:** In real historiography, such time–effect interpretation is lost, and reconstruction of the past turns into a narrative based on the experience of the interpreter. Such a narrative is to be distinguished from sci-fi only by the veto of sources (black dots).

deposits, point towards historical scenarios that offer more sense than other ones. In other words, history cannot be computed according to some formula, it is the subject of interpretation based in a narrative—i.e., storytelling. This fact will distinguish historiography, of which evolutionary biology is a part, from 'pure' science: it represents a different kind of knowing, open forever for different interpretations based on new findings or conjectures. Open to whom? Obviously, to researchers from relevant fields and for their audience. However, here we will stress that a similar 'narrative' is being weaved by *every* living being. Hence, biological evolution has two types of historiographers: learned scholars on one side, and—before all—all living beings on the other.

Historiographers of any kind are usually not able to say anything about singularities—origins of given historical entities. Here we also proceed—from prebiotic conditions discussed in the previous chapter—directly towards cells and biosphere, leaving the very origin of life in abyss. The singularity of cellular origins left behind, we can rely in our narrative on *scientifically* proven cues—yet the narrative itself depends on our interpretation of such cues.

As already stated, cues to the characters of the primeval biosphere are to be found mostly as mere palimpsests preserved on the background of extant life—such cues may, of course, raise multifarious and often contradictory explanations. As a further complication, paleontological findings and geological proxies are very scarce and raise many controversies, as shown in the next chapter.

5

Concepts of Heredity and Theories of Evolution

On the theory of descent with modification, the great law of the long enduring, but not immutable, succession of the same types within the same areas, is at once explained; for the inhabitants of each quarter of the world will obviously tend to leave in that quarter, during the next succeeding period of time, closely allied...

—C. Darwin 1876, 311

The conscious memory of man dies with his death; but the unconscious memory of nature is faithful and indestructible. Whoever has succeeded in impressing the vestiges of his work upon it, will be remembered forever.

—E. Hering 1897, 32

Descent with Modification

'Descent with modification' is the most profoundly simple summary of a phenomenon that has struck human thought from time immemorial. That progeny displays the variation on parental traits, with siblings at the same time slightly dissimilar from each other is long known; yet only C. Darwin (1809–1882) made it the centerpiece of his theory. The consequences are as follows: as resources are not infinite, siblings compete among themselves and with the rest of their kin, as well as all other inhabitants of the biosphere.[13, 14] Moreover, the physical environment may leave a different stamp on each of them. Only those who successfully conjoin—or, fit into the network will be favored by natural selection to sire next generation of descendants, who, under certain conditions, may maintain the advantageous traits of

[13] Darwin can rightfully be considered as the founder of ecology.

[14] Of course, this idea was, as usually in science, reflected in works of other thinkers like E. Blyth, P. Matthew and mainly A.R. Wallace. Both Darwin and Wallace were inspired by the book *Principles of Geology* by C. Lyell (1830) and an essay by T. Malthus (1798).

their parents. As for the traits, Darwin focused on the *appearances* of living beings, mostly animals.

Let us put Darwin's statements above into the context of evolution as viewed in his time. If we do not consider those who presupposed actions and interferences of Providence (however defined), three factors come to the fore: origin, teleology, and mechanism (for more details see Markoš 2002).

(i) The title. *Origin of species* is somewhat misleading: for our times at least, *Transformation of species* would fit better. Darwin does not deal with origins, least of all with the origins of life. Fifty years before Darwin, Lamarck combines Aristotelian *generatio spontanea* with contemporary physics to propose a speculative assertion about how, under specific circumstances, the lowest forms of life could have and still may emerge. We get from Lamarck a concept of crystallization of a kind, which takes place whenever conditions will allow it. In the middle of the 19th century, the status of microorganisms was still not clear: recall Pasteur's efforts to dismiss spontaneous generation of the newly discovered bacteria, Haeckel's 'discovery' of *Eozoon,* or the hard opposition against Virchow's slogan *Omnis cellula e cellula.* Crystallization is a physical, lawful process leading *always* to the same set of the appearances of life as its outcome.

(ii) Generatio spontanea. It is obvious that higher organisms do not emerge spontaneously; they reproduce, and the progeny resembles its parents. Evolution is apparent, yet again—as a purely physical process, physical causality, i.e., teleology, cannot be reflexively excluded as the driving force (a kind of sophisticated crystallization again). Long after Darwin, F. Galton (1822–1911) speculated about the spontaneous overturning (as if recrystallizing) of such 'crystal lattices' into another form (out of many possible, but not endless forms; Gould 2002) as driving force of new appearances in evolution. Moreover, aberrations from neat physical processes would be possible if parents transmit to the progeny some of their traits acquired during their lifetime, or even transmit *knowledge of the need* to undergo such transformation. Again, Lamarck tried to put Aristotle into a scientific robe—just to be interpreted later as a mystic and vitalist. The physical basis of Lamarckian teleology comes to the fore when he speaks about the sequences of events: all forms of, say animals, strive towards, or better, are predestined to give rise—after generations—to human beings. In short, for Lamarck, if all mammals would go extinct, they will 'grow' again from lower forms of life, reconstituting previous forms including humans. A variation to the theme was the theory of endless cycles coined by F. Engels (1820–1895)—who stressed the lawfulness of life and human appearance in any part of the universe, whenever conditions would allow it, or its reappearance after catastrophes (Engels 1972).

Most biologists of Darwin's time were proponents and seekers of Platonic laws of form, as per J.W. Goethe's formulations (Sapp 2003), hence their strong opposition towards Darwin's theory. Attempts to expel historicity from evolution continue to our days. In the 1980s, biological structuralism made an attempt to revive interest in life's forms (Sibatani 1987, Webster and Goodwin 1996). S. Conway Morris (2003), a paleontologist, revived the deterministic view of evolution; as to the place of Platonism, see Monod (1972) below, and chapter two. Contemporary theories of evolution are largely derivatives of Darwin's historical view, while at the same time declaring their scientific foundations (in the sense 'natural sciences'). 'Laws of form' recently entered by the backdoor with explanations of the structure of protein molecules (see Chapter 6).

(iii) Heredity. Darwin seemingly did not (and could not) have any clear-cut opinion on the issue. Later scholars rightfully point to the fact that Darwin—viewed by our times—can rightly be labeled as a Lamarckian concerning heredity issues (see below, section J.B. Lamarck). In his time, however, such labeling would not make much sense. Heredity and epigenesis were acknowledged, of course, but how memory and developmental programs are transmitted, remained elusive. Mendel's contemporary metaphysical attempt to explain heredity remained unattended and waited for rediscovery in the 20th century. In sum, the theory of descent with modification remained at the phenomenological level, and received genetic support only much later.

Darwin's principal contribution to the debate lies in the fact that he introduced history into the evolutionary process—with all its idiosyncrasies, contingences, trends but not determinations, unpredictable catastrophes, etc. The fact is, as we believe, the main point causing clashes with both religion and the 'true' sciences—neither could accept the burst of history into their realm. Recall that even the history of human culture was considered by many of Darwin's contemporaries (Hegel, Marx) to be a lawful, deterministic, and 'progressive' teleological process.

Science, as many scientists stress and maintain up to our times, longs after the existence of a Platonic element in nature, the basic, atemporal, level of description. One of the founders of molecular biology, J. Monod, states: "In science there is and will remain a Platonic element which could not be taken away without ruining it. Among the infinite diversity of singular phenomena science can only look for invariants." (Monod 1972) Many similar voices can be found in books containing interviews with famous scientists of the past century (Brockman 1995, Horgan 1998). One can find therein their multiple concerns that the destiny of contemporary science will not always remain true to the ideal of atemporal laws as formulated by its founders. Despite all the previous development in both society and biological science, dependence on the historical contingencies of evolution were unpalatable to most biologists of that time, as well as the general public.

Darwin's essential contribution thus lies in *not* being faithful to any such Platonic ideal: history definitely is not Platonic. In his treatise on the history of biology, Emanuel Rádl (1909, English translation 1930) perceived the problem clearly: "Since about 1890 practically no advance has been made in our knowledge of the origin of living organisms. Can this, by any chance, be due to the fact that our methods have been too exclusively the methods of science? It would be quite possible to treat the history of the world of living organisms as an historical problem, and to attempt to solve the problem by historical methods. It would then be analogous to the study of the origin of religions or of the various European states. The more I, personally, think on the subject, the more I believe in this method of attack." (Rádl 1930) Later on, he addresses the contribution of Darwin as follows:

"Darwin sought to make his presentation historical. It is based on the thought that the history of the world is full of events; that every transformation of one organism into another presents the world with a new fact; and that there have been an infinite number of such facts. [In contrast] Goethe, Fichte, Schelling, Oken, Hegel, Braun, von Baer—all those thinkers paid attention to the events by which evolution proceeds, merely to deduce therefrom a law of development, an idea." (p. 3) Elsewhere: "[Darwin] taught us to regard organisms as structures with a history, but as the product, not of an idea, but of a series of events in the real world. This was a great thought. We may study an animal as thoroughly as we please; we may compare it in the most detailed manner with other forms; *but this will not enable us fully to understand its nature, for it bears within it the traces of the past, traces which can only be revealed by historical study.*" (Rádl 1930, 174, italics ours)

The heritability of traits in macroscopic organisms has been an enigma from the days of yore. With the possible exception of plants propagating vegetatively through cuttings, all reproduction of living beings is accompanied by a drastic transformation of body appearances; even worse, *no* appearances are recognizable during the act of begetting. Many correlations are known, of course, between molecular and cellular processes, and body patterns; but how the developing body attains the species-specific likeness remains enigmatic. For more than two millennia, there were no techniques available to push our knowledge beyond Aristotle's ingenious insight.

With the advent of genetics and molecular biology, the causal connection between the script and body became clearer, and a century of struggles between 'Lamarckian' and 'Mendelian' models of inheritance followed. At its heights, biologists fell into an opposite extreme, claiming that it is solely the script that counts. Fictions such as Jurassic Park epitomize such a shift in a common sense. In recent years, the interplay between the body and virtual information stored in genes leads us towards a better understanding of hereditary processes. Below, we try to give a brief history of development in last two centuries.

J.B. Lamarck

Before Darwin, the idea of evolution as species change was formulated by J.B. Lamarck (1744–1829), who was a proponent of the idea of species transformation (unlike Cuvier, the most famous biologist of Lamarck's time), i.e., species developed out of previous species by the accumulation of inherited traits. In Lamarckian philosophy, an organism's behavior and habits—its ways of relating to its environmental circumstances, are the major constituents of the organism's appearance and behavior. So long as the environment remains stable, the species is likewise unchanged. However, if an organism encounters new environmental challenges, these lead to new experiences and thus the appearance of new organs and habits. Likewise, the old organs (and habits), for which the animal no longer has use, disappear. In the Lamarckian view, environmentally created habits are a major cause of species change. Moreover, changes of physiology such as musculature growth, organ losses or behavioral novelty, can be passed on to subsequent generations, i.e., they can become hereditary (Gissis and Jablonka 2011, Lamarck 2012 [1809]).

Lamarck did not introduce a specific, elaborated theory of the inheritance of acquired characters. He explained the variability of organisms on a behavioral basis, as part of the law of use and disuse (used organs become enhanced in the progeny, whereas those not used disappear). On physiological ground, he explained the variability by the dynamics of fluids and solids in the organismal bodies. Liquids were of a dual kind, either those that could fill various tunnels, tubules and cavities in the body as would gas or water, or those that were able to pass throughout the body, such as electricity or heat. In organic matter, fluids would be channeled by an inner life force, which represented the physical necessity for augmenting organismal complexity.

This immanent physical force represents a second major feature of Lamarckian philosophy. It drives living organisms to constant improvement. Because of the postulation of such power ('pouvoir de la nature'), Lamarck has often been interpreted as a vitalist, though he considered such force to be physical (e.g., heat, electricity, magnetism). He argued that the tendency to see some goal is merely a factor of human interpretation and not the real character of such power (Por 2011). Lamarck considered such power to be inherent in living organisms, an idea also expressed by Newman and Bhat (2009), who specifically refer to Lamarck.

Although today we intuitively (but mistakenly) see Lamarckian and Darwinian theory as mutually incompatible *en bloc*, it was quite common for many proponents of Haeckel's recapitulation theory (as Haeckel himself) to be great admirers of Darwin's work, though these thinkers clearly preferred Lamarck's law of heredity of acquired characters to Darwin's law of natural selection. This was due to not only the teleological character of Lamarckian theory of evolution but also because the emphasis on individual contribution to development is a good fit for any theory that involved development through the accumulation of traits (Otis 1994).

Lamarckism was thus quite popular in the mainstream biology of the 19th century. Even Darwin, in reaction to criticism of his *Origin of species* (1859), formulated his own conception of the heredity of acquired characters in a discussion of the problem of atavism (Darwin 1868). This theory, called pangenesis, supposed the existence of gemmules, small cellular particles that were capable of accumulating and distributing the organism's experience. Gemmules reflected the hidden variability of the germ cells, as they contain much more than is manifested in the characters of the body. In consonance with his pangenesis theory, the germ cell is an assembly of particles not only from the cells of an organism, but also from cells of earlier stages of development—and occasionally, even from ancestors (Otis 1994). Such particles are attracted to the germ cell from every part of the body, assembling in the germ cell primordia and participating in forming mature germ cells or new organs. They could remain, for many generations, in a dormant state, just to create a sudden developmental atavism (Darwin 1868). Others also introduced similar elementary units of heredity: H. Spencer (1864), W. Roux (1883), K. von Nägeli (1884) and A. Weismann (1893). Their definitions varied, but each reflected the contemporaneous research of cells and the development of microscopy.

After Lamarck: Organic Memory

The basic assumption was that if memory is responsible for the preservation of an individual's experiences, it is also responsible for the preservation of hereditary characters. Neolamarckians thus attributed a memory faculty to all organic matter and spoke of cerebral memory as a special case of this faculty. All traits, hereditary as well as acquired, were transferred as memory traces to future generations, via learning and repetition, and lead—through repetition—to instinctive and automatized behavior.

The physiologist E. Hering (1834–1918) was the first to explicitly attribute the faculty of memory to all organic matter in terms of analogy between heredity and memory. Hering was also the first to work with the term memory without involving, or requiring, consciousness. For him, memory is more a faculty of unconscious life than it is of conscious thought (Hering 1897). The human, cerebral memory represents only one concrete example of this general faculty of organic matter. Hering concluded that organic matter is able to store all kinds of stimuli even at the cellular level, and his organic memory concept was the intellectual background for his research of color vision and spatial perception (Hering 1897). A new skill, instinct or memory are related to the ability of living tissue to be modified or altered by sensory impressions (as referred in Otis 1994) and to store these alterations for future generations. Hering was not able to identify the process of inheritance of such traits, and referred only to the material trace. In this sense, every organism is a sum of material traces, accumulated from the beginning of its life span.

Accepting the inheritance of acquired characters as one of the most important processes of evolution (and as a force at least equal to natural selection), E. Haeckel (1834–1919) tried to create a theory of heredity that would support Lamarckism and, naturally, his own biogenetic law. Nevertheless, after reading Darwin's Origin, Haeckel also became a supporter of the Darwinian view on evolution, and sought for evidence of evolution through his study of embryology. Inspired by Darwinian teaching, Haeckel (1874) further developed von Baer's ontogenetic law (1828) by asserting that ontogeny recapitulates phylogeny (The Basic Biological Law—Biologisches Grundgesetz, as he called it). Ontogenesis not only recapitulates, but also *causes* phylogeny; it is the mechanistic cause (die mechanische Ursache) of phylogeny. Such recapitulation happens in an abbreviated and rapid way, and explains why development sometimes has some strange turnouts (when the embryos of terrestrial mammals have pharyngeal arches, etc.) This means, for example, that a human being passes through the same pathways as did its lineage, repeating phylogeny from an ancient unicellular organism, through the stages of coelenterate, planarian, fish, saurian, primitive mammal and ape, up to species-specific differences that appear at later stages of development (Švorcová 2012). Haeckelian recapitulation was an influential idea during the 19th century, but was later disproved in the same century, thanks to the emerging research of experimental embryology. Darwin's own opinion on the theory of recapitulation remains a dispute for contemporary historians (see Sapp 2003). On the contrary, von Baer's ontogenetic law is still recognized today; from the perspective of comparative embryology, it states only that early stages of development are more similar because of their homogeneity (not because of the supposed law of recapitulation) and that specialized traits come later. It is, as Gould stated (1977), development proceeds from undifferentiated homogeneity to differentiated heterogeneity (see also Jägersten 1972, Chapter 8).

The Lamarckian perspective is evident in Haeckel's teaching concerning heredity, wherein he draws an analogy between heredity and memory. He tried to explain the similarity of heredity and memory by postulating wavelike motions of 'plastidules', postulated as basic units of living matter (Haeckel 1876b). As the smallest protoplasmic particles, plastidules possess a capacity for memory and heredity, and their wave pattern is specifically shaped by history and environment (Otis 1994). Haeckel derived the motion of the particles from molecular motion: as every atom disposes of an inherent quantity of energy, so too, through the motion of the plastidule particles, the cell is able to remember its past and pass it on to the next generation. In this sense, every organism represents a unique pattern of wave motions.

Again, as in the case of Hering, Haeckel's memory concept of living matter was not based on any driving, creative force of conscious mind. Plastidules have psychic characters such as sensation, will and memory, but such traits are all expressed unconsciously (Haeckel 1876b, Bowler 1983). Haeckel's theory is rather mechanistic: the response of the organism to its environment is predetermined by this environment, as well as by the laws of living matter in the organism—but memory enables it to pass along acquired characters to future generations. Only later, Lamarckism began to support the idea that organisms have the freedom to shape their own evolution. Haeckel's whole teaching was based on the idea of mechanistic causality, which can lead to the development of specific organs, but all other physical processes in nature follow a similar logic. Darwinian theory, according to Haeckel, allowed the extending of mechanistic and materialistic models of Newtonian physics to all natural phenomena. This unifying principle led Haeckel to his monistic conception of nature, everything having its own natural cause and effect, denying any superior power or mystical interpretations (Haeckel 1876a [1868], Sapp 2003).

At the beginning of his interest in evolution, S. Butler (1835–1902) was a true admirer of Charles Darwin, but he later turned to the Lamarckian philosophy and placed the memory analogy at the foundation of his near-vitalist interpretation of Lamarckism (Butler 2014 [1910]). According to Butler, organisms are capable of remembering their experiences and incorporate them phenotypically, but unconsciously. This idea of an organism remembering its history perfectly correlates with Haeckel's biogenetic law. In *Life and Habit* (2009 [1877]), Butler describes phylogeny as a way of getting used to new activities which, via frequent repetition, become automatic and therefore unconscious. Butler believed that something becomes a genuine habit by becoming an unconscious instinct. Every intelligent action eventually becomes instinctive and inherited (due to physical changes in the organism). Via analogy, it is also possible to explain embryonic processes as a never-ending repetition of originally conscious activities.

There were, likewise, other intellectuals, who proposed an idea of organic memory. The French philosopher and psychologist T. Ribot (1839–1916) argued (similarly to Hering) that conscious memory is only one aspect of a more fundamental phenomenon: he proposed thinking about memory in terms of physics—as heredity is but one possible version of the law of energy conservation. Ribot considered the nerve impulses electrical and suggested that the term of organic memory should be explained only in terms of work, force, energy and particle motion (Otis 1994). In his physics-based memory theory, Ribot is symptomatic of the Lamarckian tradition in France as rather materialistic philosophy, inspired and influenced by the Cartesian heritage (see Bowler 1983).

Still others considered organic memory theory important. A. Forel (1848–1931), a Swiss clinical psychologist, wondered how an embryo could remember which adult organism it is to develop into. He saw heredity and memory as essentially the same, and aligned heredity as a species-specific memory that carries individual memory within it. German zoologist R. Semon (1859–1918) unified the concepts of memory and heredity into one single term called *mneme* (1904). The *mneme* represented the basic capacity of organic material to maintain an after-effect of stimulation as a stable modification of irritable organic tissue, creating 'engrams' in which hereditary information was somehow inscribed (Semon 1904).

Butler's friend F. Darwin, a botanist and son of Charles Darwin, also postulated that a kind of unconscious memory exists even at the level of plants. He described the germ cell as telegraphically communicating with the whole of the plant body. The idea of organic memory was also central to E. Rignano (1870–1930), an Italian philosopher, who introduced a very similar doctrine as that of Butler, describing phylogeny as the habitual process of gradual improvement.

A. Weismann (1834–1914), a big fan of natural selection and the author of the theory of germ plasm, was never a supporter of the Lamarckian organic memory concept (Weismann 1893). As he did not approve of the heredity of acquired characters, he did not see the point of comparing heredity and memory.

Although today we intuitively (but mistakenly) see Lamarckian and Darwinian theory as mutually incompatible *en bloc*, it was quite common for many proponents of Haeckel's recapitulatin theory (as Haeckel himself) to be great admirers of Darwin's work, though these thinkers clearly preferred Lamarck's law of heredity of acquired characters to Darwin's law of natural selection. This was due to not only the teleological character of Lamarckian theory of evolution but also because the emphasis on individual contribution to development is a good fit for any theory that involved development through the accumulation of traits (Otis 1994).

The 19th century notion of memory reflected attempts to explain evolution as a teleological continuum: every organism serving as a link between past and future, promising an eternal life of some sort, when instinct, habit and memory represented manifestations of a single underlying capacity of organic memory to store experiences (Otis 1994). This idea of organic memory influenced disciplines other than biology, including criminology, psychoanalysis and literature.

For example, B. Morel (1809–1873) used it as the basis for his theory of degradation, explaining high incidences of alcoholism, syphilis, epilepsy, criminality or idiocy as a consequence of an accumulated hereditary burden of the sins of fathers (Otis 1994) (this was years before both Hering and Haeckel's theories of organic memory). In a close parallel to the framework of Haeckel's biogenetic law, C. Lombroso (1835–1909) described so-called born criminals as atavisms, as if criminality resulted from lower stages of phylogenetic development. S. Freud (1856–1939), and his disciple, C.G. Jung (1875–1961), were also influenced by the idea of organic memory. Freud believed that the individual unconsciously memorized not only his or her infancy but also the experiences of his/her ancestors (Freud was a fan of Hering's theory as, according to both, most memory is unconscious). Long after most biologists had abandoned Lamarck's and Haeckel's theories, Freud continued to defend them (Gould 1987, Freud 1991, Otis 1994). Freud believed that every person recapitulates stages of human evolution: as perverts and neurotics represent earlier stages of phylogeny, everyone is, at some point in their development, perverted and neurotic.

The organic memory concept was rooted throughout the intellectual discourse of 19th century, its deep social connotations were also reflected in literature. In his series of novels from *The fortune of the Rougons* (1871) to *Dr. Pascal* (1893), E. Zola (1840–1902), knowing contemporary treatises in natural science, presents heredity as a force that builds up and threatens to explode. Zola's novels are stories about accumulating pressure that will eventually produce violent movement (Otis 1994). As Laura Otis (1994) brilliantly demonstrates in her book *Organic memory: History and the Body in the Late Nineteenth and Early Twentieth Centuries*, this idea reflected the need to explain the heredity of human culture on levels both individual and collective, as well as the relationship between them—because acquired traits are transmitted to the following generations on both levels. Rather than a scientific theory, the organic memory idea was a way of thinking motivated by nationalism and a perceived need to explain cultural evolution.

19th century biology was deeply influenced by both Haeckel's idea of recapitulation and Weismann's theory of germ plasm. Lamarckians mostly identified with the first: for development to continue (after the organism recapitulated its ancestors), new traits would have to be acquired and become heritable. In addition, the fact of recapitulation was often taken as proof that Lamarckism must be the primary explanatory mechanism of evolution (Bowler 1983). Yet, Lamarckism struggled with Weismann's theory of germ plasm for the whole of its existence (Weismann 1893). By definition, germ plasm is completely separate from the somatic line, so characters acquired during the life span of an organism can never, in any way, get to the germinal line. As Peter Bowler writes in his *Eclipse of Darwinism* (1983), the main failure of Lamarckian philosophy was its impotence at finding an adequate explanation for how acquired traits might become incorporated into the germ plasm and what are the 'material units' of heredity. Later, at the beginning of 20th century after the discovery of Mendelian laws, it also failed to come with any alternative and plausible explanation for how the traits are transmitted to progeny. A renewed interest in Lamarckism came with C.H. Waddington, and with newer findings in molecular biology (Rutherford and Lindquist 1998, Taipale et al. 2010, Fig. 5.2 and Chapter 8).

Nevertheless, Darwinism experienced something of an eclipse, as Julian Huxley called it (Huxley 1974), from 1860 until 1930. At this time, Darwinian natural selection had previously been criticized as insufficient to explain the evolution of natural forms. Among several evidentiary arguments that repeated from the 1860s was the lack of evidence from fossil records, the geologic age of the Earth, and the existence of non-adaptive traits or orthogenetic interpretations of evolution (Bowler 1983, Sapp 2003). Late in the 19th century, new experiments and methods led to a development of a new biological tradition, which added yet another blow to Darwinism: embryology.

Experimental Embryology

At the beginning of its time, embryology was called *Entwicklungsmechanik*, the name reflecting the idea that organisms are Cartesian automata or machines, but further development of experimental embryology led rather to exactly opposite statements. The emphasis put on elementary units by Darwin or Weismann was too speculative for embryologists (Sapp 2003), and their experiments actually revealed the creative potential of embryonic parts. When separated from the embryo, individual cells were often able to create a whole individual. It somehow looked as though the cells are 'aware' of the whole, as shown primarily by experiments of Haeckel's student, Hans Driesch. His experiments on sea urchins led to an important turnover in embryology. In 1892, Driesch separated two blastomeres of a sea urchin, which led to two complete, yet smaller, individuals (Driesch 1914, Markoš 2002, Sapp 2003). Based on contemporary knowledge previously presented by Weismann and Roux (another student of Haeckel), one would expect that the blastomeres would develop abnormally. Unlike Roux, however, Driesch did not kill the second blastomere, but just separated them completely, which was supposed to lead to different results as in case of Roux in 1888. Driesch's experiments were successfully repeated on other organisms and it turned out that the position in the whole is what is decisive for the specific cells. They showed that even when disrupted, the development always leads to the same result and shape is something like an attractor (irrespective of size, the amount of matter and the initial conditions). The morphology exists *in potentiam*. Driesh called this ability of developing embryo to take a specific shape as *entelechia*, a term which brought Driesh into the swamps of vitalism, though wrongly, because entelecheia does not represent some mystic force like *vis vitalis*, but rather some internal order of bodily shapes (Neubauer 1989). However, it cannot be defined in physical terms.

Inspired by Roux and Driesch, H. Spemann (1869–1941) and H. Mangold (1898–1924) later demonstrated (1924) that not only the position in the whole but also the degree of differentiation of the cell is decisive for the development. Up to a certain stage of development, cells are not differentiated, and every cell contains the 'germ plasm' of the whole. Beyond this, embryologists had difficulty explaining how cells are able to orchestrate their development into a complex organism, as well as how a single cell can maintain its structure. In order to solve this, Spemann introduced the notion of an organizer, i.e., *Organisationsfeld* (Spemann 1938), and thus re-introduced the idea of morphogenetic fields.

A.G. Gurwitsch (1904) first introduced the idea of a morphogenetic field that was geometrical and temporal. This morphological field has a physical substrate on which is shape is realized—each shape following the next, emphasizing the temporal character of the field. T. Boveri (1910), P. Weiss (1926) and B.C. Goodwin (1985) further developed this idea. In general, such fields represent a 'physical' force that led the developing cells through the differentiation processes and helped them assume their role within the whole. The field was an interconnected whole, so it defined itself against the mechanistic and preformationist notions of morphogenesis. An analogy was drawn between this morphogenetic field and an electromagnetic field: it had its physical substrate, a not fully explored biochemical nature, and was not dependent on genetic information (Vecchi and Hernández 2014). The morphogenetic field was not a quantitative notion; its 'powers' were undifferentiated potentialities taking shape in a morphological space.

Another conceptual approach emphasizing the developmental processes was organicism, as represented by L. von Bertalanffy (1932) and his followers. He avoids the morphogenetic field with its presumed vitalistic connotations, but emphasizes the wholeness of organism, irreducible to its parts. The term *entelechia* was replaced by the term *organization*—the main characteristic of which is a hierarchy that enables downward causation. However, all such processes are convertible to microscopic structures, which can be described by physics and chemistry or by constellation of powers (von Bertalanffy 1932). This approach is one of holistic physicalism (Neubauer 1989); I. Prigogine and his physics of non-equilibrium systems as well as C.H. Waddington, who will be discussed later, exemplify the further development of this idea.

With the concept of positional information, Wolpert (1969) offered another important contribution in this direction; this idea assumes that cellular differentiation is position-dependent and that cells interpret their coordinations within the developing embryo (thanks to diffusing gradients of morphogens or via cell to cell communication). Together with the French flag model (Fig. 7.1, p. 120—a model that explains how the same cell can respond differently to various concentrations of diffusing chemicals), Wolpert's concept of positional information has become a theoretical framework for the definition of spatial regulation in the developing embryo (Vecchi and Hernández 2014). Later on, Wolpert and Lewis (1975) established computational embryology through arguing that the egg internally contains all that it needs to develop, and that all such resources are genetic. As Vecchi and Hernández (2014) stress, Wolpert's conception reduces cell behavior to differential gene expression, making the cell itself, and the whole concept of morphogenetic fields, causally redundant. Cells represented automata that compute the developmental outputs based on genetic switching.

Units of Heredity

During the 19th century, biological theories of heredity were rather speculative, the most successful of which was by August Weismann who hypothesized a germ plasm containing the basic particles—biophores—of the living thing, and consequently, the idea of what is known as Weismannian barrier. The behavior of and relations within the germ plasm became more and more complicated, as Weismann tried to answer criticisms of his heredity theory. The basic presupposition is again Platonic: speculations about eternal, virtual units lying in the background of life. Such line of thought prepared a foundation for acceptance of the Mendelian laws of heredity, which were formulated at about the same time. The concept of biophores was confronted, and successfully blended with the older theory of pangenes held by Darwin and de Vries. Unlike Darwin, however, de Vries rejected the notion that units of heredity are able to travel throughout the organisms, and placed them within the cell nucleus. For him, differences among individuals came from the differential representation (what we now call the expression) of pangenes in different parts of body. Such particles were held as unchangeable in order to provide stability for transgenerational heredity, and could be in latent or active state. At the beginning of 20th century, genetics came to acquire these contours.

Together with Carl Correns and Erich von Tschermak (1900), Hugo De Vries rediscovered the Mendelian laws of heredity, which show that the hereditary units are independently segregated so there is no such thing as 'blending inheritance'. At the beginning of the 20th century, Mendelian factors were known, but their material nature remained a mystery. T.H. Boveri and W.S. Sutton had already introduced the chromosome theory of inheritance in 1902–1903 (Portin and Wilkins 2017), and W. Johannsen (1909) coined the term *gene*, which was an analytical unit suitable for evaluation of hybridization results, and yet no specific material attributes of genes had been defined. Unlike Johannsen, many geneticists imagined genes as material units of some sort, but specific connections between genes and traits remained unknown. Speculations about the behavior and function of genes often involved the connection of gene activity and enzymes. Morgan was able to place units of heredity on chromatin and knew about their linear ordering, but was still quite cautious while describing their nature (Morgan 1919). Thanks to the work of British botanist R.H. Lock and Belgian cytologist F.A. Jannsens, as well as his own work on fruit fly, Morgan was able to explain the phenomena of genetic recombination, crossing over and related gene linkage (Portin and Wilkins 2017). The atomic character of gene units was later disproved by the work of A.H. Sturtevant (1925). Sturtevant was working on *Bar* mutation, which causes small and slit-like eyes in males and homozygous females of *Drosophila*. This mutation is quite unstable, reverting easily to wild-type. Sturtevant's study of *Bar* mutation led to the discovery of unequal crossing over, and the effect of gene position on the phenotype: the phenotypic expression of *Bar* alleles was dependent on the relative cis/trans position of their repeat units (a female with three units of *Bar* on one X chromosome and one unit on the other X has a more severe Bar phenotype than a female who has two units on each X chromosome, Wolfner and Miller 2016).

It was also discovered that the relation 'one gene = one trait' does not hold absolutely because of epistatic interactions and pleiotropy, yet experiments were still designed as if it were true. A great deal of work in terms of gene mapping and the nature of gene mutation was done by Morgan's student H.J. Muller (1922, 1926), who defined genes as self-reproductive, heterocatalytic entities who can mutate. He defined genes as the basis of life and evolution (Portin and Wilkins 2017).

Finally, in 1941 G. Beadle and E.L. Tatum hypothesized that genes influence the enzyme synthesis and other metabolic reactions (Beadle and Tatum 1941), and coined the rule 'one gene = one enzyme'—not the organism's morphology, but the protein molecule became the trait corresponding to the gene. The idea of genes being connected to enzymes was present in genetics since its beginnings in the works of Onslow, Haldane, Muller or Morgan (Sapp 2003), but only Beadle's and Tatum's experiments with *Neurospora crassa* confirmed it in an experimental model.

Embryology and topics emphasizing the environmental influence on organismal phenotype were rather forgotten. As mentioned below, the parallel formulation of the Modern Synthesis put into focus the adult individual able to prove its fitness; how the adult state was attained was not that important. Lamarckism was almost completely abandoned—with the unfortunate exception of Lysenkoist biology in the Soviet Union (for more details, see Medvedev 1969, Soyfer 1994, Markoš 2002). Due to its unifying role, genetics quickly raised the popularity of the work on simple model organisms such as *Drosophila*, *Neurospora*, yeasts, or bacteria, wherein experimental mutations could be induced rather simply through X-ray or radioactive irradiation.

Modern Synthesis

With the advent of Modern Synthesis, embryology was put in the background. The role of embryology in the first half of 20th century is connected with the name Thomas Hunt Morgan, who was originally an embryologist, but switched to genetics and whose studies on *Drosophila* strongly contributed to the rise of Modern Synthesis (Morgan 1919, 1932a, b). Yet, he took classical embryology as an old metaphysics (this labelling has been very common until today). Two years later, he wrote a book *Embryology and genetics* (Morgan 1934), which tried unsuccessfully to connect these two scientific branches. The merger that really took place was that between genetics and neo-Darwinism, known today as the Modern Synthesis; embryology was shifted aside.

The Modern Synthesis began, perhaps, with Ronald A. Fisher's iconic article: *The correlation between relatives on the supposition of Mendelian inheritance* (Fisher 1918, Gould 2002) and culminated between 1936 and 1947 (Lamb 2011). The basic theorem is the idea that gene frequencies (or better, proportion of alleles of a particular gene) will change in a population due to natural selection and other processes such as random genetic drift, migration and gene flow. It supposes, moreover, that such changes can be quantified and even predicted statistically. Classical genetic concepts suddenly fit into the neo-Darwinian theory of evolution. As presented in the books of the proponents of Modern Synthesis (Fisher 1930), complex mathematics became a powerful instrument for a biology striving to become an exact science comparable to physics.

In classical genetics, the Mendelian unit of heredity (the gene) was taken as a predictably behaving, stable entity, which cannot be influenced by the environment, by behavior of the organism, or by some internal power (as in orthogenetic theories); only external and not quite defined forces could cause its mutation. This became important for evolutionary biologists when they realized that the proportion (frequencies) of alleles of such a gene in a population can—in principle—be quantified mathematically, depending on the fitness which such a unit provides to the organism. By continuously comparing bearers of particular alleles of a given gene, natural selection remains the primary mechanism of evolution and evolution continued by gradual change. Evolutionary thinking shifted its focus from individuals towards whole populations. From events on the genetic level, the Modern Synthesis explained macroevolution from microevolution, i.e., phenotypes of bearers of particular alleles, and their fitness.

The Modern Synthesis brought some crucial new concepts into evolutionary biology. During its first period (1918–1932), R.A. Fisher defined organismal fitness as the number of offspring of an individual, and proved mathematically what later became known as the Red Queen hypothesis (Van Valen 1973). In brief, species must constantly evolve (and organisms reproduce) in order to sustain their fitness comparable to their competitors. S. Wright confirmed Fisher's mathematical models in his laboratory experiments with rats and, most importantly, introduced the concept of genetic drift into evolutionary thinking (Wright 1932). The question as to whether evolution happens faster in small or in large populations separated Fisher and Wright for many years (Sapp 2003). Wright also defined the coefficient of the relationship, which led to definitions of inclusive fitness and kin selection.

During the second period of the Modern Synthesis, the theoretical models were integrated with experimental work on real populations (Lamb 2011). The main figures of this period are Theodosius Dobzhansky, whose interest in studies of natural populations helped establish population genetics as an empirical research agenda (Pigliucci and Müller 2010), and Ernst Mayr, who coined the definition of species and defined allopatric speciation based on geographic isolation (he did not believe that sympatric speciation was possible). G.G. Simpson, who integrated population genetics with fossil records (1944), and G.L. Stebbins, who integrated (1950) the plant biology into the theory of Modern Synthesis (Ginsburg and Jablonka 2010, Sapp 2003), represent the third and fourth periods of the Modern synthesis.

The hegemony of the Modern Synthesis started fading away in the 1990s, with progress in study of molecular mechanisms in development. Perhaps the milestone of such development is the manifesto by Gilbert et al. (1996a, b); see also the discussion that followed (Lipshitz 1996, Gilbert et al. 1996b).

Scientists outside the Modern synthesis. There always were persons and their researches that were kept to the sidelines. During the periods of Modern synthesis, many experimental embryologists were not interested in genetic factors, but rather in shapes and structures of developing embryos. They also emphasized the role of cytoplasm, unlike the geneticists who denied its role in either evolution or development. They gathered evidence as to how, in different taxa, cytoplasm is differentially distributed in early development, which influences cell polarity, symmetry and the formation of body axis and basic germ layers (Sapp 2003).

For example, Richard Goldschmidt (1940) believed in the role of macromutations and proposed that evolution progresses through changes in genes responsible for development (Gilbert and Epel 2015). This was in contrast to proponents of the Modern Synthesis such as G.G. Simpson, who did not acknowledge the existence of macromutations: macroevolution in their eyes is merely the extension of microevolution.

So too, C. Darlington (1903–1981), who studied hybridization, polyploidisation and different system of reproduction in plants (later also cytoplasmic inheritance), presented a similar critique, namely that evolution cannot proceed only through slow, gradual changes in genetic information (Lamb 2011). Sympatric speciation is quite common in plants through hybridization and both auto- and allo-polyploidization; so plants easily escape species definition. However, proponents of Modern Synthesis considered such examples to be mere exceptions (Pigliucci and Müller 2010).

One of the most important figures to stand detached from the mainstream synthesis was C.H. Waddington (1905–1975), who tried to include embryology within evolutionary theory, and emphasized the role of the environment. He also believed that evolution cannot proceed only through gradual changes in hereditary information, as such small changes can explain neither the emergence of big taxa nor the differing rates of evolution apparent in the paleontological record (Waddington 1953). Waddington also described a process of genetic assimilation, which explains how environmentally induced traits can be inherited (key papers in the Waddington re-edition, 1975). About the same time, a similar process was formulated by I.I. Schmalhausen (1949), who coined the term stabilizing selection. Both processes had similar scenarios: first, environmental change can induce a phenotypic change, which is further maintained through the environmental conditions (not yet fixed in the genome). The phenotypic trait has to provide a higher fitness and there must be sufficient genetic variability, which eventually fixes such a trait genetically. He coined the concept of genetic assimilation—a process of transformation of induced traits into ones expressed even without such induction (Fig. 5.1).

Waddington introduced the term 'epigenotype', which reflected the developmental system as a sum of interactions among genes, and among genes and the environment (Waddington 1940), thus bringing forth the concept of gene regulation as well as an even more important idea: the emphasis on interactions among the genes themselves (or genes as autonomous units as Dawkins sees them,

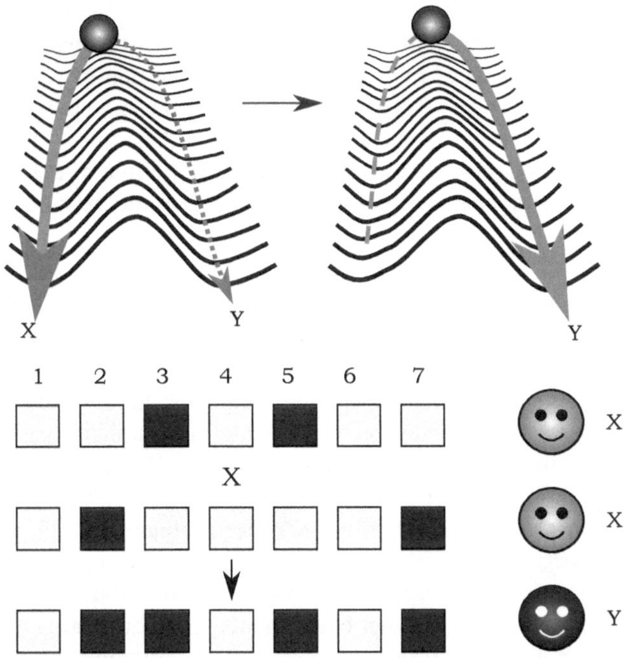

Fig. 5.1: Genetic assimilation. Above: Epigenetic landscape. Development proceeds towards the trait X, unless environmental and/or epigenetic effects force it towards the trait Y. When the effect fades away, development is steered towards X again. Such forcing, if long-lasting, may eventually fix the trend towards Y genetically. (After Waddington (1975).) **Below:** The genetic capacitor model. If the trait X is polygenic (boxes), the system may be able to buffer—to some extent—deleterious effects of mutations (black). Breeding such mutants may lead to surpassing the threshold and switch the trait permanently towards Y. (After Rutherford and Lindquist (1998).)

Portin and Wilkins 2017), and thus anticipating the genetic regulatory networks by Davidson (2006) and others.

Waddington's famous metaphor of *epigenetic landscape* (1940) depicts the space of morphogenetic states as a continuous, undulating terrain. The state which is momentarily taken by cell/tissue/organ/organism is symbolized by a marble in a specific position on the landscape and its pathway illustrates the course of morphogenesis. Such developmental pathway is stabilized in some direction and resistant to perturbations. Pathways can merge (which means that embryo can get to the same goal while taking several paths, as was discovered by Driesch) or they can branch out. Such pathways do not necessarily always lead 'downhill' or may not be direct. However, it is advantageous to use the physico-chemical processes or concentration gradients for orientation: such processes are landmarks, not causes. These things are often confused: the polarity of future embryo is not caused by the earth gravity or the tissue differentiation is not caused by a specific signal. These are only orientational points in epigenetic landscape. Thus, a homebox mutation is then an orientational disorder of the organism in its own landscape. The developing organism must 'walk' through the landscape, not just swing down through it (Neubauer 1989).

Finally, yet importantly, we would like to mention two personalities whose work had already slipped into history by the time of the Modern Synthesis, but deserves to be mentioned. T.M. Sonneborn (1905–1981) is known for his experiments with *Paramecium* (1964), which demonstrated an example of non-genetic inheritance of ciliar structures (Beisson and Sonneborn 1965; Fig. 6.6, p. 113). Non-Mendelian processes in evolution were in the focus of another rather forgotten insect biologist, R. Matsuda (1920–1986), who emphasized the role of the environment in developmental changes, embracing the idea of panenviromentalism (Matsuda 1987, Hall et al. 2003).

The Age of DNA

In his book *What is life* (1944), E. Schrödinger (1887–1961) compared the structure of molecules of heredity to that of aperiodic crystals; their structure enables the transfer of an almost infinite number of possible informational variants to future generations. Schrödinger popularized the ideas of Max Delbrück, who claimed that a gene is a polymeric entity of chained atomic structures. P. Leven and W. Jacobs first discovered that genes are actually DNA, not proteins in 1929, with the claim that DNA is a repeating tetramer. M. Chase and A. Hershey made another important step with experiments on bacteria that demonstrated that the carrier of hereditary information is actually DNA. Further experiments with chromatography conducted by E. Chargaff revealed that the amount of adenine and thymine in nucleus is always the same. X-ray crystallography experiments conducted by W.H. Bragg, W.T. Atsbury, J.D. Bernal—and most importantly R. Franklin, M. Wilkins and L. Pauling, significantly contributed to the discovery of DNA structure by Watson and Crick in 1953 (Watson and Crick 1953a, b, Sapp 2003).

By that time, molecular biology had come to dominate the study of biology, and with Watson and Crick, Modern Synthesis grew even stronger in its grounds, because the material unit of heredity, its structure and its behavior were finally uncovered. Issues of soft inheritance or the influence of development on evolution were marginalized; they were studied, but not included within the synthesis (Gissis and Jablonka 2011). Lamarckism became connected with intentional changes in DNA molecule. For instance, R. Fisher (1930), although not knowing the character of DNA molecule yet, mocked Lamarckism as an ability to produce mutations through the mental states of a given organism. Especially nowadays, such interpretations are considered as historical anachronisms. Lamarck did not know the material unit of heredity and such tendencies should be at least called Neolamarckian.

After 1958, when F. Crick formulated the central dogma of molecular biology (1958, 1970), biological study has focused primarily on the DNA molecule and related phenomena such as genetic code, transcription and translation. The notions of DNA as a carrier of instructions for proteins and phenotypes, together with localized view on causation and primacy of endogenous factors in biology were brought into biology (Vecchi and Hernández 2014).

The discovery of the model of gene regulation by F. Jacob and J. Monod (1961a, b) solved the puzzle of cell differentiation. Their model, *Lac* operon, is a simple feedback circuit switched by a repressor or an inductor. Monod simply extrapolates this knowledge from bacterium *E. coli* to eukaryotic and multicellular organisms. Monod's operon model (Fig. 5.2) began a series of analogies between software-hardware and genotype-phenotype. Such views are still very common today (Davidson 2006) and Monod himself, further developing the Cartesian metaphor, was convinced that organisms are chemical, cybernetic machines operated by means of the program written in DNA molecules (Monod 1972). The program metaphor of life was also supported by Alan Turing's thought experiment of a hypothetical machine able to rewrite any program written in binary code. A biological Turing machine should be able to transform a finite input string of DNA into output strings of RNA and proteins (Neuman 2008).

The gene was defined as a chromosome section that encodes for a functional product (either RNA or polypeptide) and Crick (1963) introduced his hypothesis of 'one cistron = one polypeptide'. The discovery of hnRNA splicing complicated the relationship one cistron = one RNA-one polypeptide: it has been shown that up to 90% of genes have different splicing variants, and therefore the definition of gene got even more complicated than before. It was about to get worse: with the discovery of *trans*-splicing, genes were found to be overlapping or, on the contrary, found in pieces that can be located on different chromosomes (Portin and Wilkins 2017). Additionally, genes were found *not* to be a singular level of heredity.

Molecular biology helped to establish another branch of evolutionary biology: neutral evolution. Introduced by Japanese evolutionary biologist M. Kimura (1983), this theory was applied to the molecular level of evolution, where most mutations are actually selectively neutral, thus do not affect organismal fitness. Therefore, not natural selection, but genetic drift is responsible for their propagation. After R. Dawkins published his popularization of Hamilton's work in *Selfish gene* (1976), genetic determinism was completed through the metaphor of discrete units of selfish genes which 'care' only about their transfer into future generations. Organisms are interpreted as mere carriers of selfish genes, molded by natural selection and random

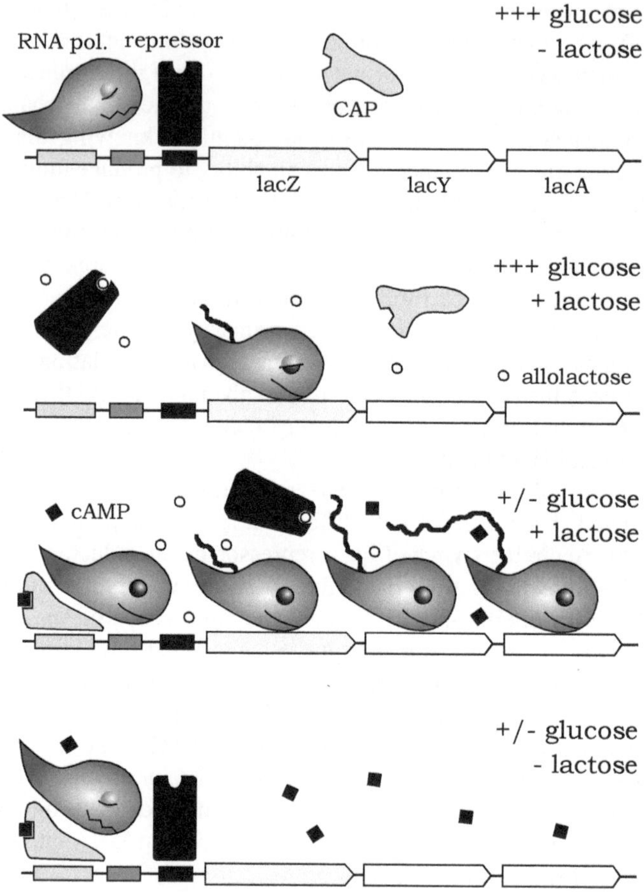

Fig. 5.2: The *Lac* operon model in *Escherichia coli*. Tight interaction between cellular organization (presence of enzymes and signals), genetic memory, and environmental cues (presence of substrate). **Rows 1 and 4:** In the absence of lactose, the operon is in a repressed state. The presence of lactose is signaled by allolactose that removes the repressor: transcription is possible, yet very slow if glucose is in plenty (**row 2**). Only if glucose is limited, the stress signal cAMP activates the transcription enhancer (protein CAP), and the transcription proceeds at full speed (**row 3**). (After Jacob and Monod (1961a, b).)

mutations. Such genetic determinism had many critics among them S.J. Gould, R. Lewontin, B. Goodwin, S. Newman, as well as W. Gottlieb or S. Oyama (Gilbert and Epel 2015).

The Human genome project began in 1990, and between 1990–2003, it sequenced 92% of the human genome. On one hand, this was comparable to such achievements as the Moon landing; on the other hand, the project brought certain disappointment because neither did it answer the great question, 'who we really are?', nor did it confirm the expected causality of gene-disease relationship. It has been shown that only about 2% of human DNA differs from that of the chimpanzee, and only about 7% of the genome is vertebrate specific (Marks 2003). As to the content of genes, the human genome is also much smaller than previously expected, containing only about 22,300 genes (and the number will get even smaller after ENCODE project, see below) similar to the genomes of flies, or some plants. Clearly, there is no correlation between the amount of coding DNA and the morphological complexity of the organism (Davidson 2006). In addition, the expected causality between a specific gene and a specific disease was proved as not that simple as expected (Sapp 2003).

Evolutionary Developmental Biology

As previously mentioned, embryology and developmental biology were not part of the 20th century Modern Synthesis, although examples of development influencing evolution were known. In the first half of 20th century, biological science was too concentrated on mutations, model organisms and the study of genes, such that processes of soft inheritance or changes in developmental sequence were considered to be exceptional and unimportant for evolutionary change. Development was considered a black box between genotype and phenotype. The return of embryology is usually associated with the study of genes which regulate development because what evolutionary developmental biology (evo-devo) showed in particular is that the evolution of endless forms is connected to the evolution of regulatory networks of ancient genes, rather than the origin and elaboration of new genes (Carroll 2005). Evo-devo still involves comparative embryology and morphology (within new approaches: concepts of phenotypic morphospace and quantifications of shape transformation); yet it also allows studies of developmental genetics, experimental epigenetics as well as theoretical and computational approaches (Müller 2007). Except for changes in genetic developmental modules, evo-devo seeks for an understanding of homology, homoplasy, and the emergence of body plans and the generation of phenotypic novelty.

The classic example of ancient genes engaged in development genes is toolkit genes (Carroll 2005, Carroll et al. 2006) such as *Hox* (or homeotic) genes in animals, and analogical genes in plants (Figs. 5.3, and 7.1, p. 120). Most of the developmental genes encode for transcription factors regulating the activity of other genes, often through complex interaction with other transcription factors and proteins, binding to enhancers of genes active in embryogenesis. Hox genes were found in all studied animals, and play a crucial role in the ontogenetic diversification of individual body parts. Moreover, short regulatory sequences flanking each such 'genes for' appear to be even more important than the tools themselves. The shuffling of such elements across the genome may easily recruit old tools to new and unexpected space or time contexts (heterotopy and heterochrony), thus influencing both appearance and function of the organism.

Hox genes can activate or repress groups of downstream genes by binding to DNA sequences of Hox-response enhancers; they can similarly control other executive genes or also regulate their own expression, using transcription factors, morphogens or signals that they encode (Pearson et al. 2005; see also Chapter 7). *Hox* genes determine the main embryonic regions along the anterior–posterior body axis, and specify the particular identity and relative position of given structures (Slack et al. 1993). Expression begins just beyond the place of the future brain (head formation takes place in a different way, with a different toolkit). Later in development, the

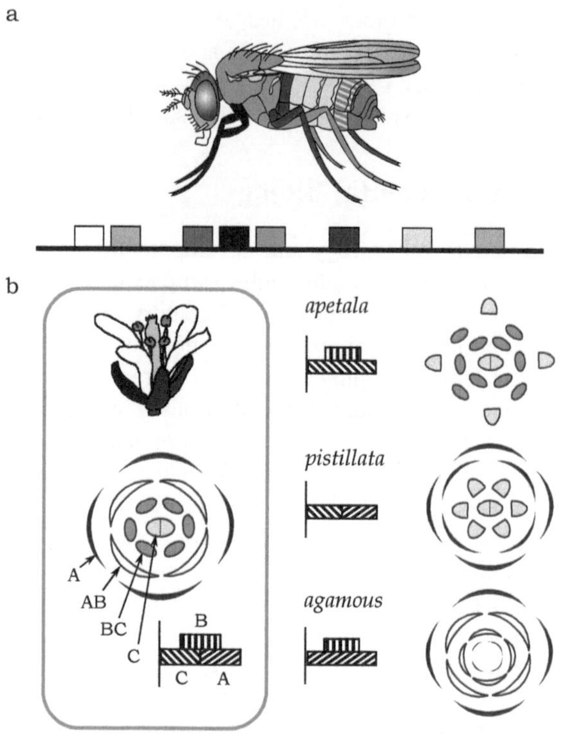

Fig. 5.3: Homeosis—improper setting of body modules. The body plan is defined by establishing genetic regulatory modules in early morphogenesis. **Above:** in animals, the anteroposterior axis is set by expression of homeotic and related proteins. Their expression in an improper module leads to homeotic mutations, e.g., legs in place of antennae or genital appendices. **Below:** The basic plan of flowers consists of four rings of organs (calyx K, corona Co, anthers An, and stamens P), fixed by three cassettes of homeotic genes (in settings A, AB, BC, and C). Examples of mutations: misfunction of A-cassette (*apetela*, C, BC, BC, C) leads to a mirror phenotype PAnAnP; mutation in B (*pistillata*; A, A, C, C) gives rows KKPP; and deletion in C (*agamous*, A, AB, AB, A) leads to a sterile, mutilated flower.

products of *Hox* genes also act as region-specific determinants in diverse structures and tissues (Carroll et al. 2006). The sites of their expression often overlap, and they inhibit or activate each other by binding to the *cis*-regulatory sequence, thus forming complex regulatory network modules in time and space. For example, the Hox protein encoded by the *Ubx* gene (*Ultrabithorax*) and the hox protein Abd-A (Abdominal-A) act as repressors for the *Dll* gene (*Distal-less*) in the abdominal part of the Drosophila epidermis but as activators of the *decapentaplegic* gene (*dpp*) in visceral mesoderm.

Drosophila melanogaster and its *Bithorax* mutation, discovered by C. Bridges in 1915 (McGinnis et al. 1984), has contributed greatly to what we know about homeotic genes. The function of the homeotic gene is experimentally detectable by homeotic transformation (homeosis), as in Bithorax in which morphological structures arise on the wrong part of the body. This mutation is manifested by the transformation of the third thoracic element into the second, and therefore, instead of halteres, the fly develops a further pair of wings and stomach. Another notorious mutation is the so-called *Antennapedia* mutation, where fly poles are replaced by limbs.

Although animal homeotic genes have specific effects on the morphological structures of the organisms of each species, they all share a certain functional property, which is a DNA sequence called homeobox (183 base pairs that encode the 61 amino acids of the homeodomain). The transcription factor, i.e., the Hox protein, that then binds the cis-regulatory sequence of the DNA using the homeodomain is highly conservative module of a Hox protein. Later it was found that not all protein products of homeotic genes have such homeodomain (especially in case of plants) and vice versa; there are some homeotic genes' products whose mutations do not appear as homeosis. Homeotic gene products act as genetic switches and thus influence the expression of other genes. Their regulatory network is quite conservative: for example, the gene *eyeless* is a key gene in eye development in *Drosophila* (and other animals as well). When the mouse variant of this gene was transferred into the genome of blind *Drosophila* mutant, the induction resulted in the reversion to the norm. Of course, the new structure was an eye of a fly, not of a mouse. Such plesiomorphy suggests the common ancestry of all animals.

The discovery that the same series of genes control identical functions in the formation and pattern of body areas and parts (even those with very a different design) in insects, vertebrates and other animals has meant a complete rethinking of natural diversity, animal history, and the origins of structures (Carroll 2005).

Homologous genes also appear to play a role in processes such as cell division, cell adhesion or cell movement (Pearson et al. 2005). Enhancers, to which the *Hox* gene products bind, are tissue-specific. For example in *Drosophila*, the enhancer of the *dpp* gene dependent on the UBX protein is active only in visceral mesoderm, CNS and somatic mesoderm (and inactive in the epidermal mesoderm). Mutations in cis-regulatory sequences are probably responsible for the evolution of morphologies. However, a duplication of homeotic gene clusters (one set carries 8 orthologs) also contributes to morphological diversification, as well as mutations and the resulting changes in Hox proteins themselves (Pearson et al. 2005).

Genes such as *Hox* genes create complex gene regulatory networks (GRN), which Eric H. Davidson describes in his volume *The Regulatory Genome* (2006) using examples of classical model organisms. He mentions the development of specific structures such as the skeleton of sea urchin larvae or the heart of the *Drosophila*. Davidson's interpretation of GRN is a reductive model that treats developing embryos as computational machines. Inspired by computer networking, Davidson uses terms such as web architecture, nodes, or peripherals to describe genes as serially hardwired, conservative 'batteries' that are shared across taxa. The genome represents a set of programs and subprograms, corresponding to the hierarchical activities of the cells. Davidson compares genes to communication protocols: the nodes of the network control the information-processing units. It is a precise, mechanistic description of spaciotemporal regulations in development, yet the time sequence, which is decisive in these processes, cannot be grasped from the schemes themselves.

Phylotypic stage. The conservative toolkit genes are also connected to one important developmental stage, which is phylotypic stage (sometimes called the pharyngula, tailbud or, in insects, germband stage; Ballard 1981, Sander 1983, Slack et al. 1993; Fig. 8.1, p. 136). It is that specific developmental stage wherein all the members of the same phylum most resemble each other. Developmental pathways leading to this stage are usually quite different even among closely related taxa (usually differences in cleavage). After gastrulation, the morphological divergence of all members belonging to the same phylum is highly reduced; this marks their entrance into the phylotypic stage, after which the developmental pathways start to differ again. The vertebrate phylotypic stage is usually morphologically characterized by the presence of the neural tube, notochord and somites, the head with pharyngeal pouches, heart and optic anlagen, and a tailbud (Richardson 1995, Richardson et al. 1997, Švorcová 2012; see Chapter 8 and Fig. 8.1).

Ernst Haeckel first drew attention to this conservative developmental stage (1874), but his depiction was later criticized as overly simplifying. Some authors recognize conservation on the genetic level, but not on the morphological one (Richardson 1995, 1997, Bininda-Emonds et al. 2003), due to the many morphological differences caused by heterochrony or somite numbers. Slack et al. (1993) were the first to recognize that during the phylotypic stage, the most conservative, evolutionarily oldest, orthologous genes (such as the *Hox* genes) are activated (Domazet-Lošo and Tautz 2010). Also, during this period there exists an intense web of interactions among organ primordia wherein the whole phylotype acts as a singular developmental module with a high degree of interactivity (Galis and Metz 2001, Irmler et al. 2004). Only later, after phylotypic stage, development operates on the level of semi-autonomous modules within the body; this modularity ensures higher phenotypic stability during development and prevents the pleiotropic effects of possible environmental perturbations (Wagner et al. 2005).

Modularity. The concept of modularity was introduced and emphasized within developmental biology by, e.g., Riedl, Lewontin, Bonner, and Raff (Nelson 2004, Wimsatt and Schank 2004). Modularity can be recognized on many levels of biological organization including genetic, developmental, anatomical and behavioral (Müller 2007). A module is generally defined by a high degree of internal and a low degree of external interactions (i.e., interconnectedness with other modules; see hairball in Chapter 7); hence, the pleiotropic effects tend to be restricted to behavior within a specific module. Modules are evolutionarily persistent and often reused in new contexts. They can be autonomous, which means they are insensitive to perturbation of the context in which they are embedded (the module is able to maintain the same function in abnormal tissue environments such as when transferred to a different location within the embryo, Schlosser 2004).

Modularity is usually considered advantageous in terms of higher phenotypic stability during development (canalization), which is an ability to facilitate adaptation (the effects of the environment are not pleiotropic, but selective on specific modules) and possibly an escape from adaptive constraints (Wagner et al. 2005). As Müller demonstrates in his review on the evo-devo agenda (2007), the genetic background need not remain the same in cases of the same anatomical modules, i.e., the change in developmental sequence still leads to the same conserved phenotype. In such cases, it appears that the bodily level is more important than the gene regulation networks—

the whole body may often be taken for a single module thoroughly isolated from the surroundings (see Chapter 8).

The Rise of Epigenetics

After the 1980s, knowledge of epigenetic modifications such as DNA methylation, histone modification or different types of small RNA molecules accumulated, and epigenetics came to be taken for granted (Chapters 6–8). At first, the emphasis placed by the Modern Synthesis on genes and gene mutation being the ultimate cause for the construction of an individuum, as well as the specter of Lamarckism (along with an unhealthy pinch of Lysenkoism, though this is not part of our reading), made epigenetic discoveries hard to swallow. However, by 2009, Jablonka and Raz were able to publish a review of more than 100 cases of epigenetic inheritance, which demonstrate that epigenetic causation is responsible not only for cellular differentiation and DNA packaging, but also represents another level of hereditary information.

ENCODE. Following the Human Genome Project, another Project called ENCODE was launched. Unlike the human genome project, ENCODE researched not the genetic script itself, but the biochemical activity of the genome. It began in 2003 and its first results were published in 2012 in *Nature* and *Journal of Molecular Biology*, and the effort has since continued. More than 450 scientists from 30 institutions took part in this project, and studied the biochemical activity of the genome in 147 human cell types. Their results demonstrate that the amount of protein coding sequences (i.e., genes) is even smaller than previously thought, comprising only about 1.5% of the genome (about 20,000 genes). They found that more than 80% of the genome has some biochemical function, and that there is a lot regulatory sequences which serve as silencers, enhancers or promoters (there are about 500,000 regulatory sequences in total, of which 400,000 are enhancers), i.e., they have a protein binding function. Moreover, the majority of biochemically active DNA sequences are coding for different types of regulatory RNAs, which interact among themselves, with the chromatin and other proteins (see Chapter 7).

The Encode project further complicated the definition of the gene such that the authors prefer to use the term 'functional element' for a sequence provided with biochemical function. In addition, protein coding sequences often overlap each other or are discontinuous. Encode also revealed a great variability in epigenetic markings, finding about 3 million nucleosomes with differently modified histones (Fig. 3.2). However, only about 3700 types were shared in all cell types (The ENCODE Project Consortium 2012).

Epigenetic inheritance. As defined by Jablonka and Raz (2009), the term epigenetic inheritance represents, in a broad sense, "the inheritance of developmental variations that do not stem from differences in the sequence of DNA or from persistent inducing signals in the present environment". In Weismannian terms, this involves soma to soma transmission between mother and offspring. In a narrower sense, epigenetic inheritance usually represents the inheritance of chromatin marks and RNA molecules. Jablonka and Raz (2009) also include within this category the transmission of self-sustaining metabolic feedback loops, the cellular inheritance of 3D structures including cilia (Sonneborn 1964, Beisson and Sonneborn 1965) or the self-propagation of prions, and also the heredity of cellular membranes.

Epigenetic modifications on chromatin can be induced randomly, but also environmentally, and then transferred from one generation to another (when we consider only eukaryotic organisms). For example, in mammals, stressful experiences such as malnutrition, insufficient maternal care, and exposure to toxins or fear can thus influence the progeny's health, sexual preference or appearance (Meaney 2001, Waterland and Jirtle 2003, Weaver et al. 2004, 2005, Meaney and Szyf 2005, Anway et al. 2006, Morgan and Whitelaw 2008, Diaz and Ressler 2014). The reactions are quite plastic: the epigenetic markings can be erased, when the initial stimulus is no longer present in the environment (Chapter 8). There are also suggestions that epigenetic induction might eventually determine the evolution of instinctive behavior (Ginsburg and Jablonka 2010, Robinson and Barron 2017). Environmental factors that lead to changes in epigenetic markings can, together with environmental signals processed by neuroendocrine system (followed by hormonal production), be a primary source of gene expression alteration, usually associated with phenotypic plasticity.

Other cases of extragenetic inheritance. Extragenetic inheritance is a subject of studies of extended evolutionary synthesis (see below), which usually involves epigenetic inheritance, ecological inheritance and social transmission. Extra-genetic inheritance has become important for theories that attempt to extend the concept of inheritance beyond genetics. Ecological inheritance involves the inheritance of modified environments, i.e., niche inheritance (Odling-Smee et al. 2003). Such an inheritance system does not rely on the reproductive behavior of discrete replicators and can be also transmitted externally and horizontally, among various organisms, within and across generations (such organisms do not have to be related). Although cultural transmission is usually held as referring to humans, it also applies to non-human animals (e.g., dogs) and basically denotes the transmission of information via social learning (observation, imitation, teaching, memes, etc.). The sum of such information is a culture, which according to Richerson and Boyd (2006), only in case of humans is of a cumulative nature. We aim to broaden such definition of culture also to other organismal communities (see gratuity, and Chapter 7).

Extended Evolutionary Synthesis

Phenotypic plasticity is an ability to develop multiple phenotypes above a single genotype; usually, the variations are caused by environmental factors, such as temperature, stress and nutrition, physical effects such as pressure, gravity or light, and the presence or absence of predators, or even members of the same species (for further reading, see West-Eberhard 2003, Piersma and van Gils 2011, Gilbert and Epel 2015).

For proponents of Extended Evolutionary Synthesis (EES), who originally came from the circles around evolutionary developmental biology, ecology and behavioral sciences (Pigliucci 2009, Pigliucci and Müller 2010), phenotypic plasticity is often viewed as a primary evolutionary condition. Organisms are inherently plastic (plasticity is the first step in adaptive evolution), whereas genetic fixation of the phenotype comes later (West-Eberhard 2003, Newman et al. 2006), depending on environmental conditions, developmental bias, and selection (see also genetic assimilation above, and Fig. 5.1). For classical theory, the environment selects for genetically fixed variability expressed as phenotypic plasticity, whereas in the scenario of phenotypic plasticity, the environment creates phenotypic variants and then selects them.

Developmental bias is another key concept of EES; it emphasizes that some morphological forms emerge more easily than others (Laland et al. 2015). Thus, some examples of convergent evolution such as the development of eyes in distant taxa could indicate developmental bias and selection work in tandem. Developmental bias can be seen, for example, in the even number of segments in more than 3000 species of centipedes, or the parallel evolution of cichlid fishes in African Rift valley lakes (Johnson et al. 1996, Parker and Kornfield 1997), or parallel speciation of sticklebacks in Canadian lakes (Braithwaite and Odling-Smee 1999). The diversification of cichlids apparently started from a single pair of fish, to end, after several thousands of years, in tens of species. A comparison of genomes should reveal whether the speciation event is based in mutation or alteration of the genome, or whether the whole event is epigenetic.

In general, for EES proponents the variation generated by mutations, recombination and subsequent selection are not sufficient to explain adaptive evolution (Pigliucci 2009, Pigliucci and Müller 2010, Laland et al. 2015).

Organismic System Approach. The questions of developmental bias and evolutionary novelty, together with the evolution of form and causal processes that lead to phenotypic organization, are the main topics of organismic system approach (OSA) represented by S. Newmann, G. Müller and R. Bhat. Whereas neo-Darwinism interprets phenotypic plasticity as an evolved adaptation resulting from selection pressures and changes in genetic program, OSA sees plasticity as an inherent and primitive characteristic of living matter, i.e., plasticity is not programmed, but stems from the physical properties and the ability of the organisms to react to environment. The genetic program is not a precondition for or cause of evolution, but a product of evolution (Newman and Müller 2000, Newman et al. 2006, Newman and Bhat 2009).

For OSA, the primary morphological determinants in evolution are epigenetic processes, not in the modern sense of epigenetic modifications, which are responsible for cell differentiation, but rather in the Aristotelian sense of epigenesis, the emergence of form. These primitive epigenetic processes are the interactions of cell metabolisms with and within the outside world and the cell, the interaction of primitive cell masses with the environment and these primitive tissues among themselves.

The examples of phenotypic plasticity as we observe them today (Piersma and van Gils 2011) may represent the remnants of the original state of affairs, when epigenetic processes were not yet tied up by genetic hardwiring. Thus, the whole concept is based on two phases of evolution: pre-Mendelian and Mendelian. During the pre-Mendelian period, before the Cambrian explosion, organisms were extremely plastic and did not have the integrated phenotype of modern organisms. They were strongly influenced by the physical and chemical properties of the environment and of themselves. The authors assume the existence of primitive cell aggregates that are able to react to electrical, chemical or mechanical impulses, which led to the emergence of primitive cell masses. The interactions of these masses with the inner and outer environment led to primitive organ forms and body plans (segmented, multilayered, etc.).

The oldest known and most reliable multicellular fossils, from the time between Ediacaran and Cambrian, are from 700 Ma ago, the most studied are from the period 580–542 Ma (although some authors date them even earlier as mentioned in Chapter 4, Zhu et al. 2016 or Bengtson et al. 2017). They were simple, soft and flat animals without body cavities (the body sheets emerged only in Cambrian); there are no obvious similarities to contemporary organisms and, mainly due to the lack of body envelopes, the fossils are rather rare (Newman et al. 2006). By 550–530 Ma, during a period of the Cambrian explosion, all Metazoan body plans known today emerged. Classical neo-Darwinism is not able to explain this sudden emergence of body plans, but OSA is (Müller and Newman 2005).

On the genetic level, all Metazoa apparently use a conservative genetic toolkit for their evolution and diversification, without actually changing the genetic script (see example of Hox genes above). These toolkit genes are responsible for development of functionally similar, but morphologically different and taxonomically distant structures.

Many other conservative regulation networks exist beside the Hox system. Another example in animals is the *Pax-, Six-, Eya-, Dach-*gene regulatory network functioning as a multiply deployed module in vertebrate as well as insect eye development (Wagner 2007). Homologous regulation network in *D. melanogaster* is formed by *ey* (*eyeless*) gene, *sine oculis* (*so*; with homology to the *six* gene family in vertebrates), and two transcriptional cofactors, *eyes absent* (*eya*) and *dachshund* (*dac*). Such regulatory networks are very complex (Chapter 7), but what we need to emphasize here is the varied redeployment of the same conservative genetic script, Pax being only a single, however, notorious example.

Newman and Bhat (2009) believe that toolkit genes existed in evolution even before the emergence of Metazoan organisms. They divide the toolkit genes into two basic categories: developmental transcription factors (DTF) and dynamical patterning modules (DPM). DTF were already mentioned above and they involve the group of Hox genes, and families like Hes, Tbox, Dlx, Pbx and others. They determine the cellular type differentiation and specific organ regions formation. Davidson (2006) infers from the existence of homologous pathways of DTF that they played a major role in evolution during Cambrian explosion; however, Newman and Bhat (2009) do not believe it is a sufficient condition for the evolution of morphogenesis.

They also emphasize the role of DPM, which involves the genes for cadherins, Notch, Wnt pathways and their ligands, TGF-beta/BMP, FGF, Hedgehog and their receptors. DPM are crucial for cell adhesion, differential adhesion, the properties of cellular surfaces, supracellular gradients, etc. They had to play a major role in the emergence of Metazoa. For example, in the presence of Ca^{2+}, cadherins bound to cell can gain the property of adhesion, which was the basic cell property for multicellularity and morphogenesis. Their homologs are already present in the closest non-metazoan relatives to Metazoa, choanoflagellates (Abedin and King 2008, see Newman and Bhat 2009), which are able to form colonies. The appearance of cadherins led to the emergence of cell types and tissue layers. Of course, the heredity of such states was dependent on the stability of the environment (the amount of Ca^{2+} in the environment). Differential adhesion has led to different cell islands, to the lumen formation, different types of tissues and tissue layering or different types of gastrulation, etc. (see also Chapter 7).

Newman and Bhat (2009) also take into account the processes of biochemical oscillations that can regulate the surrounding cells within the cell aggregates, whether the regulation is juxtracrine, as in case of Notch pathway or paracrine, as in case of Wnt pathway. Furthermore, they also mention lateral inhibition, where the pathway Notch-Delta plays a crucial role in causing the cell to choose between two possible states (thus maintain the cell mixture in more than one biochemical state). The authors further unwind the role of Wnt pathway in determining apical-basal and planar cell polarity, the role of DPM in synchronized biochemical oscillations, and the morphogen gradients in the formation of spots, stripes or different boundary patterns (Chapter 8). Newman and Bhat believe that generic, epigenetic processes based on DPM created all types of extant body plans, which involved major transformations such as the creation of cell mases and mixtures, tissue masses and layers, body segments and cavities, cell polarities, etc.

Only later, marking the formation of the Mendelian world, these processes of physical morphogenesis were fixed by genetic information, but before this world began, the organismal forms were very plastic, and were based only on the physical properties of the environment and the properties of their cells. One genome corresponded to many phenotypes because the genetic hardwiring was not as decisive as environmental conditions (variations above the Norm, as we called it in Chapter 1). Later on, phenotypes became progressively more coopted—bound by genomic regulatory networks through stabilizing selection, genetic assimilation, mutation and duplication. Eventually, organisms became more robust to environmental conditions and perturbations, and physical conditions no longer had the primary role in morphology. Yet correlation between the genotype and phenotype is only a derivative of these morphogenetic processes, rather than an original state of affairs or even an evolutionary cause.

Interestingly, in one of their publications, Newman and Bhat (2011) are not afraid to wake the ghost of Lamarck by referring to an internal, physical power that leads to the cumulative improvement of organismal forms. Lamarck himself considered this power to be purely physical, a natural necessity of some sort. In his conception of internal power and by emphasis of physical processes in organic matter, Lamarck anticipated the ideas of the organismic system approach to evolution: an approach based on generic, physical, epigenetic processes (Newman and Bhat 2011).

The 'Phenotype First' Approach. As mentioned above, in some evolutionary scenarios the developmental plasticity can be viewed as an inherent property of living organisms responsive to environmental changes, whereas in the classic Modern Synthesis scenario, genes are usually seen as a sufficient explanation of evolution—and developmental plasticity is seen as a result of such evolution. The 'phenotype first' approach, which integrates Waddington's genetic accommodation (Waddington 1942), the Baldwin effect and epigenetic inheritance systems, offers a completely reverse scenario, where genetic fixation is not the prerequisite, but a result of evolutionary scenario (West-Eberhard 2003, Piersma and van Gils 2011).

J.M. Baldwin published his idea in 1896, emphasizing the role of (mainly learned) behavior in evolution when an organism faces new environmental conditions. Such new behavior (or other functions) can, in some cases, facilitate selection, i.e., it can be adaptive (phenotypes with the new behavior can have better reproduction success or other advantageous properties). Although Lamarckism was still popular during its eclipse, Baldwin himself did not consider this process to be Lamarckian. Some modern interpretations of the Baldwin effect (Hall 2001) claim that a mutation has to occur to fix such phenotypic change; such a mutation (or mutations) can then spread throughout the population by being favored by selection. Compare this with the genetic assimilation hypothesis, where the mutation is not required because the existing (yet to the moment unexpressed) genetic variability comes to be expressed in an assimilated phenotype, and is subsequently favored by selection.

The idea to merge genetic accommodation and the Baldwin effect came from Mary Jane West-Eberhard (2003) who developed a modern conception of genetic and phenotypic accommodation. She claims that genes are followers, not leaders in evolutionary change. In this scenario, the plastic, phenotypic change is a first step towards evolutionary change. Thanks to stable environmental conditions and to the modular organization of the phenotype, such phenotypic change can be adaptive and thus accommodated phenotypically. This means that change in the regulatory gene architecture is sufficient for phenotypic accommodation.

According to West-Eberhard (2003) and Newman et al. (2006), organisms are primarily plastic, so they can respond to environmental changes. The phenotypic response, often adaptive, can become integrated phenotypically (thus, we speak about phenotypic accommodation). A new phenotype (or trait) does not have to be fixed genetically; it can become fixed epigenetically, or through changes in the modular organization of the phenotype. In other words, the evolution of a new phenotypic trait does not require the evolution of new gene complexes, but only changes in the regulatory architecture of the old ones.

However, if environmental conditions persist and if such a new phenotype becomes advantageous, natural selection can fix it through the changes in the genotype; thus, the traits also come to be genetically accommodated. Such genetic change may either wait on some random mutation, with subsequent fixation of a phenotypic trait (Baldwin effect) by deployment of undiscovered genetic variability (Waddingtonian assimilation), or on some quantitative change in gene frequencies (genetic accommodation). In such cases, individual adaptation is a sufficient cause of population adaptation.

Such a scenario can also explain the dual inheritance hypothesis of biocultural evolution (Richerson and Boyd 2006), i.e., such cases wherein cultural adaptation can lead to genetic changes, such as the spread of the genetic ability to digest milk when humans started dairy farming. The cultural adaptation of dairy farming increased the relative fitness of genes for lactose tolerance in the human populations with access to fresh milk. Laland et al. (2010) shows whole sets of genes that have recently been subjected to rapid selection due to various cultural selection pressures.

Niche Construction. The last EES concept we mention is a crucial concept of niche construction, which integrates the evolution of culture, ecology, genetics and evo-devo. The niche construction concept emphasizes the dynamic and reciprocal relationships between organisms and their environment. In the framework of classical neo-Darwinism, the environment is merely a background for evolution—however important the selection pressure it generates. By contrast, the niche construction concept emphasizes that organisms themselves are also affected by changes in the environment that they co-create. This idea comes from R. Lewontin, who often criticized genetic determinism and described organisms as constructors of their environment and niches as constructions of organisms. M.W. Feldman, K.N. Laland and F.J. Odling-Smee furthered Lewontin's concept under the term niche construction (Odling-Smee 2003).

Niche construction is that process by which the organism alters the immediate environment in which it and neighboring species live. Successful niche construction increases a species' chances of survival, and thus directs its own evolution—and also that of other species. Hence, not only does the environment generate selective pressures on individual members of a population, but also individuals of the species change the environment through the process of niche construction. Thus, biotic relationships represent a constant feedback between the processes of natural selection and niche construction wherein as the organism affects the environment and also influences the selective pressures upon itself.

An example of niche construction par excellence is the above-mentioned influence of dairy farming on our genes, but people are not unique in their ability to change their environment. Birds build their nests, earthworms chemically change the soil in which they live, and thus affect plants and other organisms associated with it. Beavers build dams, and thereby change the watershed and provide a source of water for other organisms, as well as alter the forest: roots, soil structure, and all. Algae and plants alter the amount of oxygen in the atmosphere, which affects the flow of energy and nutrients. Fungi and bacteria decompose organic matter; bacteria also produce substances of various kinds that change the environment, bacteria that dwell in our body cavities influence our physical and even psychical health or development (Chapter 10). The list of examples could be almost infinite.

A quite popular example of niche construction concerns the ant species *Myrmelachist schumanni*, which lives in Peru's forests and uses their formic acid as herbicide to the detriment of most trees, with the exception of *Duroia hirsuta*—the tree in which these ants live.[15] Ants not only protect the tree from uninvited visitors such as caterpillars or grasshoppers (who the ants harass collectively until visitors leave), they also inject formic acid into the leaves of other (competing)

[15] BBC documentary of David Attenborough from the Cycle Life in the Undergrowth, episode Intimate relations.

plants. Such injections usually kill the plant within a few days. These actions result in drastic changes in the tree biodiversity of the habitat, leading to tree monocultures that Peruvian natives call the Devil's Gardens. The largest observed Devil's Garden contains over 320 trees of *Duroia hirsuta*; it is about 1300 m^2 and probably over 800 years old.

Constructed niches are often heritable, and so niche construction theory involves the concept of ecological heredity. To sum up, organisms can change their and other species' environment through their own activity (the ecology calls such process usually as ecosystem engineering). The development and existence of certain traits really depend on the construction of the environment by these or other species inhabiting the locality. This means that modifications of the environment by organisms (niche construction) and inheritance of the modified environment (ecological inheritance) are processes not only of ecological but also of evolutionary nature. The constructors change the selection pressures by changing the environment through niche construction. Such selective pressures can be directed by organisms themselves or had been directed by their ancestors, and influence at least one population in the ecosystem (Matthews et al. 2014).

Philosophical Approaches

Moreover, in relation to EES, we need to mention the philosophical approach called Developmental Systems Theory (DST; Oyama 2000, Oyama et al. 2001). This theoretical approach emphasizes epigenetic processes in evolution, as does our book. It denies any basic level of description for evolutionary and developmental processes, and considers these as existing in mutual influence and unity. It also stresses the structural aspect of the living, wherein the higher structure cannot be reduced to the lower one. DST considers the ontogeny of an organism as a contingent cycle of interactions among various resources of variation, not only the DNA molecule, but also the whole structure of the cell/organism, as well as interaction with the environment. It discusses all the aforementioned topics including extended heredity, niche construction, contingency of evolution, etc.

Process biology offers yet another approach within the ontology of biological phenomena. In order to understand process biology, we need to explain the tradition of process philosophy. Unlike traditional substance-based ontology, process philosophy emphasizes processes against substances. It dates back to Heraclitus who perceived stability as an illusion, and understood reality as undergoing constant change. The most famous process philosophers are A.N. Whitehead (1861–1947), and H. Bergson (1849–1951), but G.W. Leibniz (1946–1716) is also often conceived as a process philosopher, as are the American pragmatists including J. Dewey (1859–1952), C.S. Peirce (1839–1914), W. James (1842–1910), and many others.

Bergson is a process philosopher primarily because of his concept of *durré* (duration). Time cannot be defined as a homogeneous quantity, which can be modeled on space, as Newton perceives it (as time points, one after another). Duration, i.e., the way we experience time, is something that unites past and present into an organic whole. Both past and present permeate each other as well as the totality of a living organism, whose parts penetrate one another, although they are theoretically distinguishable. Bergson speaks about *succession without distinction*, i.e., we are able to distinguish the individual events only thanks to our ability of abstraction. Duration is an aspect of our mind; it cannot be understood in terms of quantity or substance. Also, organisms are definable as duration wherein the whole past of an organism exists in actuality (Bergson 2001, 2004, 2012). As Merleau-Ponty (2003) emphasizes, Bergson defines living by history. For him, organisms are outside of every comparison with the physical system because they are primarily temporal beings.

However, these days, process philosophy is generally related to A.N. Whitehead. His book *Process and Reality* (1929) rejects both Aristotle's metaphysics of substance as well as of materialism, that is, that the world consists of extended substances. In contrast, Whitehead's process philosophy claims that reality consists of actual events that are experienced through actions of *prehension*. This experiential activity enables one actual event to relate to the past events, which we integrate into one organic whole through the process of *concrescense*. It is this process of synthesis that lets all past events collapse into a single actual experience. The world is thus a chain of events and experiences rather than composed of material substances. What we describe as permanent things are in fact processes, and only our ability to abstract enables us to talk about them as spatially and temporally extended objects. Whitehead calls this phenomenon a 'fallacy of misplaced concreteness': the error of mistaking the abstract for the concrete (Whitehead 1997). This fallacy also concerns situations when we grant scientific models ontological status. So far as organisms are concerned, Whitehead defines them as intrinsically active, interdependent (as actual entities), and self-related subjects.

Recently, a group around the English philosopher J. Dupré has elaborated a natural metaphysics providing an ontology for current biology (Nicholson and Dupré 2018). Every science, whether or not the scientists are able to accept it, has a metaphysic background. Contemporary science has a substantial one because it models and defines phenomena as things, countable, stable, autonomous, and atemporal, whether they are genes in biology or atomic particles in physics. Such metaphysics is connected with mechanicism because it explains living organisms as composed of mechanically interacting things and parts. Process biologists replace the substance metaphysics with a process one. They have several reasons for this: (i) It is difficult to find anything that does not change. Change is observable on many levels of organization. Galaxies, stars, planets, ecosystems, organisms, cells, DNA, molecules and even atoms change. (ii) Substance has no effect without process. Inert substances without dynamic interactions are irrelevant. For a process biologist, the dynamics of interaction are more important than the entities in interaction. The idea applies, for example, to toolkit genes where spatio-temporal interpretation of the same genetic script is a decisive factor in the resulting morphology. (iii) There are many processes that cannot be described as substances such as storms, fire, meaning attribution, scent, etc.

A reader may notice that even within a process approach, it is hard to avoid terms such as *things* or *entities*. Process biologists argue that what we observe as things are actually well stabilized processes. Change is omnipresent; so what needs to be explained is why some things remain stable. They define process as a coordinated family of change/event, which can be functionally or causally interconnected. Processes are composed of events, contingencies and occurrences—none of which are things.

This line of thinking also interprets organisms as processes. Again, process biologists provide some arguments: (i) Organisms do not exist in equilibrium; they are open systems constantly exchanging matter and canalizing energy. (ii) They are defined by causal continuity, not by the stabilization of their properties. Stability must be maintained by the organism; it is not a primary characteristic, but a thing that has to be achieved. (iii) Organisms exhibit life and developmental cycles, i.e., they undergo many phenotypic changes during their life, passing through different stages, yet remain the same organism. This feature distinguishes them from machines. (iv) Organisms exist as ecologically interdependent (see niche construction in this chapter, or Chapter 10 on holobiotic interactions). This means that they are not independent things, but interdependent processes (Nicholson and Dupré 2018).

This ontological approach of process philosophy helps overcome certain conceptual difficulties that pertain to defining a biological individuality. For example, as every living thing is an interconnected, processual whole, organisms cannot be defined without their relation both to other organisms and to the environment. Sometimes it is difficult to demarcate the exact boundaries of an organism, as it is with fungi, plants or holobionts. In addition, the concept of the organism as a process defines itself against the mechanistic and reductionist approach prevalent in contemporary biology. Organisms are perceived as self-stabilizing processes (even genomes are considered structures which are maintained and stabilized by the cell, a view isomorphic with the phenotype first approach mentioned above). Moreover, organisms can be described neither from a single genetic level of description (antireductionism) nor as machine-like entities, precisely because they are not mechanically interacting entities made up of their parts, but processes of varying degrees of wholenesses.

Processual philosophy is important for our book in many ways: we emphasize the community into which every individual is born, as well as interorganismal relationships that have evolved through the mutual understandings of the Norm and the being together in the Heideggerian sense. We also emphasize that the evolution of living beings is dependent on the experience and the memory of the lineage. Unlike in traditional theories, which seek for laws and invariants, these issues of life can be grasped as essentially processual.

Philosophical Turnoff on Causality

In terms of causality in biology, classical evolutionary theory was strongly influenced by Mayr's paper *Cause and effect in biology* (1961). Mayr was not original in such distinction, it was more or less used by R. Owen, C. Whitman, and E. Conklin (Beatty 1994), but Mayr was the first to elaborate this distinction as such, and introduce it into evolutionary thinking. According to Mayr, there are two complementary ways of asking in biology—'why' questions and 'how' questions.

When we ask HOW question, we are asking for mechanical influence on a trait: for example, how nutrition influences metabolism, how exposure to some toxin influences sexual selection, etc. These questions are usually associated with ontogeny and are concerned with proximate causation, i.e., causes dealing with, as Mayr puts it, *all aspects of the activation and operation of genetic programs within an individual lifetime* (Mayr 1961).

On the other hand, WHY questions concern ultimate, i.e., evolutionary causation, which deals with evolutionary changes of such genetic programs in time. These causes present historical explanation why an organism has some specific trait rather than another. In his dissertation thesis, Mayr tried to answer an ultimate question: why do birds migrate? (Beatty 1994) Thus, every biological phenomenon can have both ultimate and proximate explanations.

The distinction led to a division of biology into that of proximate causation (functional biology) and ultimate causation (evolutionary biology). In Mayr's view, developmental processes were only proximate causes; they do not have any evolutionary impact. Mayr argues that these explanations are sometimes wrongly considered as competing accounts, as if one could rule out the other. In Mayr's conception, the very notion is absurd, for questions of how and why are complementary (Mayr 1993).

Based on EES as well as his own conception of causality, one proponent of EES, K. Laland, criticizes Mayr's distinction 50 years later. He thinks that not only is Mayr's definition of ultimate causation not clear, but it can also easily misguide a scientist in asking questions. Mayr's distinction can channel thinking in several ways (Laland et al. 2011). First, some scientists interpret ultimate causation as a function that basically links evolution and function under a single term of ultimate causation. But functions are not causes. Function is merely a description of what an organ or character does, but function cannot determine the occurrence of the character. Rather, the occurrence of a function is historical—it happens to occur, or not, based on the evolutionary history. Second, the proximate and ultimate distinction strongly supported the separation of evolution and development, with the presumption that proximate (developmental) causes cannot interfere with evolutionary ones.

Since proponents of EES suggest that a big portion of phenotypic variability is actually caused by changes in development, it means that development can direct selection, it can bias the action of selection, or even decide which genes will be exposed to selection (Laland et al. 2011). Thus, Mayr cannot be right in arguing that ontogenetic processes do not affect ultimate explanations as they belong only to proximate causes. Rather, Laland argues, such processes cannot be considered as *only* proximate.

Third, this dichotomy leads to channeled thinking about evolutionary causality in unidirectional linear terms, where phenomena such as coevolution (i.e., also niche construction) or sexual selection cannot be fully explained. EES prefers notions of reciprocal causation, where not only A is a cause of B, but also B is a cause of A.

We also think that Mayr's dissociation of ontogeny versus phylogeny breaks down in cases of acquired epigenetic traits, or rather, acquired states of developmental regulation (Portin 2012). Such states do more than influence the ontogeny of an individual; they can also serve as sources of novel variation and direct selection. The physical embodiment of proximate versus ultimate causation is represented by the Weismannian separation of soma and germ plasm. However, in many organisms germ plasm is not even segregated from the soma, and even in the case of mammals (where they are segregated quite early in the development), it has been shown that their boundaries are more penetrable than previously thought.

Evolution as History

On one hand, further development has acknowledged Darwin's revolutionary breakthrough; on the other, however, it has seemingly attempted to frame it in ways more palatable for science. As for Darwin, organisms essentially meant black boxes, opening them and uncovering their elements and processes appeared a great challenge and task for science. Mendelian genetics introduced a highly successful Platonic element to this 'normalizing' of Darwin, bringing with it notions of genotype—phenotype duality and free combinations of genes (alleles). This became combined with notions of genetic immortality (in the sense in which atoms are immortal in chemical systems), and of occasional (random and rare) mutations representing the only element of contingency, uncertainty. The neo-Darwinian synthesis, i.e., the fusion of Darwinism with genetics and statistics, soon followed by the advent of molecular genetics, became a powerful *scientific* explanatory platform for the evolution of life. Everything is determined by genes (more precisely, alleles) competing for their transfer into the next generation via phenotypes. This model of evolution became the Norm taught in textbooks and discoursed in eloquent books such as those by R. Dawkins (1976, 1982). This was a very successful model: many casual students of biology even took to mixing it together with ontology.

Our aim here is to point towards three directions that—in our opinion—were neglected, even ridiculed, because they were impossible to reconcile with requirements of natural science. We return here to the concept of Descent with modification, in its historical dimension: the keywords for our quest are historiography, the absence of any basic level of description, and—most importantly—the 'agency' of living beings (see also Chapter 11).

1. Evolution is a historical process analogical to that of cultural history: in short, neither past or future development can be calculated, and the past can only be reconstructed, interpreted. All Darwinian biologists of whatever orientation are thus historiographers constructing a narrative based in *scientifically* proven sources. Comparative morphology, paleontology, or sequence comparison in molecular biology[16] provide examples of scientific background for such narratives, but it is narratives that count, and their interpretation may substantially differ in various scientific schools.

2. As already mentioned above, the idea of descent with modification is not rooted in any basic level of description: evolution proceeds on and across many levels—structural, morphological, genetic, ecological, etc.—all mutually

[16] In construction of trees, e.g., of *E. coli, E. maximus, E. helioscopia*, the appearance and role of that particular living form can be disregarded.

interconnected and dependent, but not one being the 'one ring to rule them all'. With no convenient word at hand, we shall call all such processes 'epigenetic'. Members of a group (species, population, biome) are not atoms; they are unique, not interchangeable. The group, on the other hand, is not a mere pile of individuals. As a result, the composition of the group as well as the character of its members will change with time—with all descendants bearing the whole of the consequence.

3. Living beings are not passive entities exposed to the push-and-pull of external forces: they, as heirs of their lineage, are seat of the memory and experience of that lineage; they are born to a community that is also an heir of that same memory and experience; and, they go through individual experience during ontogeny. All this enables both cells and multicellular assemblages to interpret their situation here and now (Kauffman 2000), as they originate *their own* narrative of their own world—or umwelt. In our way of understanding, each living being is a historiographer.

6

New Dimensions of Diversity

In ragas, or solo jazz play, sounds are limited to a restricted sphere, within which a gigantic range of inventiveness opens up. If you have all the colors available, you are sometimes almost too free. With one dimension constrained, play becomes freer in other dimensions.

—S. Nachmanovitch 1990, 85

For what animals bequeath to their descendants is only their genetic systems, and all biological evolutionary changed that affects the properties of animals is ultimately change in DNA.

—R.A. Cameron et al. 1998, 610–611

In the introductory chapters of this book, we followed the history of prebiotic evolution, the establishment of life with its 'Norms', and the promiscuous exchange of life's achievements within the framework of such a Norm. We stressed that a substantial part of the Norm included, not only basic metabolic and genetic processes (those that could have been established in the prebiotic phase), but also a principal novelty that could be introduced by cellular life alone—the exchange and processing of information, both within the cells and from outside. In this chapter, we discuss basic components of information networks that may have been present in the first biosphere; in Chapter 7, we offer examples of its diversification in emerging lineages of life. Of course, a similar description could also be provided in parallel for the realms of metabolic, morphological, or ecological diversification. Everywhere the same pattern appears: as principles come to be established in both cellular and biospheric LUCA, different versions of life's appearances maintain conspicuous analogies, their properties always overlap to certain extent, enabling the meaningful communication of distant lineages.

On Digital

One often encounters statements claiming a digital character of this or that trait of the living. For example, R. Dawkins (1995) proclaims: "Life is just bytes and bytes and bytes of digital information. Genes are pure information—information that can be encoded, recoded and decoded, without any degradation of change of meaning. Pure information can be copied and, since it is a digital information, the fidelity of copying can be immense." What escapes too many biologists is the fact that DNA is simply not a medium to which information is 'saved' (as a blank sheet of paper, CD, or hard drive): the DNA 'medium' itself must be copied hand in hand with the information—and the medium is 'carnal', i.e., not digital.

Let us, for the rest of this book, put aside the shortcut 'life = genes = (implicitly) DNA = digital information' because there is a flaw in the usage of the word 'digital'. Digitality is exclusively the property of virtual entities such as natural numbers, alphabets, logical syllogisms, or information theory. It follows that things in the physical world can never attain digital properties; the most they can be is discrete (Markoš and Švorcová 2009). Discreetness takes a lot of effort to achieve even when necessary as it is in speech, information technology, and genetic processes such as transcription or replication. All these, of course, can be modeled *as if* digital (and the 'information content' of such a reduced model calculated), but the actual ontological status of such models should be observed with vigilance. Take, for example, the speech transcription into the digital *alphabetical* inscription, by observing the rules of grammar and orthography. What we get is a *model* of speech, not speech itself (Harris 2009, Cowley 2016).

The motif of parsing the world into discrete chunks (units) with blurred boundaries will recur with various modifications throughout this text. This may constitute the basic principle of tacit orientation by living beings within their individual 'here and now': things (concepts, patterns) are recognized with a fair dependability, while the blurring allows (reciprocal) adjustments within the instant (imminent) situation.

Recurrently, this principle can easily be recognized at different levels of living, e.g., Darwin's principle of descent with modification, or in parsing of morphogenetic gradiens (Chapters 4 and 7). Actually, *all* copying (repetition) means not the producing of identical copies of something previous, but the creating of variations—this holds for rituals, refrains, interpretation of rules and laws, and even the replication of DNA: the fidelity of such 'copying' is appropriate only as it concerns some established goals within some momentary conditions.

G. Deleuze states rather poetically, as if answering Dawkins' quoted above: "It is therefore true that God makes the world by calculating, but his calculations never work out exactly, and this inexactitude or injustice in the result, this irreducible inequality, forms the condition of the world. The world 'happens' while God calculates; if the calculation were exact, there would be no world. The world can be regarded as a 'remainder'." (Deleuze 1994) The central message of this book is epitomized here: in the opposition of these two quotes (Dawkins and Deleuze)—there is always tension between our exact models of the world, and the endless shimmering in the world around. Most illustrations in this book should be looked at in just this way: somewhat out of focus, with edges blurred.

Shaping DNA

Very long molecules of DNA (often tens of centimeters), plus the letter notation of nucleotide sequences, plus the basic knowledge in molecular genetics—all this gives many students the impression that DNA is a linear molecule, essentially characterized by the symmetrical double helix. From this perspective, the string of DNA is rightly portrayed by 'ACGT' notation. A large amount of *information* (sensu information theory) can be extracted from such a string frozen in this reduced way, but substantial knowledge is also lost. Whereas numbers and other kind of *virtual* digits can be written by arbitrary notation (alphabetical, Morse, ASCI, or other coding), molecules of water, adenine, serine, etc. are not mere symbols, and cannot be replaced by different kinds of molecules. Their shapes and therefore their properties are not 'frozen' (digital), but 'embodied', hence somewhat blurred by (i) internal or external thermal noise as well as, before all else, (ii) dependence on the context wherein they find themselves; the same holds for more complicated molecules such as strings of DNA (Fig. 6.1).

What in textbooks looks as if it were a completely symmetrical double helix, is in fact a structured shape depending on the sequence and placement of nucleotides. Note that in no means does a DNA molecule represent a thread of 'letters': each base-pair is bulky—about 1500 daltons. The contours of the string depend on the specificity of bases (i.e., GC is different from AT, TA, or CG), as well as on their neighbors in the chain. At any particular site, the proteins bound to that site and/or

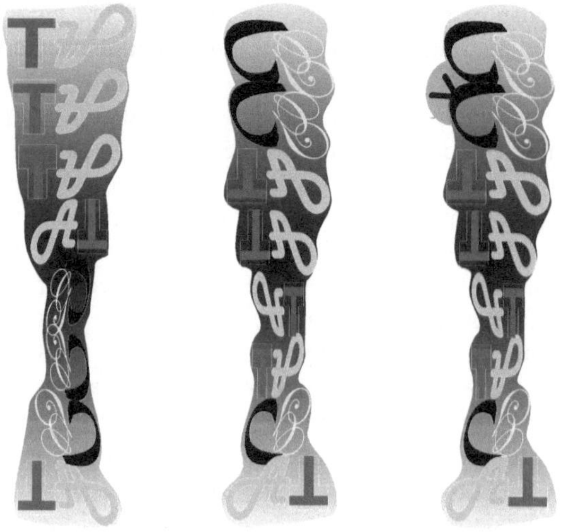

TTTAGCCA　　CCTTAAGT　　CČTTAAGT

Fig. 6.1: The shape of DNA. Left and middle strings have different shapes because they differ in sequence. The middle and right strings differ not in basic sequence, but in (reversible) covalent modification (epimutation C→Č), resulting in change in shape of the string. Twisting or bending of the string will also change its shape, as will change binding of proteins with an affinity to the string. Such modifications may result in binding or repelling of proteins with a specific affinity to a newly attained shape.

its vicinity also determine the context—and hence modify the shape. The ecosystem of proteins in interaction determines the momentary shape of the DNA string, thus opening the possibilities as to what proteins RNAs, etc. can bind—as well as when and where.

Moreover, bases can be chemically modified (e.g., methylation of cytosine or adenine), thus further increasing the repertoire of available base pairs—and of shapes of the string. The external shape of the string also depends on whether it is bent, stretched, or twisted by external forces—another factor contributing to context-dependency.

It follows that what *is* recognized by 'reading' the DNA script is the surface properties: the DNA-binding proteins feel the overall shape by 'running their fingers' over a particular section of the string. In fact, the sequence is read—'cover to cover', or base upon base—only during replication or transcription of the string. On such occasions, however, special measures are taken to maintain the discreetness of the sequence *close* to digitality, but again, the 'machinery' involved works by palpation, not reading.

We now understand why mutations (*irreversible* substitutions, deletions, insertions) of the string sequence may often bring about drastic consequences—as to the availability or information content of the string; we also understand why such consequences may occur throughout the string—even outside the protein-coding regions. The cell, in contrast to the string, reigns with a repertoire of means to avoid, minimize or repair such events. These buffers include heat shock proteins and redundancy of inscribed information. If, however, there is a need for change at a particular locus of the string, it is much safer to introduce—by action of the protein ecosystem surrounding the string—*reversible* epimutations that can influence, when needed, the overall shape as vigorously as mutations, yet allow the *Norm* to remain untouched. Controlled *irreversible* mutations of the genetic text are reserved to specialized somatic cells: as in complete deleting during erythropoiesis, or special tinkering in differentiation of lymphocytes. At another level, deep irreversible encroachments into existing DNA shaping are committed by lateral gene transfer, which is the transformation of cells by foreign DNA.

We will later return to DNA, but let us first turn to the molecular shaping of proteins.

Proteins

We urge the reader to recall (Chapter 2) that the universe of polypeptide chains repeatedly collapses into much smaller sets of shapes. Here we will develop the image further. First, the clear and distinct 'crystal' structures seen in Fig. 2.4, p. 17, represents an idealization: as explained above, the thermal noise somewhat blurs the structures. Second, the requirement 'provided that external conditions remain the same' should be kept in mind all the time. Minute changes of temperature, acidity, salinity, and other environmental factors as well as, before all else, the ever-changing presence or absence of other proteins may shift protein conformation substantially. Third, as in case in DNA, a mutation in the sequence of amino acyls in the chain may shift its 'collapsing' into a substantially different structural class. Fourth, the very function of a protein molecule depends on its *ability to shift* between its conformations—in a controlled manner—upon specific impulses such as ligand binding. Fifth, a successful protein molecule cannot serve as a matrix for the production of its copies (compare with the replication of nucleic acids). Even if abiotic production of random peptide polymers may have thrived in the age of prebiotic evolution, the subsequent 'normalization' that occurred at the dawn of life led to the establishment of genetic coding that safeguards the controlled, repeatable synthesis of proteins *tested* (by natural selection) for a given task. Saving the information about an amino acid sequence into a medium that is *not* an amino acyl polymer but can be copied with high reliability is undoubtedly one of the major breakthroughs that accompanied cell appearance. It is also a paradigmatic example of code systems present in living forms.

At this point, we expect the reader to know the principle (the Norm) of transcription and translation. We start with the simple model 'one gene—one protein'. 'Gene' is—here and throughout—to be understood as that part of DNA that will be transcribed to RNA which, upon being processed to mRNA, will serve as a direction for translation. Our second assumption is that each gene will 'produce' only one species of mRNA, and each mRNA will be translated into a set of identical native proteins. Note that

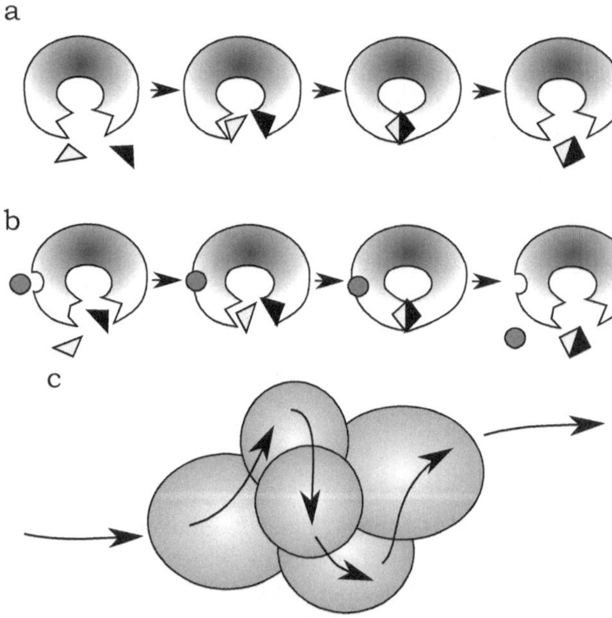

Fig. 6.2: Workings of proteins. a. An enzyme with two binding sites for its *ligands*; ligands are recognized and bound with a high affinity. Binding evokes conformation switch bringing both ligands together and allowing their ligation; the final compound is released and the protein is ready for a new cycle. **b.** Allosteric regulation: the protein has a binding site for the regulator (ringlet); in the presence of regulator, the protein will change its conformation which will result in a substantial change of its activity (activation in our example); **c.** The metabolon: arrows mark out the advance of a hypothetical metabolic sequence.

the diction of molecular biology often abounds with expressions suggesting an active role of genes in cellular processes: thus genes 'express' themselves, 'control' or 'regulate' this or that function, 'produce' some effects, etc. As a result, many researchers believe that "the plans for complex multicellular plants and animals are encoded in DNA" (*pars pro toto*, Stadler and Eisen 2017; see also processual biology, Chapter 5). In fact, all such functions come true only via ensembles of proteins.

It follows that we mostly do not encounter the ideal state described above. A single gene may give rise to a plethora of different RNAs (Pelechano et al. 2013), with some giving rise to mRNAs coding for proteins (Gilbert et al. 2016); finally, the translation of identical mRNAs may lead to a set of different proteins. The set of such 'native' proteins enters the ecosystem of proteins already present in the cell. In such an assembly, further shifts (in some cases drastic) in protein shape (hence its function) may be introduced, either passively—simply by a 'mass action' of the molecules involved, or by specific, often reciprocal, active interaction and modification.

(i) As in the case of DNA, fluctuations of the physical environment, especially extremes in temperature, acidity, salinity, concentration of heavy metals, etc., may play a decisive role. If a given organism is not adapted for such harsh environments, and if such extremes are but transitory, special proteins—chaperones—may help attaining and/or maintaining the conformation of proteins responsible for performing vital tasks. Some native proteins may need such conformation-nursing even in 'normal' environment (Hartl et al. 2011), or they can even interact—with very high affinity—while remaining in such an internaly disordered state of the molecule. There is a prevalent idea in molecular biology that from a DNA sequence, the protein shape and, thus, its function can be predicted. However, even the interaction of ligand and an enzyme does not work as a key and lock. It is not a precise pairing, it is rather a negotiation or construction of a niche: both sites are adapting to each other. Recently, intrinsically disordered protein were discovered which never completely attain a 3D structure and can rapidly change the form depending on the momentary context and function they need to provide. They often work as chaperone proteins, stabilizing other proteins (Borgia et al. 2018).

(ii) The presence of ligands that can be bound specifically to a given protein is decisive: the property to bind ligand(s) is the basic requirement for most proteins. Small molecules, macromolecules such as proteins, DNA, polysaccharides or specific surfaces, can also act as ligands, both constitutive (e.g., substrates of enzymes) and regulative (e.g., second messengers). For example, specific regions on other proteins (*epitopes*) may function as a ligand for a given protein; such interaction, as in the case of any other ligand—will change the shape of both interacting partners and often lead to the establishment of multiprotein complexes of tens of proteins (see *hairball* below; Figs. 6.2c, 6.3–6.6).

(iii) Mutation in the protein-coding sequence of DNA or RNA may lead to an *irreversible* change in the amino acyl sequence of the resulting protein, causing (usually deleterious) conformation shifts in one or more contexts listed above. As in (i), the set of chaperones may—to some extent—help buffer the damage (see capacitors, Chapter 8), or natural selection may eliminate the bearer of the mutation.

(iv) *Epimutations* cause *chemical* modification of some amino acyl residue(s) in the protein, hence augmenting the set of amino acyls in the chain from the standard 20 to (often much) greater numbers (Fig. 6.4). The outcome is, at the first sight, comparable to a mutation at a particular site. Often causing serious damage, mutations may occur either spontaneously or by the action of noxious agents, either chemical or physical. The remedy may come, again, from chaperones, or through the removal of the damaged protein. If, for whatever reason, such adjustment is impossible, the damaged protein may accumulate and cause serious malfunction—as in the case of protein plaques in the brain, which cause dementia. In contrast to such haphazard events, our focus here is the introduction of specific, thoroughly *controlled* and mostly *reversible* epimutations into protein molecules, carried out by specific sets of enzymes. In other words, the cells thoroughly keep the universality of the code (the norm), yet broaden the sustainable epigenetic modifications in order to increase or modify some set of given features (for protein methylations see, Biggar and Li 2015, Lu et al. 2016).

Such adjustments are required for very practical reasons that can be explained here on enzymes, but the explanation holds also in other types of proteins. Briefly, the performance (activity) of an enzyme is a function (trade-off) of two constants that can never come to a mutual agreement in evolution. The Michaelis constant (k_M, expressed in units of concentration) is the measure of the enzyme's affinity (binding specificity) towards the ligand. Catalytic constant (k_{cat}, expressed as cycles per second) is a measure of the turnover number, the flexibility or 'agility' with which the enzyme performs its task in a unit of time. The lower the value of k_M, the more accurate the recognition and binding of the ligand; an absolute value tending to zero would mean crystal-like, rigid structure of the enzyme-ligand complex, incapable of further conformation changes. The higher the k_{cat}, the swifter the conformation changes (performance), but at the expense of the accuracy of ligand binding. Should the cell need different binding and/or performance at different places or times, it may harbor many similar genes coding for different settings of both constants. Such is the solution in case of globins, parts of hemoglobin complexes. Different versions of hemoglobin in mammalian development match different requirements for the transportation of oxygen in embryo, fetus and air-exposed, born individual. An improper insertion of the globin variant may cause developmental disorders or cancer. Similarly, the core histones that are most frequent in nucleosomes (H2A, H2B, H3.1 or 3.2, and H4) are, in some developmental phases, replaced by some of minor variants (20 in humans; Maze et al. 2014). Our final example points to receptors to the same ligand that differ in sensitivity and/or distribution that can work simultaneously (Fig. 7.2, p. 121).

An alternative solution of how to accomplish protein's tasks under different contexts is through epigenetic modulation. First, however, we proceed from a single protein to protein complexes.

The Hairball

It has been known for some decades that the proteins in the cell do not work 'on their own', freely and randomly floating in the 'cytosol'.[17] Usually, they are organized into tightly coupled multiprotein complexes, assemblages even of tens of different proteins often containing lipids, sugars, or RNA. Since the 1970s, examples have abounded in the literature, which include metabolons (e.g., pyruvate dehydrogenase complex, citric acid cycle), membrane rafts, transcription initiation factors, ribosome, nucleosome (Fig. 3.2, p. 45), signal particles, cytoskeleton and extracellular scaffolds, etc. But with the advent of proteomics, there came a fuller insight into this cellular 'small world' or, more poetically, this 'hairball' of multiprotein complexes interconnected by somewhat looser bonds with other such complexes (Kauffman 1993, Lander 2010, Mo et al. 2017; Fig. 6.3). Modern analyses allow us to map the networking of thousands of protein interaction networks; for both normal cells and some genetic diseases, see, e.g., Drew et al. (2017), or outputs of programs like ENCODE (Chapter 5).

What is important is that adding or removing a single protein, or even changing the conformation of this or that constituent of the complex, may change (often profoundly, in an 'on/off' way) the structure, function or composition of such complexes, or even of the whole 'super-hairball' which is the cell. Equipped with this knowledge, we proceed from the rather unpalatable hairball metaphor to a more pleasant metaphor of 'hairdo'—from tangled mess to *controlled* (targeted) epimutations.

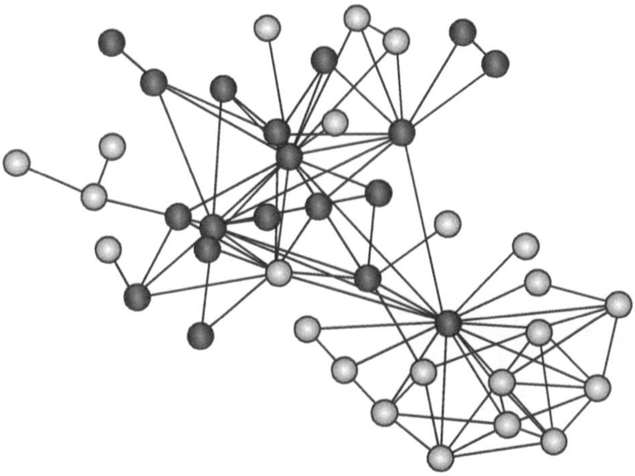

Fig. 6.3: The hairball. Tightly coupled multiprotein complexes serving metabolism (metabolon), chromatin binding, transcription and translation (elongation complex, ribosome) or signal particles. Two such complexes are shown, interconnected only loosely.

[17] The quotes are apt: the original meaning of 'cytosol' is 'unsedimented part of highly diluted cell homogenate after centrifugation at 100,000 g for 1 hour.' It contains a suspension of whatever has escaped the cytoplasm or organelles. This technical term passed through laboratory jargon into textbooks, to designate what was previously called cytoplasm. Ironically, it was simultaneously clear that cytoplasm is not homogeneous, i.e., no soap or 'sol'.

Targeted Epimutations

The reversible derivatization of amino acyl residues in proteins is a common tool for regulating the properties of a given protein (Fig. 6.4 and Table 6.1 below give some examples). Take a paradigmatic, textbook example of epimutation, one of the first that was discovered. In animal cells, glycogen is a polymer of glucose, synthesized by repetitive additions of glucose moieties by the enzyme glycogen synthase. The reverse process—glycogen decay to glucose monomers—is catalyzed by the enzyme phosphorylase. Both processes, of course, cannot run simultaneously in a vicious circle, only one of the two enzymes is ever in an active state. Both enzymes are therefore subject to derivatization by *protein kinase A*, an enzyme that adds a

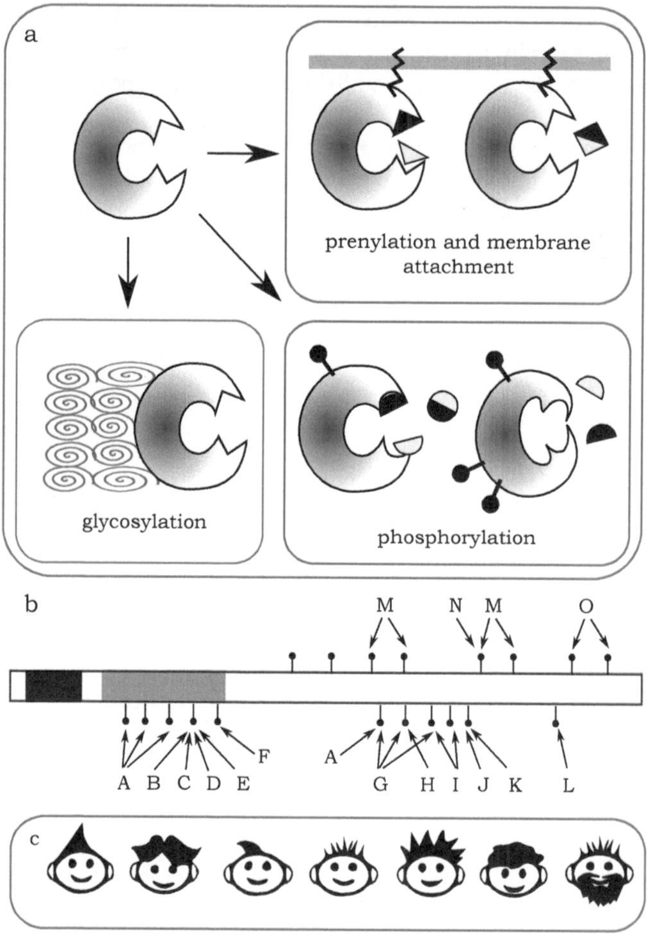

Fig. 6.4: Epimutations. a. Epimutations of native proteins leading to their different localization, shape, or function. **b.** Insulin receptor substrate is the protein hub integrating signal inputs from different receptors, and passing the processed information further. Each 'barcode' leads to a unique conformation (shape) of the protein, enabling or impeding binding of proteins A–O. **c.** Depending on the composition, the resulting 'hairball' will assume a different 'hairdo', with often far-reaching consequences for cell structures and functions. See also Fig. 3.2, p. 45.

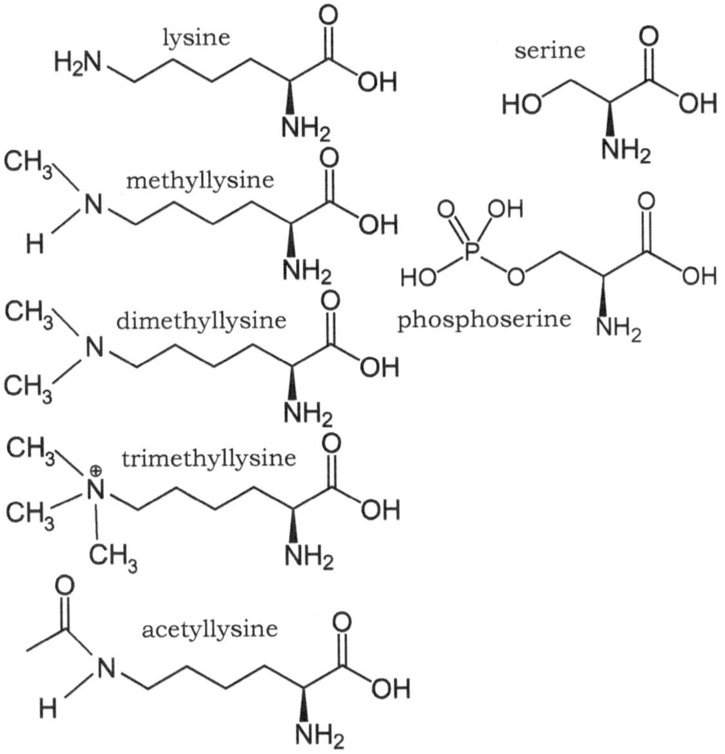

Fig. 6.5: Epimutations of amino acids. Transformations of lysine (1–3 methylations or acetylation) or serine (phosphorylation) shown as examples. See also Table 6.1.

Table 6.1: Examples of amino acyl modifications in proteins.

Amino acyl	**Modification**
Serine, Threonine, Tyrosine, Histidine, Aspartate, Glutamate	phosphorylation; acetylation; glycosylation
Cysteine	long-chain fatty acylation; geranylation; farnesylation; carboxymethylation; dimerization (S-S)
Lysine	hydroxylation (and sugars attached through OH group) methylation; dimethylation; trimethylation; ubiquitiation; sumoylation; biotinylation
Arginine	methylation; dimethylation; mono- and poly-ADP-ribosylation
Proline	isomerization; hydroxylation (and sugars attached through OH group)

phosphate group to a particular serine residue in their molecule, changing, thus mutating the serin moiety to a moiety of phosphoserine.

Note that though there are many serine residues in the molecule of both proteins, only single one, occurring in a context of a specific shape—epitope, is accessible to the kinase. Such epimutation—the substitution of serine with phosporyl serine in the string—will switch glycogen synthase to the off state, while in contrast activating the phosphorylase. The counterpart of protein kinase—*protein phosphatase 1*—will remove the phosphoryl group from both phosphorylated epitopes, thus reverting both enzymes towards glycogen synthesis.[18] In other cases of such regulations, the simple on/off switch may also be finely attenuated to states in between, according to the state of the cell and the setup of regulatory tools. It is obvious that genetic tools cannot provide such regulation.

Our second example is the protein called insulin receptor substrate (IRS) that serves as a hub that receives inputs from many upstream signal-processing pathways, integrates them by assuming some of the many possible conformations and, through the specifics of the conformation change, broadcast the information downstream (Taniguchi et al. 2006). The scheme in Fig. 6.4b shows that the protein can be phosphorylated on multiple serine and threonine moieties with different outcomes as to its activity. The task is accomplished by the set of protein kinases and phosphatases, which is as regulated by internal crosstalk as it is by the plethora of inputs received from elsewhere (see also next chapter). Through such modification, the IRS protein is prone to changes of its very barcode, and thus adjusts differently to inputs, and changes its activities in the adjacent time interval. IRS—along with its sets of upstream and downstream connections—is but one item of the endless life of cellular and intercellular networks.

The third example concerns two nuclear proteins—lamins and histones. The vertebrate protein family of lamins (Dittmer and Misteli 2011) forms a complex meshwork that lines the nuclear lamina (Mahamid et al. 2016). Among other modification, they also undergo multiple phosphorylations (there are more than 30 conserved phosphorylation sites in lamin A) via some set of different kinases/phosphatases. Moreover, the phosphorylation pattern changes distinctively depending on the cell cycle (assembly/disassembly of nuclear lamina), tissue differentiation and/or homeostasis, maintenance of tissue specificity, regulation of chromatin organization, cancer, etc.

We discussed the second set of nuclear proteins—histones—in Chapter 3 and will further that discussion below. Note that whereas histone genes are very conservative (Chapter 3 and Fig. 3.2, p. 45) and the set of histone comprises only some 20 genes, the gene family of lamins is extremely heterogenous, with many paralogs; moreover, quickly changing in time: their homology across different lineages is blurred.

A parable comes to mind at this place: one recurrent motif in works by U. Eco is the contrast between the dictionary and the encyclopedia (Eco 1994a). The crystal-clear classification contained in dictionaries, textbooks, lists of enzymes or other items is opposed to the actual known world (i.e., a cultural construct) with its endless mutual references and (re)adjustments according to contexts—comparable to an endless browsing of an encyclopedia. Understanding encyclopedic entries allows an interpretation of one's situation in the world, here and now, so as to get along living one's life. The adding of new entries or dragging existing ones into new contexts is evolution. Lamins and histones represent different strategies of 'browsing the cellular

[18] As usual, the scheme is not as simple as shown here: other proteins and regulators also control the activity of the kinase/phosphatase switch, via different mechanism(s).

encyclopedia' in mutual synergy. Table 6.1 above may represent dictionary entries, as does the list of possible protein modifications. The encyclopedic whirling use of such entries, among others, is life—both in our and (we feel) Eco's understanding.

Above, we gave examples of but one type of reversible epimutations of proteins—phosphorylation. The tool (or better—set of tools comprising of a cohort of kinases and phosphatases, themselves regulated) introduces changes with deep consequences not only in the structure and function of the proteins regulated, but also of cells, tissues, and whole organisms. Other tools introduce or remove many different epimutations—each example in Table 6.1 involves, or evokes, complicated events in the cell in concord with other similar processes.

Now take a fresh look at Fig. 3.2, p. 45, which shows the scheme of a nucleosome. Its core is composed of a histone octamer (4 homodimers, i.e., only four kinds of proteins in terms of native proteins). What is conspicuous is that histone genes are amazingly conservative as if it were extremely important to stick to Norm set at the level of native proteins: for example, the human histone H3 differs from its homologue in corn (*Zea mays*) merely in two amino acid positions (Fig. 3.2, p. 45). This extreme conservativism of native sequences is counterbalanced by a rich spectrum of epigenetic modifications of the histone molecules. Each molecule offers multiple targets (barcode sites) for editing by big set of regulatory proteins contained in chromatin (Dann et al. 2017). The reason behind the extreme conservatism is perhaps that histone mutants in editable positions may lead to serious malfunction or cancer (Pengelly et al. 2013, Herz et al. 2014). From simple combinatorics it follows that human chromatin may contain tens or even hundreds of millions of unique nucleosome shapes, with patterns characteristic for given cell types in a given time period. Each elementary nucleosome 'hairball' thus will get a specific 'hairdo' that is appropriate for a given structure at a given time.

As discussed above, lamins stay in a sharp contrast to conservative histones. Before all, they are restricted to the animal kingdom, and gained brand new functions only in vertebrates—apparently connected with complicated tasks in morphogenesis and cell function. This holds especially for lamin A. Several genes code for different versions, and even more splice variants exist. Protein variants (about 700 amino acids long) occur in different assemblies in different tissues, thus contributing to structurally unique nuclear laminas, weaving a complicated meshwork of lamin polymers, with components epigenetically modified in many more ways than mentioned above. By tens of interactions, they bind especially to heterochromatic parts of the chromatin, thus structuring its domains. They also sequester transcription factors and signals, and canalize outputs of such assemblies to nucleus insides. Especially when compared with thorough conservativism of histones, it is conspicuous that lamin A is a protein that harbors in humans perhaps the highest frequency of mutations. About 1000 alleles are known, with about 350 of mutations connected with serious inborn diseases—laminopaties. These affect, before all, tissues of mesenchymal origin (such as cardiomyopathy, muscular distrophies, etc.) and cause premature aging syndrome (Dittmer and Misteli 2011, Maraldi 2018). Of course, other eukaryotes must also guarantee nuclear lamina functions—and do so by other families of proteins. Perhaps all such families started from a common ancestor, but their mutual similarity became blurred due to high mutation rates in each eukaryotic lineage.

Let us now extend the hairball-hairdo metaphor to complexes comprising tens to thousands of different proteins in a dynamic network. Any epimutation may cause profound changes in the form of a different 'coiffure', which causes a different functioning of such assemblies. Chromatin served above as an example of such a complex reshaping; a similar example can be found, e.g., in the ribosome (Natchiar et al. 2017), or in signal particles mentioned in Chapter 7; the IRS protein discussed above and in Fig. 6.4 is but one component of such a signal processing assembly.

At the level of nucleosomes and their assemblies, substantial progress in understanding has been achieved: for example, similar structures can be found near, and on, the basal lamina. Likewise, it has been shown that enhancers are topologically close to regulated genes, etc. Yet the overall structure of the intact nucleus remains elusive, "revealing that it is a structurally disordered 5- to 24-nm-diameter granular chain" differing in density across the nuclear volume (Ou et al. 2017). In other words, the hierarchical packing model propagated in textbooks over the last two decades (derived from *in vitro* studies) probably needs revising. Undoubtedly, there exist regions with active and inactive chromatin, characterized as areas of topologically associated domains or chromosome contacts (Stevens et al. 2017). Yet, in spite of enormous effort of consortia such as ENCODE (Nature, 6 Sep 2012; see Chapter 5) or 4D Nucleome Network (Dekker et al. 2017), little is known of how they are organized in space and time, as well as the correlation between such structures and function. Chromatin loops, held together by special proteins, apparently play an important role (Terakawa et al. 2017), as does nucleoskeleton, and epigenetic modifications of histones and other regulatory proteins that take charge at various given times and places (Dann et al. 2017, Nagano et al. 2017). Larson et al. (2017) argue that the interior of a nucleus may have the character of immiscible droplets (as in 'lava lamps'), with a separation of chromatin elements between its various phases. The phase itself emerges upon phosphorylation of heterochromatin protein 1 (only in the presence of DNA), which, when in non-phosphorylated and thus 'soluble' state, spreads across vast regions of heterochromatin, as an organizer of the heterochromatic state. Phase separation, hampering entry of transcription and other factors, may represent a safeguard against the reactivation of such regions (Larson et al. 2017).

What was said above mainly concerns the interphase nucleus; the dynamics of chromatin during the cell cycle is another problem that became tractable only through new techniques of observing the process *in vivo* in single cells—for first details, see Nagano et al. (2017). All such global chromatin modifications have their local counterparts in groups of protein remodelers (Dann et al. 2017).

Combing the Shapes of Nucleic Acids

Let us extend our hairdo metaphor to DNA. As we have seen above (Fig. 6.1), the double helix is about 2 nm thick, and this allows recognizing not only overall structures such as major and minor grooves, but also much more sophisticated surface parameters.[19] In no ways is the DNA molecule a monotonous ever self-same structure. This shaping gets another dimension with epigenetic shifts in the sequence (mutations) as well as derivatization of nucleotides (epimutations)—an analogy of what we have seen in the protein realm.

Troubles with propagating oil palm (*Elaeis guineensis*) may serve as a case illustration of the importance of the fact (Ong-Abdullah et al. 2015). Seedlings of elite hybrids intended for plantations are produced not from seeds, but from tissue cultures of somatic cells—as is the case in many other domestic plants. In plants, it is quite easy to produce embryos and seedlings from an undifferentiated cell mass grown in a laboratory culture; hence, theoretically, all should be genetically identical. A substantial proportion of palm seedlings, however, shows severe aberrations leading to drastic reduction of the yield of seeds, and thus of oil. No mutation of the DNA string has been found, and the culprit is an epimutation of a short sequence of non-coding DNA. As in other eukaryotes, the palm genome is punctuated with retrotransposons (LINE sequences) of retroviral origin. One LINE sequence is inserted into an intron of one specific gene: **...exon1/intron.LINE.intron/exon2...** Such intron should theoretically be deletable during the processing of the transcript (without regard to the presence or absence of LINE), leaving no traces in the resulting mRNA and protein. Yet the final phenotype depends on how the insert is methylated: any hypomethylation of this tiny part of the string leads to a far-reaching rebuilding not only of this particular area, but of the whole chromosomal locus containing many genes—which results in the above-mentioned change in the phenotype towards negligible yields.

Such somatic epimutations are common in somatic cells, both in plants (Baker et al. 2011) and animals—see Chapter 8. We shall return to the phenomenon and its consequence, here it serves only to illustrate how the overall shape of a DNA molecule—as well as that of its parts—will markedly influence the state of the molecular ecosystem of the cell. As mentioned above, the shape of a DNA molecule is undoubtedly (even primarily) due to the sequence of nucleotides, but can also be influenced from outside—e.g., by bending or twisting the chain, or by the binding of proteins of different kinds. In addition, the very sequence can be changed by epimutations (in case of oil palm by methylation of cytosine) in a way similar to what we have seen in proteins (see above). The goal of such processing is also similar: protein machineries that ensure the origin of replication or transcription, safeguarding repairs and/or regulation of gene expression, etc., do not read the sequence of nucleotides but recognize specific shapes on DNA's body (or the body of DNA plus associated proteins). Mispairings during DNA replication and/or spontaneous mutations of bases will create 'blobs' on the surface, easily recognizable by proofreading and other repair systems.

However, what is even more important to the context of this book is that covalent (and reversible) modifications of bases change and/or create new docking

---

[19] To remind the reader: an average amino acid has a molecular weight of around 100. In contrast, the m. w. of a triplet coding for an amino acid is about 1500; in the DNA double helix, it is 3000. Add the ribosome, tRNA and aminoacyl-tRNA synthases, and you get quite an unusual 'letter'-code indeed.

sites for DNA-binding proteins. The most common modification is methylation/demethylation of cytosine at specific CG-sites, creating a 'bar code' on the DNA molecule recognized by a large cohort of regulatory proteins that establish the code itself and thus define the overall state of a given region of DNA molecule (for cytosine modifications, see Yin et al. 2017). Less frequent modifications are comprised of, e.g., methylation(s) of adenine, glycosylation, hydroxymethylation, etc., of different bases. Such subtle changes (thousands of derivatizations per cell) will enormously amplify the cell variability in some organs, e.g., the brain (Luo et al. 2017). Derivatization of nucleotide moieties adds another dimension to the state space of attainable patterns. In concordance with our combing metaphor, very little changes may influence the coiffure of the whole chromatin—see the oil palm story above; see also Zhang et al. (2013).

The lesson to take from this survey is that local changes in DNA shape need not be attained only by drastic and irreversible, irreparable, events such as Mendelian mutations and/or insertions and deletions of bases. Much more frequent changes come from epimutations, that create bases not belonging to the original set of four, reversible and targeted at very specific sites and causing changes in and of protein, RNA, and the structure networks responsible for setting the overall state of the cell—at the same time being created, modified, and obeyed by that network.

Even more creative is the modification process in RNAs (thousands of modifications per cell, with a broad spectrum of possible alterations). This story starts with high transcript heterogeneity, leading to RNA isoforms obtained or obtainable even from a single gene (Pelechano et al. 2013, Schimmel 2018). Such sets of RNA may regulate protein hairballs directly, or—after translation—deliver new kinds of proteins to such interactive networks. Here, epigenetic regulations play roles similar to those shown above in proteins. As in the case of proteins, RNAs have much greater flexibility than DNA in assuming secondary structures (Ding et al. 2014). Adenosine-to-inosine transformation requiring a hairpin structure with A/I on its tip, with the transformation being performed by a special enzyme is a frequent example (Walkley and Li 2017).

Epigenetic change is typical especially for non-coding regions of mRNA, and for RNAs designed for roles other than translation (Gilbert et al. 2016, Harcourt et al. 2017). Many classes of RNA exist that constitute ribonucleoproteins: this includes not only well-known ribosomes and spliceosomes (Harcourt et al. 2017), but also, e.g., chromatin (Hendrickson et al. 2016). They are also engaged in the fine attenuation of a plethora of genetic regulations (Hausser and Zavolan 2014, Holoch and Moazed 2015). Schimmel (2018) published an extensive review concerning but a single class of tRNAs, which reveals an unexpected complexity consisting of billions of different tRNAs and their derivatives (chemical modifications of bases and/or many ways of fragmentation), as well as many splice variants of aminoacyl-tRNA synthases. In addition to the role prescribed by the Norm, i.e., the functioning of the decoding process in translation, such derivatives and fragments have functions in subtly attenuating codon preferences, the regulation of protein (or the whole hairball) activities, the rate of translation, signal transduction, etc.; they also play a role in tumor propagation. In transgenerational inheritance mediated by sperm cells, they even modulate the translation processes in the zygote: "Transmission of an acquired male phenotype could be mimicked by injection of sperm-derived RNA fragments into fertilized, wild-type oocytes. This work clarified, with direct functional observation, an earlier study showing that RNAs had a role in epigenetic inheritance. Together with a subsequent finding that tRNA fragments are abundant in sperm, these findings led to a pair of independent studies that confirmed the significance of tRNA fragments for epigenetic inheritance" (Schimmel 2018; see also Chapter 8). Such fragments also play an important role in the assembly of ribosomes (Kim et al. 2017). The environment may play its role via influencing subtle regulatory functions of some of such variants: "cellular homeostasis balances on a tipping point, which is exquisitely sensitive to the levels of specific tRNA species" (Schimmel 2018). In bulky egg cells, such games undoubtedly are even more sophisticated. Other groups of transcripts (lnRNAs, iRNAs, etc.) play similar roles in the meshwork of any cell.

As we see, all the constituents of cellular networks are prone to chemical modification. DNA, however, stands out in a very specific way: despite all epigenetic modifications, it remains linear and allows replication of its strings. Moreover, the replication system may read the epigenetic bar-coding of the mother string, and faithfully introduce it to the daughter copy—this allows *maintaining* such modifications through generations of cells, and even through generations of sexually reproducing organisms (Chapter 8). The possibility of replicating and maintaining epigenetic bar coding (the string 'diacritics') in DNA and RNA is probably unique among constituents of the cellular network (but see also the reconstitution of histone bar-coding after each mitosis). It follows that the newly begot cells/zygotes may not receive virgin DNA without diacritics, i.e., constituted only from 4 bases A, C, G, T; it may also transmit settings established by the predecessors, hence receiving and reflecting some of the memory and experience acquired during their lifetime.

What should follow is a survey of enzymes catalyzing the introduction and/or removal of epigenetic markers, as well as a survey of DNA- or RNA- or protein-binding proteins regulated epigenetically. Lack of space does not allow this, though we will introduce some examples in the following pages.

Of course, not everything goes: the cell has a plethora of means of correcting excesses caused by not only mutations, but also epimutations, i.e., of wrong 'diacritics' of epigenetic modifications. Thus it can keep in check the presence of other cells (e.g., pathogens), or physical factors (e.g., heavy metals). For example, the system of stress proteins (chaperones aka heat shock proteins) buffers various disorders of proteins (Rutherford and Lindquist 1998, Taipale et al. 2010, Jarosz and Lindquist 2010, Rohner et al. 2013), or even mutations caused by transposons (Specchia et al. 2010). Creative powers must be kept solidly within usable frames, to prevent runaway leading to a quick extinction.

Protein Modularity

The hairball is not the only solution to building multifunctional complexes: another way is by the fusion of different genes or their parts (exon shuffling), thus creating a single gene coding for a protein with different domains. Take, for example, an enzyme catalyzing an endergonic reaction that requires coupling with an energy source such as splitting ATP (e.g., protein kinases, or membrane pumping). Such an enzyme usually contains two domains, one with a function of ATPase, the second performing its specific task. The ATPase domain is homologous with domains in a large set of similar proteins performing a plethora of different function—actually just a handful of such ATP-splitting domains can cover all of the reactions that require propulsion by ATP. It is obvious that the domains were not independently acquired in evolution, but were added to the gene through gene (exon) shuffling. The advantage of such a merger of gene sequences is obvious: a single protein is immediately able to perform its function, instead of waiting for its partner—the product of another gene.

The modularity principle, observable on many levels of biological organization (see Chapter 5), is extremely prominent in proteins of the extracellular matrix—linear assemblies of many—even tens—of such domains bearing different functions, which often come in variants due to RNA processing and splicing (Fig. 6.6). Complex structuration of such elements is achieved by a large number of enzymes and structures, and by their interconnections with the cell interior. All this suggests that in addition to mechanical function, the extracellular matrix (ECM) also provides an informational and marshalling highway for cells.

Collagen may serve as an example (Fig. 6.6). Its basic function is to secure mechanical, tensile strength of tissues: the task is fulfilled by the special helical structure of the protein, with every third amino acyl in the string being glycine. The native protein, however, is about 30% composed of the amino acyl proline and also contains a high amount of lysine. These residues become readily hydroxylated (to hydroxy proline and hydroxy lysine), and thus become prone to glycosylation by long linear polysaccharides (glycosamino glycans). The glycosylation is not uniform; it differs in various collagen variants and/or tissues—again, the idea of a barcode of some sort comes into mind. Of course, no such posttranslational modification is required for the mechanical function of collagen fibrils; it serves as a set of communication tags for the entire extracellular matrix network.

All proteins of the extracellular matrix are glycosylated in this way—especially proteoglycans, which, together with the huge polysaccharide molecule of hyaluronic acid, constitute the 'barbed' structure of an interconnected saccharide matrix. Again (as in case of the basic function of collagen), the meshwork harbors water and salts (in order to maintain the turgor of the tissue), but extensive barcoding of the proteoglycan, fibronectin, and many other proteins (some sketched in Fig. 6.6) is not aimed at this simple function: its complicated structure suggests that it also plays an information role.

We must add that cell surfaces are also 'barbed' in a manner similar to that of anchored glycoproteins, and interconnected across the membrane with the cytoskeleton.

Cells entering the structure of ECM (and at the same time building and rebuilding it) orient themselves—move and/or differentiate—by the abovementioned cues (barcodes) anchored as epigenetic markers. Obviously, there may exist a 'sugar code' (Buckeridge 2018, Gabius 2018) or other sorts of codes that are not inherited genetically even as basic constituents (as in case of proteins and RNAs). Some seeds of the structure may be present epigenetically, yet the whole contraption is being built (and its composition and function decided) by contexts introduced by

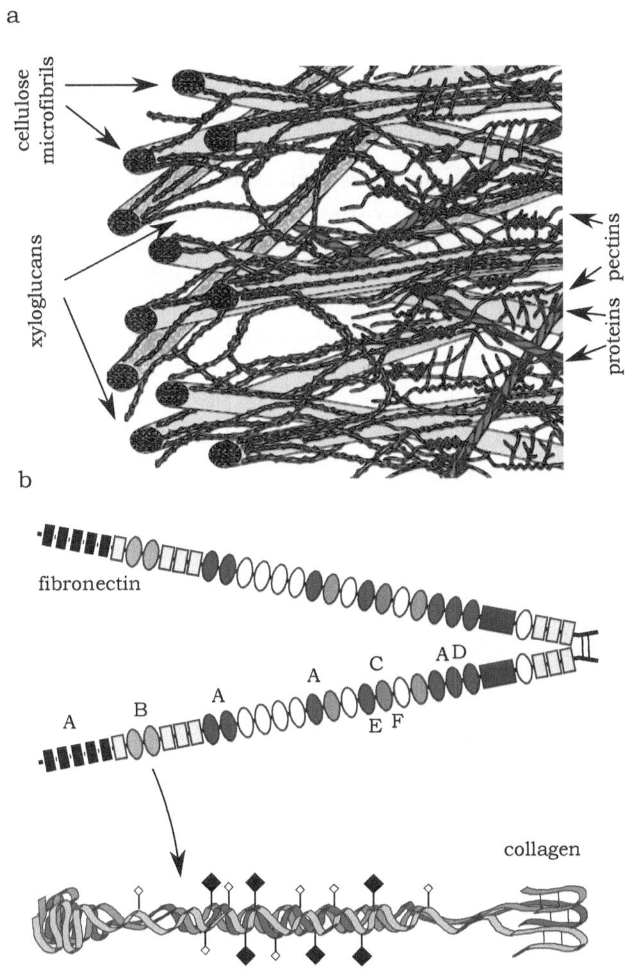

Fig. 6.6: Extracellular matrix (ECM). a. The scheme of plant cell wall as an extracellular matrix structure—a dynamic scaffold of polysaccharide fibrils and hundreds of proteins constituting and/or rebuilding the scaffold. Analogous dynamic structures are present in other lineages. The network is interconnected with cell surface glycoproteins, and through them with the cytoskeleton, or even nucleoskeleton. **b.** Scheme of two linear proteins of animal extracellular matrix—fibronectin and collagen. Fibronectin represents a row of different protein domains (modules) serving interconnections with different elements of the matrix web (e.g., domains **B** bind collagen). Three collagen proteins forming a coil, a part of higher-order collagen strings. Besides mechanical function, the amino acid residues are amply hydroxylated (small diamonds) and through such moieties, glycosylated (big diamonds). Such glycosylated barcoding serves not a mechanical purpose but interaction with other components of the extracellular matrix.

ontogenetic and/or environmental cues. For more about information processing and signal transduction in extracellular matrix, see Fig. 7.4, p. 123 and the following chapters.

So far, we have only mentioned animal solutions; in plants and fungi, analogous structures play the role of coordinating ontogenesis and tissue integrity. As cells cannot move actively, their growth and division in complex tissues and organs will depend solely on a reciprocal exchange of information, based on cell-to cell contacts and cues from the extracellular matrix. Mouw et al. (2014) speak about "specific extracellular matrix signature that is comprised of unique compositional and topographical features."

Compare now two evolutionary solutions constituting sophisticated networks of linear contraptions. On one hand, cytoskeletal structures are highly dynamic—they can quickly assemble, disassemble and/or restructure upon need. The extracellular matrix, even if far from being static, i.e., built forever, is slower in changing, and requires tens, or even hundreds of enzymes. The function of extracellular matrix is to provide a scaffold of structures and cues informing the surrounding cells of their coordinates in time and space (Chapter 7). Cells reciprocally secrete their constituents and thus keep turning over the constituents and structure of their scaffolding. On the other hand, such long-lasting functions are safeguarded better by keeping cells together in a single structure.

Not Rooted in the Norm Yet Faithful to It

As we have seen in previous sections, DNA is faithfully copied with very high precision, thus maintaining the nucleotide sequence in the string. DNA cannot do the task by itself, assemblies of proteins must steward it—and organize the many exceptions from the rules (such as the assembly of antibody genes, crossing over, exon shuffling and other recombination events, or horizontal gene transfer).

As to native proteins, the string of amino acyls will more or less depend on the information in DNA. 'More or less' because the mRNA synthesis involves extensive processing of a primary transcript—as defining exons and their splicing, or even extensive editing—see mitochondrial transcripts in trypanosomes (Chapter 3, p. 47). The ribosome, then, will define the reading frame, including the Start(s) and Stop(s) of translation, and often further idiosyncrasies. Yet, in most cases (the Norm), there usually exists an almost perfect correspondence between the sequence of amino acyls in the protein, and of nucleotides in the corresponding part of the gene. Posttranslational modifications, however, follow another rule 'negotiated' within the ecosystem of present proteins and their structures. They decide which items will enter or leave the extant assembly, how they would be processed epigenetically, and what structures they will join (recall the simple operon model mentioned in Chapter 5; Fig. 5.2, p. 78). The protein ecosystem is highly emancipated from the genetic script maintained in DNA, yet rooted in it via genes and regulatory sequences needed for DNA-binding proteins.

We now proceed to structures that are even more emancipated from the 'DNA Norm', structures that depend on information *not* encoded in DNA, but on stable structures or dynamic ensembles of proteins. These include 'eternal subcellular structures' that can best be epitomized by cortical structures of ciliates (Beisson and Sonneborn 1965; Fig. 6.7).

The cortex is the surface structure of many protists; it hosts many sub-structures and functions. In ciliates, most conspicuous cortical structures present rows of cilia

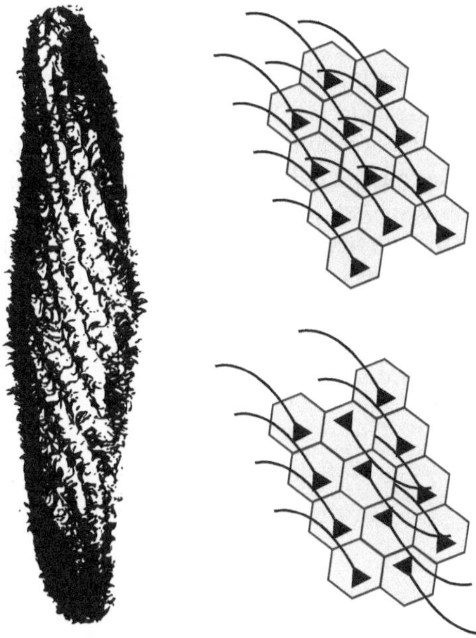

Fig. 6.7: Cortex in ciliates. The cortex contains many structures, among them rows of cilia. Before cell divides, it must duplicate all such structures by growth and division of existing ones—they cannot be assembled *de novo*. Micro-operations turning a row 180° lead to an anomaly that cannot be repaired even after hundreds of generations, or via sexual prcesses. (After Sonneborn (1964), Beisson and Sonneborn (1965).)

providing locomotion for the cell, and the oral apparatus (mouth). The synthesis of such new units requires the presence of preexisting cilia or mouth; in other words, the structure must be present to organize the assembly of a new one. Each cilium grows from an apparatus called kinetosome (basal body): the destruction of basal bodies will devoid the cell of its ability to make its cilia even if the necessary proteosynthesis is working normally. The oral apparatus is similarly dependent on the preexisting one. Rearranging the cortical components by transplantations (doubling the amount of mouths), or inverting a row of cilia, will lead to a progeny bearing—for hundreds of generations—the same anomaly. The cell has no means to fix such a crippled phenotype—even sexual processes are of no help (Sonneborn 1964, Beisson and Sonneborn 1965, Frankel 1989; Fig. 6.7). It can be speculated that tiny differences in cortical structures may lead to the reproductive isolation of sympatric protistan species (Amato et al. 2007).

In 'ordinary' cells, most, if not all, organelles come into existence only by division (e.g., mitochondria or plastids), or by transformation of preexisting organelles (e.g., Golgi, nuclear envelope, or transport vesicles); after all, the cell itself belongs to this category too—it is a replicator. Other entities such as the extracellular matrix may represent a floating structure in continuous transformation, yet never disappearing and reappearing. Obviously, the structural information resides in such structures; in many cases, it cannot be gained *de novo*—by merely mixing the necessary components: they will not appear spontaneously and require a well-coordinated service from the cell.

We have never been dead, as we are reminded by the French philosopher R. Ruyer in his *Princeton gnose* (1974): "As no living being had been dead so far, its existence extends far before its individual birth, and reaches up to beginnings of life and of the universe." The quote could stand as one of the principal themes of this book, with two reservations. First, the singularity of life's origin makes a qualitative landmark in cosmic evolution. Second, we do not cling to gnosis: for us, neither life nor the universe needs any Sovereign to rule them. As to life, we state in the title of a previous book (Markoš et al. 2009) that it is its own designer. As to the universe, it has its inherent rules, but because it is not alive, it cannot negotiate its status: it is not a conscious unity, as the Gnosis would suggest.

7

Information Boom

We don't have the luxury of sitting down to read the equation that governs the universe; we just observe data and make an assumption about what the real process might be, and 'calculate' by adjusting our equation in accordance with additional information. [...] What I am talking about is opacity, incompleteness of information, the invisibility of the generator of the world. History does not reveal its mind to us—we need to guess what is inside of it.

—N.N. Taleb 2008, 268

To create is to inhibit automatic or learned solutions; to act is to inhibit all the actions that we do not take. The refusal to lose ourselves in complexity is an attitude.

—A. Berthoz 2012, 13

Quorum Sensing

In the book *The limits of growth* (Meadows et al. 1972), we read the following riddle: "Suppose you own a pond on which a water lily is growing. The lily plant doubles in size each day. If the lily were allowed to grow unchecked, it would completely cover the pond in 30 days, choking off the other forms of life in the water. For a long time the lily plant seems small, and so you decide not to worry about cutting it back until it covers half the pond. On what day will that be? On the twenty-ninth day, of course. You have one day to save your pond." As an analogy we offer a laboratory bacterial culture, growing in a suspension, in a nutrient-rich medium. Bacteria grow exponentially, as does the water lily in our illustration, but the possible outcomes would be more gruesome for bacteria: if, during the exponential growth, the nutrients are exhausted or noxious wastes accumulated, most bacteria would die. The fiercely dividing cells must—before this happens—switch to a stationary regime of survival, to sporulation, or to a different metabolic regime that allows growth on some other component of the cultivation medium. Some fraction of cells may even thrive on their kin dying all around them, or they actively change their surroundings (e.g., by foaming) to facilitate their dispersal to more favorable localities. In other words, to avoid the 'tragedy of the commons' (Hardin 1968), it is expedient to sniff the peril in advance and stop growing *before* the onset of harsh conditions. There are even more dangerous perils, e.g., antibiotics that must be detected so that appropriate measures can be undertaken in time (for review concerning resistance to antibiotics, see Harms et al. 2016).

To sniff means to know—for individual cells—the density of the suspension in which they grow, and to confront the knowledge with internal settings based on whatever the critical parameters of medium composition may be. Note that such a quorum is not easy to guess from the concentration of nutrients. The bacteria may still occur in a fairly rich medium (at half concentration or so), or nutrients may fluctuate independent of cell density. A more dependable signal molecule is one that is *not* a metabolite: instead the cells release continually some signal compounds, and simultaneously monitor the level of that very compound in the medium (quorum sensing). Cells develop special pathways for producing such signaling compounds, as well as other pathways for signal detection and amplification; together, these end in behavioral decisions (for review, see Whiteley et al. 2017).

The reader should keep in mind that our model experiment with an exponentially growing bacterial monoculture is extremely rare in nature. First, as a rule, many different bacterial species will share common space. Second, bacteria tend to adhere to surfaces by creating mats, plaques, holobionts and similar structures: in such a case, the distribution of the signal may attain the complicated spatiotemporal dynamics of cells and extracellular matrices, this time created by many species. As to complexity, such signals may match those encountered in multicellular animals and/or embryos. Third, signals tend to be, if not universal, then, shared across many species ('lineages'). Fourth, there exist many different quorum signals that may comprise ever-changing cocktails. Fifth, quorum signals represent only a single instance of much broader palette of signal molecules, including antibiotics, attractants and repellents, gaseous pheromones, cues dependent on the structure of extracellular matrix and cell surfaces, signals of eukaryotic origin, etc. Sixth, members of the community bring to the community differently orchestrated adjustments of releasing and sensing the components of the entire mix (cocktail), depending on species, composition, and the history of the entire ensemble (ecosystem).

The result of all this teeming is extensive crosstalk, within the community, as well as with other communities (Ben Jacob et al. 2004, Pátková et al. 2012, Čepl et al. 2010, 2014). Each member of such a consortium (often composed of tens or even hundreds of different species) applies its own strategy of how to deal with circumstances. For example, one species may produce high amounts of quorum molecules, thus inhibiting the growth of its neighbors and, thanks to such cheating, thrive far longer on the available nutrients. This brings to the fore an image of sophisticated communities strikingly similar to macroscopic ecosystems (as forests, meadows, clonal multicellular organisms, holobionts, etc.), or even to the econosphere of modern society (Kauffman 2000). Such communities are, of course, forced by flows of energy and nutrients through the community, but are also capable of subtle adjusting, canalizing, and even foreseeing alterations of these flows. This chapter and Chapters 8–10 offer examples of such gaming.

A short note before we proceed further: Gardner and West (2004, West et al. 2007) recognize four types of interactions within such communities: mutualism, competition, altruism and spite. But despite the emotionally loaded words, their players of the game in such models do not actually *play*: they interact as programmed robots that do not 'want' to compete, defend, or so forth. However, vernacular meaning of all these words points toward *understanding* and we stick to such meaning in what follows (Švorcová et al. 2018). We will return to the topic in connection with so called insurance effect (Chapter 3, p. 44, Chapter 9, p. 191). As we proceed, we emphasize that the view above concerning bacterial (and archaeal) consortia largely holds in eukaryotes—especially with the invention of multicellularity.

Signal Transduction

We now turn to a summary of some general features of signal transduction, amplification and assessment, followed by a resetting of the pathway to the state sensitive for a new signal. The process often involves tens or even hundreds of proteins, adjusted into a network by previous history of the community and its players: events, ontogenetic stage, season, genetic memory, etc. Note also that the absence of a signal may also represent a message for the receiver.

(i) Transformation of gradients. There is a tendency, whenever possible, to transform (transcode) scalar information or gradients (quantities of whatever sort, including concentrations, light intensity, thermal differences, etc.) into discrete units. The distribution, comparison and processing (amplification or suppression) of the resulting discrete value (pulse) in space and/or time is much easier to register than working with continuous time- or concentration-curves, interfering waves, patterning of 'fields', or assessing of 'colors' or 'bouquets' of complex mixtures (Figs. 7.1 and 7.2). Such parsing of the surrounding world into manageable symbolic chunks (and their mutual proportions) is common to all domains of life—from cellular to intellectual. Here we deal only with signals at the cellular level, assuming implicitly that similar rules also hold at the level of multicellular sensors, e.g., eyes.

(ii) Receptors coupled to amplification pathways play a principal role in securing the recognition, assessment and amplification of the percept. Transmembrane receptors—proteins or protein complexes responsible for recognizing extracellular signals and passing the message into the cell—will serve here as paradigmatic example of sensors. Many receptors, however, also reside intracellularly and may not be localized to membranes. Intracellular hubs for signal processing, as the IRS protein in the previous chapter (Fig. 6.4, p. 104), also work by similar principles.

Also for simplicity, we only discuss molecular signals such as, e.g., hormone, pheromone, elements of extracellular matrix, or cell-cell contacts, but antennae receiving electromagnetic, heat, or tactile inputs work on similar principles. Usually, all such signals are easy to distinguish from common nutrients or metabolic intermediates. They usually exert their influence at very low, nanomolar concentrations (as 'first messengers', such as quorum signals discussed above, or adrenaline or acetylcholine in animals). The binding site of the receptor must recognize and bind the signal out of complicated background noise, and with high specificity and affinity. Conformational change of the complex signal-receptor will launch a cascade of responses on the cytoplasmic side of the membrane (Fig. 7.2).

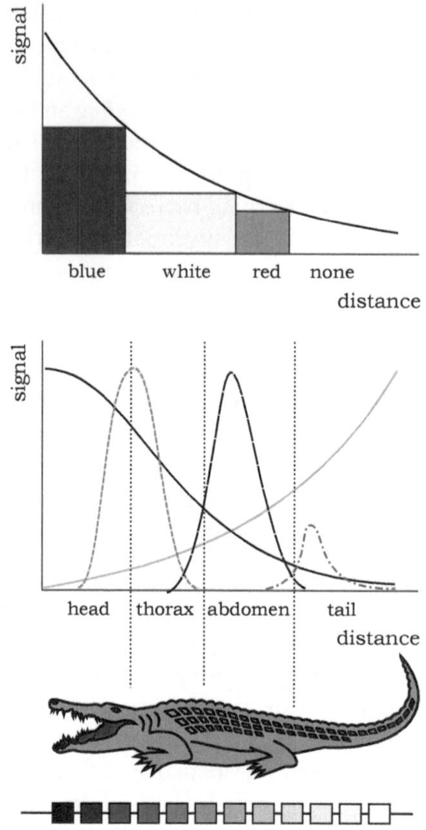

Fig. 7.1: The 'French flag' model and its transforming discrete modules. a. Parsing the morphospace with a single morphogen. Cells along the gradient carry three types of receptors for the signals, with different affinity, increasing from left to right. In the 'blue' zone, all three receptors will be activated, whereas in the 'red' one only that with highest affinity; zero activation is also a signal. **b.** The transient cocktail of morphogen at a given coordinate will be translated into a putative body map, and endorsed by 'hardwired' casettes of gene/protein complexes. See also Fig. 5.3, p. 80. (After Wolpert (1969).)

To fulfill the abovementioned requirement of parsing the signal into discrete units, the receptor must be able to attain three states: (1) inactive (empty) and receptive to the signal, (2) active, with signal molecule bound, and (3) empty but refractive, i.e., unable of binding the ligand after the removal of the previous one. All three 'strokes' are prone to regulation as to duration (time constant), affinity towards the ligand, amplifying efficiency or intensity, and a quality of forcing exerted either towards other pathways and structures or received (feedback) from them. Such thorough adjustment can be attained by the plethora of factors dealt with in the previous chapter—i.e., the genetic settings, epigenetic modifications both through ligands or epimutations, and the putting of proteins into different 'hairball/hairdo' contexts.

(iii) Amplification. Essentially, three downstream amplification mechanisms, or combinations thereof, work both at binding and during the binding of the signal. In

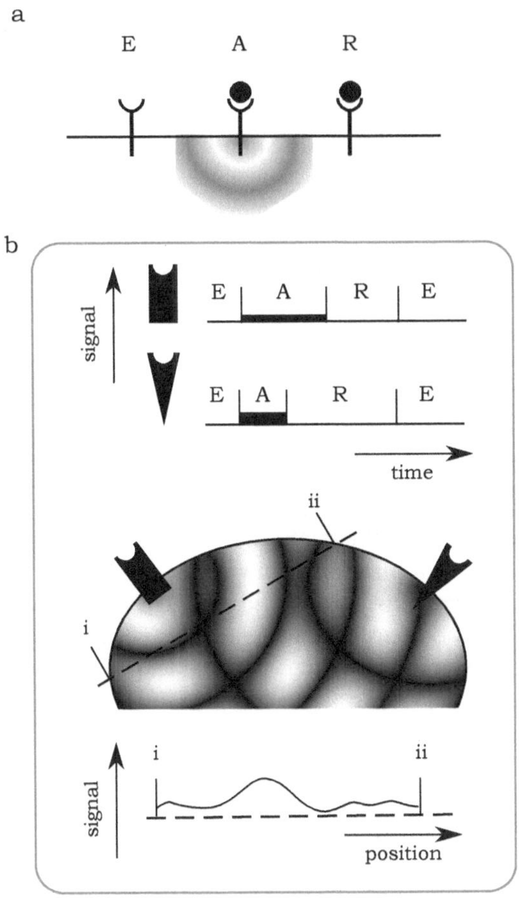

Fig. 7.2: Signal transduction and amplification in cells. a. Three phases of receptor functioning: E—empty and inactive; A—occupied and active; R—refractive, occupied or not. **b. Above:** Two receptors for the same external ligand: the time constants of A and R phases may differ, and so may differ the affinity and amplification power in A phase (portrayed as line thickness); the setting is subject to regulation. **Middle:** Each receptor sends pulses of amplified signals of different amplitude and frequency, mutually interfering. Even if the cytoplasm were homogeneous, the 'concentration' of signal would vary at each point in space and time (**below** shown for a single space transect and time point).

one case, it triggers the activity of enzyme(s) producing small molecules ('second messengers'). They allosterically activate (or inhibit) a class of enzymes, usually protein kinases or protein phosphatases (or a cascade thereof) that introduce epimutations to their target—downstream (e.g., transcription regulators) or upstream (crosstalk with other pathways). In the second case, the complex may open membrane gates, allowing the passage of some compounds (as, e.g., ions Ca^{2+}, Mg^{2+}, Na^{+}, or K^{+}) —they, again, function as the second messengers mentioned above. Whereas in the first case the cell needs pathways synthesizing and controlling production/destruction of second messenger, here it must maintain its (electro)chemical potential across a membrane, and secure its quick pumping away to return to the refractive/receptive state. In both cases, the amplification is powerful: at each step, the original ligand, bound to its receptor, will arouse the production of hundreds to thousands of such second messengers, in a single pulse, and at each step of the amplification pathway.

In the third case, the activated receptor becomes self-phosphorylated, and as such attracts and herds proteins into a multiprotein complex—a 'signal particle' (Langeberg and Scott 2015). This hairball works as a protein kinase/phosphatase factory: one or more cascades of protein kinases/phosphatases become involved in cell functions. The end members of such amplification pathways influence the 'executive' parts of the cell, thus controlling metabolism, gene expression, secretion, cell cycle, cell differentiation, etc. (Fig. 7.2); one such target is regulation of the glycogen synthase/phosphatase system, presented in Chapter 6. Variants of such principles (even not necessarily homologues) are, of course, also at work in bacterial and archaeal cells.

Again, as in the case of receptors, the sensitivity and half-lives of the activated state of any member of such signal particles is controlled by the availability of components within the surrounding ecosystem of proteins and other compounds, which thus change the composition of the whole ensemble. IRS is again an example.

(iv) Crosstalk. There exist hundreds of different receptors on the cell surface, each serving or sharing many amplification pathways. All members of such pathways may be crosslinked with other pathways, exerting epigenetic control of their nodes. The pathways become networked to a very high degree, i.e., the output of the network may result from multiple regulatory inputs, integrated by the network (Fig. 7.3).

Before closing this section, we stress again that regulatory pathways tend to be universal (see the next chapter), even if, of course, little homology exists between, e.g., mammals and angiosperm plants. What is different, however, is the possible

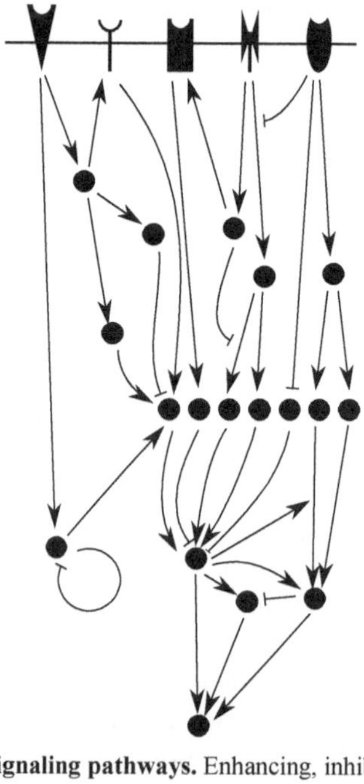

Fig. 7.3: Interrelations of signaling pathways. Enhancing, inhibiting, and crosstalk in a net.

networking of such pathways in different 'lineages' (i.e., what elements are present in the cell at what time), and, of course, inputs and outputs of such pathways. This is how symbionts, pathogens, commensals, etc. (see Chapters 9 and 10) are able to couple to their partner or host's signaling networks, and mutually adjust each other's responses (e.g., mycorrhiza or bacterial symbionts).

An intracellular web can also communicate with its surroundings 'mechanically'. Through the constituents of the extracellular matrix, cytoskeleton, and other more transient 'signal particles', it may connect the extracellular milieu with the cell interior (Mouw et al. 2014). An extracellular matrix represents a genuine 'living environment' for the cell (in contrast with the mineral scaffolding of the abiosphere). Its 'landscape' is the source (as in the case of signals in a gaseous or liquid phase) of specific cues or messages that serve to orient cells embedded or living on its surface (provided, of course, that the cell is able to perceive the particular signal). At the same time, cells incessantly work upon their environment, build it up, take care of it, or place orientation cues within it. Extracellular matrix is a genuine milieu informing cells about their 'here and now' in various contexts. It is utterly vital to microbial consortia, to morphogenesis, to maintain the function and identity of tissues (basal membrane, cartilage, cell wall, etc.), and to organize the immediate surroundings of living beings (e.g., rhizosphere, see below). In addition to providing support, it is a repository of information and memory, comparable to quipus, the 'talking knots' of old Andean cultures. Multidomain proteins of animal extracellular matrix (Fig. 6.6, p. 113) often also contain domains homologous to growth factors. Such deposit may stimulate the responses of cells that come upon (palpate) such a site, without the need for an active sender of an appropriate signal. An example of this in embryonic pilgrimage of neural crest cells is sketched in Fig. 7.4.

a

b

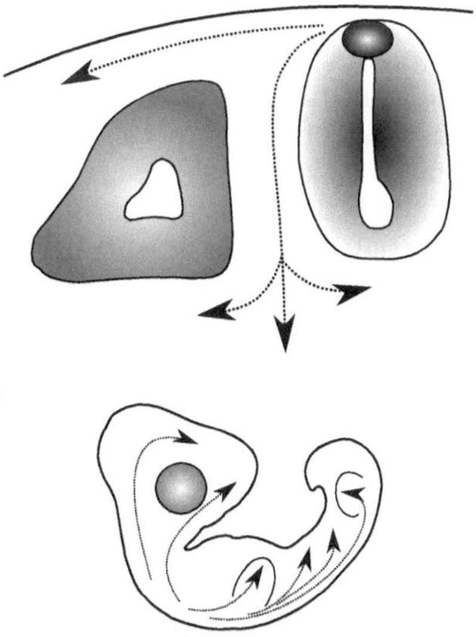

Fig. 7.4: Neural crest cells travelling in embryogenesis. a. In the trunk, cells are directed (by extracellular matrix and other signals) mainly into dermis, or towards forming vegetative neural system. **b.** In the anterior part, they give rise to structures and tissues of the branchial arches (cartilage, glands) and the facial part of the head (facial cartillage and bones, teeth, glands, etc.).

Through special receptors, e.g., integrins, extracellular matrix can also literally 'pull' extracellular matrix receptors in plasma membrane connected on the cytoplasmic side with the cytoskeleton and nucleoskeleton—and thus influence cell behavior and its gene expression. In reverse, chromatin can use this same way to readily inform the surrounding cells about its state (Perris and Perissinotto 2000). Similar liaisons with the extracellular matrix may also exist in microbial communities (Humphrey et al. 2014).

Chance, Necessity, and Gratuity

At this point, even the most casual reader is likely to understand that our views are, to a truly significant extent, broader than those allowed within the 'genocentric' theories formulated by the founders of molecular biology (see Chapter 5). In their thinking, the 'aperiodic crystal' of the genome works as a seed and represents all that is needed for the 'crystallization' of the whole cell (and recurrently the multicellular organism) out of some 'mother liquor'. Yes, at present, cytoplasm (any and all) must work as a substitute for the liquor (as demonstrated fictionally within the Jurassic park franchise), but in near future the whole process might become fully defined. Serious discussions got under way as to whether Martians, upon obtaining the sequence of tiger's genome, would be able to resurrect the beast.[20] As witness to the spirit of the 1970's, J. Monod's *Chance and necessity* stands a catechism of natural philosophy for biological acolytes of the time, yet even his book reacts to the recently discovered phenomenon of allosteric regulation of proteins that, in our opinion, marked a new chapter of epigenetics (see Fig. 5.2, p. 78). Monod copes with the new phenomenon by introducing the concept of *gratuity*: There is no chemically necessary relationship between the fact that the enzyme **E** processes the compound **C**, and the fact that the biosynthesis of **E** (i.e., the activation of its gene) is induced by the same or other compound. Physiologically useful or 'rational' as it may be, this sort of relationship is chemically arbitrary—'gratuitous', one may say. This deserves a longer quote:

"This fundamental concept of *gratuity*—i.e., the independence, chemically speaking, between the function itself and the nature of the chemical signals controlling it—applies to allosteric enzymes.[21] [...] Between the substrate of an allosteric enzyme and the ligands prompting or inhibiting its activity, there exists no chemically necessary relationship of structure or of reactivity. The specificity of the interactions, in short, has nothing to do with the structure of the ligands; it is entirely due, instead, to that of the protein in the various states it is able to adopt, a structure in its turn freely, arbitrarily *dictated* by the structure of a gene. From this it results—and we come to our essential point—that so far as regulation through allosteric interaction is concerned, *everything is possible*. An allosteric protein should be seen

[20] At present, people ask whether tigers that live for generations in zoos belong to the same species as their free-living relatives. Yes, they share the genome and can mate: but should we not take seriously the fact that reintroduction into 'original' conditions does not work?

[21] The statement also applies to epimutations however unknown at the time (AM).

as a specialized product of molecular 'engineering,' enabling an interaction, positive or negative, to come about [...] Thanks to the concept of gratuity, we see how and why these molecular regulatory interactions, foiling chemical constraints, succeed in being selectively chosen solely on the grounds of their contribution to the coherence of the system." (Monod 1972)

The same, we add, holds also for chemical modifications of macromolecules. In other words, by introducing the term gratuity, Monod thereby acknowledges rules that stand *above* the rules defined by chemistry, and depend on a cybernetic 'hardwiring' that comes free, gratis. This of course also holds for the genetic code and other conventions (Norms) developed in evolution.

But is it indeed for free? Take a parable from our everyday life: whosoever needs a knife, can afford it cheaply, almost for free. This is because uncounted generations invested their efforts in acquiring the skill of producing tools for cutting or stabbing —ever more efficiently, ever more cheaply: in this, they furthered the normative procedure of cutting. If we were to pay our ancestors royalties for such skills, nobody could afford a knife. These days, nearly none of us has any idea how to produce a knife, but all of us know the process of cutting. Were we to find ourselves in bad need, we may even 'return' to the old ways and look for a broken piece of metal or glass, or even a suitable piece of flint. Such regress would be punctuated by our memory and experience (again gratuitous): but the usage of flint knife in some extraordinary situation would not mean our return to the Stone Age. Such 'remembering' might have eased the return of mammals to the oceans; yet in no way do whales repeat either appearance or behavior of ancestral Palaeozoic fish. This is repetition with a difference, tradition with an added value of novelty, descent with modification.

With this analogy in mind, we ask the reader to regard the sophisticated web of intracellular and intercellular relationships. If we disentangle it, we find at the very kernel the 'flint'—the basic processes taught in the secondary school, including translation, operon model, transmembrane pumping, etc. The more sophisticated web is a gratuitous upgrade of those primeval processes that were negotiated in the biosphere LUCA, and a result of the effort that has taken billions of years. Moreover, cells can step back to the previous versions of the 'program' where their doings may be more similar to those of other lineages, as in our parable of recollecting the utility of flint. At such deeper levels, signals become more understandable, communicable—even more so than the tools themselves (e.g., the set of hormones in circulation); they are more or less universal for basic lineages of life (recall also *umwelt* in Chapter 4). Biologists proceed in a similar way when they breed their laboratory clones, strains or germ-free lines, and keep them under standard conditions. In laboratory strains, the manifolds of the biospheric interactions is reduced to a minimum, and on such a unified background the scientist is able to reveal more general features of the processes of life. We now return to examples of signals, pathways, and interpretations.

Regulation of gene expression (see also operon model, Chapter 5 and Fig. 5.2, p. 78). RNA polymerase II, the enzyme producing transcripts of DNA-coding genes, is bound to the promoter of the gene-to-be transcribed, as part of a pre-initiation complex. Transcription can start when activated protein kinases phosphorylate the polymerase—as well as some other members of the hairball, thus restructuring it and releasing RNA polymerase to do its job (exemplary paper: Schilbach et al. 2017). The quality and intensity of the external regulations (the 'hairdoing' of the hairball) can adjust the rate, or even the output, of transcription. Following steps (RNA processing, translation, etc.) are prone to even further regulations.

Gradients. A signal may often arrive unevenly distributed in the surrounding space, in the form of variously shaped gradients in space; such gradient may be short-lived, or may represent unevenly distributed cues in the extracellular matrix, on various surfaces, etc. Take the simplest arrangement with some transient signal generated at a single point (source), diffusing into surrounding space, and ending in a sort of a sink. The cells that occur in the space are confronted with differing intensities of the signal, and this fact may influence their behavior. Many males can find the female (the source of the pheromone signal) across very long distances; the same holds for sensing scattered food (e.g., prey or cadavers), or avoiding predators. Amoebae of cellular slime molds gather along such lines of force, diffusing from the source, to establish a multicellular fruiting body. Motile bacteria, due to their small size, cannot sense such gradients as Brownian motion will blur it at micrometer scales. They have, however, developed a sophisticated motility schedule to register the gradients, be it nutrients or noxious compounds.

Sedentary cells spread across a signal gradient (say, in an embryo) may react by parsing the field as epitomized by the famous 'French flag model' (Fig. 7.1). The signal diffuses from the source, along the row of cells. Cells seat three kinds of receptors for signals of differing affinity: in the area of highest concentration, all these will bind to it and will together mediate the 'blue' behavior, whereas at low concentrations with a 'red' answer from neighboring cells, only the most sensitive receptor (and its downstream pathway) will work. Real setups may be much more sophisticated, see, e.g., the fly embryo. As such spatiotemporal gradients usually persist only for short time, it is crucial for cells to remember the situation by activating more durable memory traces—this often results in permanent changes in gene expression and/or morphological differentiation (as in case of neural crest cells) that may be maintained in a given cell and its progeny.

Pathfinding. At the stage of neurula, the vertebrate embryo (Fig. 7.4) enters a decisive and very specific stage of development. The cells on the roof of the neural tube—the neural crest—undergo dedifferentiation and move into different parts of the embryo where they give rise to a plethora of cell types, tissues, and organs (Hall 1999, Kulesa and Gammill 2010, Trainor 2014, Kaucka et al. 2016, Bronner and Simões-Costa 2016, Martik and Bronner 2017). In the trunk, the main cell streams flow dorso-laterally, giving rise to melanocytes in the hypodermis. The second stream proceeds ventro-laterally (between neural tube and mesenchymal pouch), establishing above all, semi-autonomous neural ganglia lining the backbone in each segment; at the rear, they establish the caudal bud.

Cells that travel from the anterior part of the neural tube build—practically 'from nothing'—the facial part of the head and its branchial arcs (bones, teeth, cartilage, etc.). The coordinates on the neural axis determine some properties of the neural crest cells, so, e.g., only cells travelling to the head parts have the capacity to build cartilage. This part of the story is not our interest here: the principal question is *how* they find their definite position and *how* they build species-specific facial structures. Particular paths are laid down by the scaffolding of the extracellular matrix created by surrounding cell sheets (of neural tube, epidermis, mesenchyme, notochord, or endoderm), and their surfaces. The streaming cells recognize cues in the meshwork and change their properties along their path: they reciprocally add or remove cues, thus informing—and modifying the fate of—both the surrounding tissues and the neural crest cells streaming in their wake. As mentioned above, it is extremely difficult to study the structure and transformation of this floating world. Early transplantation experiments (ectopic grafts) confirmed the role of extracellular matrix at ectopic places of the embryo (Perris and Perissinotto 2000) and revealed (at least partially) the universality of the developmental code, by constructing quail-chick embryonal chimeras (Le Douarin 1973, Le Lievre and Le Douarin 1975, Hall 1999). Hundreds of papers have studied the developmental effects of deletions of particular genes for extracellular matrix enzymes, of sugar modifications, etc. (Banerjee et al. 2013, Martik and Bronner 2017). This research gained a new impetus when it turned out that the migration of metastatic cells takes place in very similar ways. What is still lacking, however, is a global theory of all such interactions.

The reciprocal (in)formation dynamics of both cells and extracellular matrix, as demonstrated above on neural crest cell migration, is by no ways unique: similar examples could be taken from development and/or maintenance of practically any tissue: from microbial consortia in mats, plaques, colonies, through animal or plant tissues, up to holobionts (Cohen et al. 2017). So-called single cell multi-omics are composed of genomic, transcriptomic, proteomic and epi-variants thereof, all gathered from single cells (Kelsey et al. 2017). It may turn out to be especially fruitful in the near future, to record (concordant) events at and between different 'layers' of cell organization in different times, and compare cells that share common destinies in tissues (even seemingly homogeneous), bacterial communities, suspensions, etc.

Processing and Responding

We can (very roughly) divide responses to signals (upon amplification) into four classes, but will here delve deeper into only the last two.

(i) The first group is of physiological reactions as responses to immediate stimuli, such as various reflexes as well as hormonal, nutritional and behavioral feedbacks: danger; food or prey; sexual partners; stress reaction; the presence of appropriate receptors and organs; neural transmission; sensitization and desensitization (Weber-Fechner law) of such receptors, amplification pathways, organs; proprioception, etc. Receptors, pathways and executive organs (appropriately set for the task) mediate such responses, and attenuate them to fit the context. Sexual cues, for example, may not be recognized either by an immature addressee or by mature one under improper settings (e.g., season, stress). Note the enormous amount of effort that must have been invested in advance—by ontogenetic processes, individual experience, physiological and ecological settings, evolution of the lineage (see gratuity, above), etc. From such a web, to single out one clear pathway or response is often a very demanding goal:

thorough selection of laboratory models and settings is a priority. Yet a very high fraction of results remains irreproducible, not because of fraud but due to minutiae of local laboratory protocols. Physiological reaction as reflex, avoidance of noxious inputs, etc. may serve as paradigmatic models for this group of experiments. A book by J. Sapp, with an ambiguous title *Where the truth lies* (1990), which maps the life of Franz Moewus—one of the founders of molecular biology, may be instructive in this respect. The 'in our hands' argument has plagued biological research from the very beginning; recently, however, it became a serious problem caused by the rocketing costs of repeating experiments in different laboratories.

(ii) The second group involves genetic responses, also along pre-established pathways or programs. Again, tools must be built in advance; here, the tools serve to arouse such responses and react according to genetic, ontogenetic, or environmental instructions. In the presence of an unusual (but not unknown) nutrient or other cue, cells (especially prokaryotic) can synthesize an appropriate metabolic pathway in an order of minutes: it would not be economical to keep and maintain such pathways idling in advance waiting for the opportunity to cope with uncommon situations (but see the 'insurance effect' above and in Chapters 3 and 9). The process, however, may also involve the subtle regulation of genes transcribed (or their variants if different promotors or terminators are possible), splice forms, translation speed, epigenetic markers, etc.

Recall also that new proteins entering the hairball may fundamentally change the very properties of the network in the immediate future (adjacent possible). A paradigmatic model for this group of reactions is the operon model of genetic regulation in bacteria (Fig. 5.2, p. 78).

(iii) Both environmental and/or internal cues may also lead to longer-lasting effects, some reversible (as in mating season, flowering, jumping transposons in plants, or cell cycles in protists), others irreversible (e.g., adolescence, pupation, or building permanent structures such as fruiting bodies). In the second case, the stimulus may be at work for a restricted time window only, as in the case of morphogenetic gradients (Fig. 7.1), or of imprinting in defined ontogenetic stages such as the juvenile stages of some animals or vernalization in plants. Whole cohorts and tissues are recruited, with cells determined towards a specific differentiation state. At the level of chromatin, conspicuous changes rest in its massive reshaping, such that some loci will be permanently switched on or off, with the maintained state also attained in the progeny of such cells (Chen and Dent 2014). Two protein groups (counting dozens of different proteins), which block or enhance the expression of some loci on the chromosomes, are outstanding in this respect. The Polycomb group (PcG) exerts a general blocking effect in all eukaryotes, which leads to the establishment of heterochromatin at the given locus: in some loci, however, activation is also possible. The Trithorax group (TxG) has, again in general, an activating, enhancing effect at the given locus. Besides these, many other chromatin-modulating protein groups exist, e.g., Hox genes (see Chapters 5 and 8). Allshire and Madhani (2018) give a review of mechanisms participating in heterochromatin buildup and function.

The correlation of the effects described with the epigenetic markers on DNA and nucleosomes is not known in much detail. For illustration, remember the 'bar coding' on histone proteins and its combinatory possibilities: Piunti and Shilatifard (2016) in their review refer to the fact that the well documented epigenetic marker connected with PcG is trimethylation of lysine in position 27 on the histone H3 (H3K27me3), whereas activation is typical with trimethylation of K4 on the same histone (H3K4me3).

If we concentrate on the Polycomb group, we find only one DNA-binding protein, and three principal complexes are attached through the above-mentioned H3H27me3, H3k26me2, or through an ubiquitylated residue H2AK118/119Ub (i.e., ubiquitination of lysine 118 or 119 on the histone H2A). These sites are the docking places for the main protein anchors of the Polycomb complexes, which then provide a 'crystallization core' attracting other proteins (the hairball).[22] The inhibition of gene expression at a given locus requires, first, the presence of a cis-regulatory sequence on DNA called Polycomb response element, which elicits DNA methylation in the region (PRE; Coleman and Struhl 2017). If the context of other cis-sequences (and their bound state) allows, the site will attract histone modifiers (e.g., methylase PRC2) that will methylate the lysine moiety in position 27 of histone H3 on the nearby nucleosome, thus providing the docking site. Second, the Polycomb complex will 'brick up' the site, i.e., inhibit it. Finally, complexes from different parts of chromatin will unite into a greater 3D hairball that keeps the imprisoned loci in a silent state (Jiao and Liu 2015, Entrevan et al. 2016). Long-lasting patterns of markers on DNA, chromatin, and the whole loci are well described for many types of differentiated cells, as is the silencing of one of the X chromosomes in mammalian females (Almeida et al. 2017), the silencing in plants upon external stimuli (Yang et al. 2017, Jiang and Berger 2017), and the reprogramming of the plant genomes during sexual reproduction (Kawashima and Berger 2014). Finally, epigenetic reprogramming of cancer cells may lead to the resistance to chemotherapy (Shaffer et al. 2017).

Mutations of key position K27 on histone H3 does not allow methylation of the site, which results in incapacity to build the Polycomb complex. This has far reaching consequences in embryonal development (for *Drosophila*, see Pengelly et al. 2013). For the role of Polycomb complex for determining the positional identity of neural crest cells, see Minoux et al. (2017).

Polycomb silencing is, as a rule, maintained in daughter cells, and may persist for the whole life (or life period, as in plants). The information thus survives cycles of DNA replication and mitosis: DNA methylation takes place concurrently with its replication. The attached barcoded histones of mother cell origin instruct PRC2, as well as connecting enzymes, in the labeling of the newcomer naive histones, subsequently reconstituting the previous state. Coleman and Struhl (2017) showed that the excision of PRE and/or inhibition of PRC2 leads to a 'dilution' of the labeled histones and an activation of genes at the site, over several cell generations. Examples of this include the context of PREs, level of DNA methylation, and barcodes on

[22] The enumeration of all such proteins and/or combinations is far more exhausting than the old lists of systematic biology (e.g., Entrevan et al. 2016); here and throughout our work, we try to express the main principle as simply as possible.

histones (note that we mentioned only a handful out of many that undoubtedly have meaning in other contexts)—as well as the variable composition of Polycomb complexes. All this plays a decisive role in whether genes at the given site, in a given context and in a particular organism and species, will be active or silenced, and for how long they will remain in such a state. To be clear, our brief discussion is of but a single—even if one of most important—regulatory complexes in chromatin.

From this simple example, the reader can get an idea of how complicated the 'epigenetic codes' may be (as well as the turns they take in time), and the scope of the challenge they present for those trying to decipher the original Norm (presuming there was any at all). Yet, it seems that there exist cues on DNA that may also be responsible for the reconstitution of interphase chromatin. This may be the case of *ultraconservative elements* (UCE), sequences shared in large lineages such as vertebrates without change for hundreds of millions of years. The human genome contains some 500 such UCEs, flanked from both ends by a waste area of less conservative regulatory elements (Bejerano et al. 2004, McCormack et al. 2012). Perhaps these serve as the principal hubs of the gene expression factory, the Norm that organizes the events.

But what is the meaning of the other 'bars' of the histone code that are often known for particular situations only (e.g., tumors, or in transitory situations like wound healing in the skin—Naik et al. 2017)? No comprehensive theory has yet been developed. Similarly, ample methylation of DNA is typical more for silenced loci than for active ones, but again, no definite rules can be extracted from this enormous body of data. We return to long-lasting epigenetic coding in the next chapter.

To close the discussion on long-lasting and irreversible effects of signals, we briefly mention two points. First, the roles played by many different classes of RNA in configuring chromatin and DNA structures are very inadequately understood, and we must wait for future research. Second, genome reshuffling (transposons, immunity, aneuploidy and polyploidy, cells without nuclei) may also contribute to long lasting, mostly irreversible changes in cell differentiation and ontogenesis.

(iv) Transgenerational epigenetic memory has been known as non-Mendelian inheritance for more than a century (Dauermodifikationen, i.e., lingering modifications, paramutations, etc.; Brink 1973, Martienssen 1996). Germ cells can transfer to the next generation both histone modifications and some of the DNA. As we have seen in case of RNAs (and probably many other types of signals molecules), signals may be brought along with gametes to influence and reprogram—immediately after their fusion—both pronuclei and the newly established zygotic nucleus. Such transfers of epigenetic markers to the next generation will be discussed in more details in the next chapter.

Simplexity

The tension between reductionist and holistic worldviews has amassed in biology since its very beginnings as a science, bringing with it both breakthroughs and infertile skirmishes. We are not going to plunge into the history of these rows, but it is worth mentioning the contribution of the French neurophysiologist A. Berthoz

(2012). His approach is close to that of the Uexküllian theory of *umwelt*, yet with important extensions (for umwelt, see Chapter 4, p. 57). Naturally, Berthoz refers to animals, i.e., creatures equipped with brains, but we contend that his conclusions may hold for all life. The principal question is this: how do trillions of molecules assemble into a cell or multicellular 'living matter'? Of course, they do not—a living being does not start 'from nothing' as we see in dissipative structures such as a flame or the Belousov-Zhabotinsky reaction, or what is called today 'active matter' (Sanchez et al. 2012, Popkun 2016). All organisms are indeed very complicated in their molecular parts. The organism is not simple, yet it is *simplex*, the term introduced by Berthoz. "Faced with complexity, living organisms chose specialization, modularity, separation of functions, division of labor, categorization, and distinction. The brain is like a hive, an ant colony, a termite mound, an army, a factory, society itself. It is a seat of extreme specialization that appears very complex but that, in reality, simplifies processing of information about the world and, moreover, that facilitates better control of action" (Berthoz 2012). In other words, the enormous complexity is organized: channeled according to experience, gratuity and the habits of the community. At any moment, there exist preferred—yet not hardwired—pathways of doing things, out of many possible and even, from the view of physics, equally probable. We are back to the longevity of lineage, its memory and experience that guarantees sovereignty in performing tasks as established within the lineage. For illustration, take, e.g., the Pax regulatory protein and its universal command 'Make eyes!' (Chapter 5). Fly mutants bearing deletions in both copies of the gene are blind; their progeny can be rescued by transformation of the germinal line by the Pax homologue from mice. As soon as the command again arrives, the gate is opened and the progeny will develop normal, insect-specific (not mammal-specific) eyes.

As Berthoz explains with an allusion to Uexküllian *umwelt,* yet also holding the heritage of evolution in mind, simplexity is based on three principles that are implemented successively or in parallel: inhibition, repertoire of inputs, and memory. Berthoz takes inhibition for "one of the greatest discoveries of evolution" (p. 13). It enables the separating of executive actions recognized as important and the simultaneous inhibition of primitive cognitive strategies or innate reflexes: "to create is to inhibit automatic or learned solutions; to act is to inhibit the actions that we do not take." The refusal to get lost in the complexities of body and world means 'bracketing off' unnecessary or redundant items, to focus on a limited amount of tasks. Again, we stress, Berthoz has brain function primarily in mind, but zooming out to all life is, in our opinion, easy.

The second principle is, in our opinion, the setup of sensors and actions which constitute the umwelt, or simply physiological aptitudes. As in the previous case, this constitutes channels (receptors) that, from the manifolds of world, pick out only those inputs and reactions that are relevant to the 'chosen' ways of living. We do not know how to interpret the croaking of frogs or the singing of a thrush, even if we can hear the sounds and construct hypotheses as to the meaning of such sounds. Additionally and importantly, a living thing can choose from within the frame of what it has at hand: "Deciding involves selecting from the information around us whatever is pertinent to the goal of action" (p. 14). Decision-making distinguishes Berthozian umwelt from the programmed automata presented by Uexküll.

The third principle is the ability to anticipate based on memory. Again, Berthoz applies the principle to an animal individual with memory. For us it is again easy to generalize it to the whole of life, encompassing the genetic and epigenetic memory of the individual, its neighbors in the biosphere, and also evo-devo. Anticipation based on memory allows, again, functioning in the complex world, the seeking of paths optimal for a given context (recall the insurance effects previously discussed). The world is not simple, and the answers to its challenges cannot be simple, but for making a living thing, they may be simplex.

Note that not one of the three principles is deterministic or hard-wired: there is always a surrounding halo of suboptimal, redundant solutions of any given task that can be tried in unexpected situations. Of even more importance, we add, is the possibility to generate novelty, to intrude into parts of the world not previously available (or more pungently said, not yet existing for that given form of life).

Flexibility based on the abovementioned principle leads us to another concept by Berthoz: *vicariance*—the ability for some functions to substitute for others when need demands, as a remedy, if the general way of doing things fails. 'Don't repair—replace!' is the slogan that characterizes vicariance. The world is stable until proven otherwise—and vicariance is a remedy for the Black Swan situations (Taleb 2008) that living beings continuously encounter. As S. Kauffman (1993) reminds us, the environment usually changes at a speed at which mutation-selection and adaptation simply cannot cope; vicariance, i.e., fertile tinkering, may help bridge the gap. To provide a recent biological example, Osterwalder et al. (2018) mapped regulatory elements (enhancers) which safeguard normal development of structures such as limbs, brain, or heart. They report hundreds of cases with five or more redundant enhancers flanking developmental genes, and conclude: "redundancy is a remarkably widespread feature of mammalian genomes that provides an effective regulatory buffer to prevent deleterious phenotypic consequences upon the loss of individual enhancers." All this redundancy is based on complicated and complex meshwork of information processing in cells, bodies, brains, and ecosystems, which enables living being to exist in a mode that is simplex and yet—thanks to redundancy, memory and vicariant abilities—sufficiently flexible to test out new pathways in the *struggle* for life.

8
Morphogenesis

Certainly, the documentation on how to build an aircraft does not contain the 'complete' information for the construction of an aircraft. [...] *Plans for the board kitchens, seats, bathrooms, seatbelts, etc. are not that crucial for somebody who knows what kitchens, seats, bathrooms, seatbelts, etc. mean, what they are good for.*

—A. Markoš et al. 2009, 184–185

The law is not a recognizer of persons; its judgments are applied at the end of a series of acts. With regard to individuals the law thus creates a fiction... [...] *The law visualizes the individual as a kind of actor with a role whom the court has located in the situational system of the legal code.* [...] *That the persons who stand before the bar of justice are identities that they appear to be personification of, and completely explainable by, the logic of their crimes, is the effect of a visible artifice of juridical thinking.*

—H. Rosenberg 1965, 136–7

The Tradition of the New

Such is the title of a book by H. Rosenberg (1965) containing essays from cultural studies and history. A recurrent motif throughout is narrations, how—in culture and history upon crises—communities resurrect historical narratives and/or mythology, but raise such old motifs in order to overthrow an existing way of life. Despite appearances, the result is *not* a return to a previous, idealized, 'Golden age', but the emergence of something new. To bring our own examples, such was the Renaissance (allegedly the return to the Roman age), the great Protestant movements (the return to the roots of Christianity), communist revolutions (ostensibly pointing to the Spartacus rebellion of 73 B.C.), or the uncounted nationalist upheavals striving to 're-establish' the Golden Age that never was (recall, e.g., the civil wars in former Yugoslavia and the Caucasus—rooted in just such inveterate tribal mythology). The analogy may help remind the reader that in spite of the great diversity in the biosphere, there are few 'body plans' in each major group of multicellular organisms. At the level of molecular interactions, the deep homology of tools is even more conspicuous. In plastic phases of evolution (after crises such as mass extinction), the lineage often makes a step back, before heading its evolution in a new direction. In no way, however, do we want to evoke the opinion that, say, the advent of birds was 'inspired' by pterosaurs, though both groups of active flyers coexisted in the biosphere for a long time. After all, they did not belong to the same lineage of vertebrates, so no descent with modification was at play in this case; yet, we can speculate that homologous molecular tools had been in use in parallel in both cases. Some 'remembering', however, *might* have played a role in the evolutionary quick transformation of hoofed land animals into whales and dolphins. Induction of a tooth structure with enamel in birds—after some 100 millions of years of the absence of teeth in the lineage—is an illustration of such deep memories (Kollar and Fisher 1980).

Multicellularity

"I am aware of the dangers of over-interpretation. Generalizations from a few specific cases which combine incomplete information from organisms of arguable comparability does not make the best science." This comes from the book *Fungal morphogenesis* by D. Moore (1998), and we ask the reader to keep these words in mind throughout the following text. We enter a shaky field and our current narrative covers a widely fragmented field of information.

If we take the primordial biosphere LUCA, as portrayed in Chapter 4, we see diverse cells promiscuously exchanging matter, genetic information and even cell structures, busy with sending and receiving signals, and building sophisticated structures of extracellular matrix. Incessant disturbances everywhere force transformational negotiations within the biosphere, but also hone a sense of alert wariness towards what surprises the environment may serve. From the perspective of such a scenario, multicellularity may—at least for a critical period at the beginning—look like an attempt to 'break free' from such bounds and execute undisturbed the basic embryonal program that will establish the elementary, lineage-specific scaffold for further development. If the attempt succeeds, the growing germ will organize a miniature individual (in contrast to a 'dividual' community), constituted by a clone of cooperating cells, and erecting body structures (differentiated cells, organs and extracellular matrix). Moreover, such development will be bound by 'traditions', hereditary ways that are characteristic for a given lineage, offering directions on how to cope with a given task. In eukaryotes, this went hand in hand (i) with sexual processes that keep the tradition (Norm) within a frame held by those who can, and do, interbreed (population, species); and (ii) by drastic curtailing of horizontal gene transfer, which may well violate such established routines through the introduction of 'malware'. Evolution invented the multicellular way of living independently in several tens of variants—thus enormously increasing the complexity of the biospheric web.

The importance of clonal beginnings in isolation from the biospheric web becomes apparent when compared with a second way towards multicellularity: the association of cells of the same kin (but not necessarily a clone), as in slime molds or myxobacteria. This variant is quite rare, and leads to multicellular bodies that are relatively small (about one millimeter). What is an even more serious drawback is that they represent not clones, but chimeras of cells of different origin. In constituting a multicellular body, cells of differing origin may start competing instead of cooperating as to what body part they engender: only those forming the sporangium will get the chance to spread their progeny, whereas stalk cells will die. In other words, in chimeras, parasitism of some clones of the same kin cannot be fully avoided. The same may also occur in clonal bodies (tumors being the most conspicuous example), yet, in general, the overall control prevails.

A chimeric body can be considered as an extension of what is regularly observed in microorganisms that live in multispecies consortia (mats). If, however, they happen to grow in monoculture suspensions, the clone will often differentiate its tasks (metabolism, coating, foaming) among different cells in population—as if substituting for the division of labor present in a consortium (Rosenzweig and Adams 1994). Myxobacteria or slime molds may represent a special case brought about by their predatory way of living and/or the necessity of spore dispersal (the need of fruiting bodies). However, such routes towards multicellularity appear to represent a dead end; as such, we will discuss them no further.

The precondition for multicellularity proper, thus, is a single cell (a zygote or a stem cell, a spore, a bacterium) or a small group of cells concentrated in a small volume that is insulated from all other inhabitants of the biosphere (Pátková et al. 2012). In bacteria, the creation of a multicellular body, i.e., colony (Shapiro 1998, Ben-Jacob and Levine 2005, Rieger et al. 2008, Čepl et al. 2010, Shapiro et al. 2012), is a very rare occasion—it occurs only when a cell happens to fall on a sterile semi-solid substrate (e.g., a freshly ruptured fruit), or if manages to clear and stake the surroundings via antibiotics (*Streptomyces*). In most situations, however, a bacterial cell is born into an elaborated bacterial community and remains dependent on it. Mutual interactions with other life forms will hinder any attempt towards multicellularity, and many lineages cannot advance in such direction even in those rare cases when the chance occurs. Indeed, many bacterial species cannot even survive, let alone grow, in axenic 'germ-free' cultures.

Eukaryotes developed better-elaborated strategies that allow regular safeguarding of the germ cell. This enabled them to start a brand new type of evolution: early ontogenesis taking place within a 'mini-biosphere' organized by its inhabitant(s), who represent a clone. In such an 'artificial' situation, the embryo can follow a program specific for a given species, based in cell multiplication and growth, determination, differentiation, apoptosis, movements, extracellular matrix construction and rebuilding, etc. (Fig. 8.1). Only upon accomplishing the early state (often characterized by assuming the phylotype state, see Chapter 5 and below[23]) can the germ start communicating with the outer world—often helped by its mother. As shown in the next two chapters, even in species with obligatory symbionts (extracellular or intracellular), embryogenesis usually starts with 'clean' germlines, which are only later infected—in a controlled way—by their symbionts (Bright and Burghelesi 2010).

[23] Primary larvae of some animal groups may represent an exception; see below: phylotype.

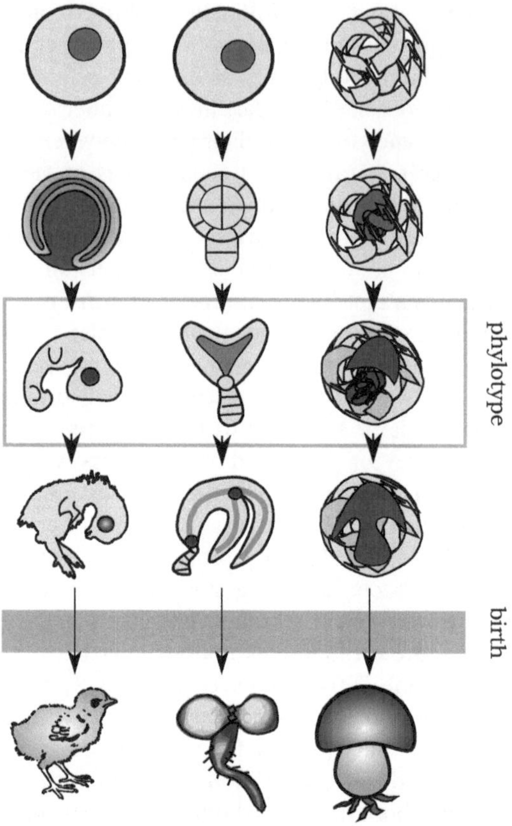

Fig. 8.1: Common features of clonal multicellular organisms. The phylotype is a common stage of ontogenesis in big lineages, as vertebrates (pharyngula stage), dicot plants (torpedo stage of the seedling), or those fungi that build fruiting bodies. Stages before phylotype are strictly insulated from the rest of the biosphere. After the phylotype stage, the insulation may loosen somewhat, but may persist up to the birth. (See also Fig. 8.2.)

Of the many multicellular lineages, we now attend to three most distinguished: 'perfect' fungi with fruiting bodies (mushrooms); animals; and vascular, particularly seed plants. The rest are multiple marginal attempts in different groups of protists, including *Volvox*, colonies of yeasts, myxomycetes, and many others: each of these have established their own way towards multicellularity, with no (or highly questionable) homology between them. Below we give some simplified, model projections of basic processes and patterns of the more distinguished lineages; needless to say, many representatives of a given phylum may differ in particular aspects of their strategies for living.

Fungi furnish the best example of the contrast—and switch—between 'communal' and 'private' ways of living, as described above (Moore 1998[24]). Imagine a species

[24] For the updated (2013) version of the book, see http://www.davidmoore.org.uk/Biog_01.htm.

living in soil and producing fruit bodies (mushrooms; Fig. 8.1). With the exception of basidia and spores, the structure of the fungal organism presents threads of hyphae, linear strings of cells interconnected by cytoplasmic contacts (as in the scheme of plant tissue in Fig. 3.1, p. 31), and in incessant contact with the soil microbiome and roots. Hyphae can grow only by divisions of the apical cell; intercalary cells, if dividing, give rise to a branch that grows again by apical growth. It follows that fungi never form true tissues such as a parenchyma or skin: if it does form 3D structures as fruit bodies, they must 'weave' them as a complex tangle of branched hyphae.

The sexual life of our model fungus is also remarkable. Haploid spores give rise to a haploid mycelium, but upon encounter ('confrontation') of mycelia of a different mating-type, the apical cells fuse. Both nuclei, however, remain separated in a common cytoplasm (a dikaryotic, but *not* diploid stage), and divide synchronously, often for hundreds of years, to occupy each daughter cell. Karyogamy takes place only in basidia, immediately followed by meiosis and the production of four haploid spores. The plasticity of the life cycle is underlined by the fact that monokaryotic (haploid) mycelia can exist indefinitely and even produce haploid spores.

Most fungal biomass is present as mycelium, which consists of hyphae, or bundles thereof (rhizomorphs) that form a web across vast areas; it is busy with decomposing organics, etching rocks, and entering into zillions of symbiotic interactions with microorganisms and plant roots (mycorrhiza). A whole forest or meadow can be thus 'hardwired' by a single mycelial web vigorously organizing flows of nutrients and information across the community (Kiers et al. 2011). Such a hyphal continuum, descending from a single spore, may have a mass of tens of tons, covering an area of many square kilometers and living for thousands of years (Smith et al. 1992 for *Armillaria*; many other reports since). In spite of all this, one hesitates to call this clonal spread an individual, even if—in contrast to other dwellers of the multispecies community—it remains interconnected thanks to the hyphae. It leads a typically *microbial* way of life and participates as a unifying element of many different consortia, but such 'mushrooming' has no shape, no pattern or individuality that would require morphogenesis; it has no 'self' (or better, no self as based on criteria defined for most animals and plants). In contrast, fruiting bodies *are* such individuals—and the mycelium must establish the preconditions necessary to isolate the early stages from the biosphere (Fig. 8.1). The process starts with a single cell that will divide many times, thus sending multiple hyphae sideways. The resulting bobble starts differentiating towards the sclerotium, with (at least) two possibilities of further development. The first outcome is a rounded structure, which aids in survival under harsh conditions (e.g., dry season); in some fungal lineages, sclerotia can give rise directly to spores (*Claviceps*). The second possibility is the development of sclerotium towards a fruiting body. In that case, the contours of a prospective fruit body (the 'embryo') become apparent at a sub-millimeter scale, and the structure starts growing and differentiating. Imagine the logistics of the process: doubling the volume of a structure that consists not of cells but of knotted threads! The hyphae involved must rely on growth with obvious tropisms, intercalations, and branching, accompanied by controlled apoptosis at places, as well as on differentiation of particular cells to form 'tissues'. At some stage of stem and cap development, hymenium—the seat of further basidia as well as spores under the cap—starts proliferating and differentiating centrifugally from the top of the stem. It is interesting that even the status of differentiated fruit body remains poised towards the 'microbial' way of life: if a severed body comes to a contact with the substratum, it again begins producing vegetative hyphae.

Animals. There is no spore or haploid body stage of animals. Gametes are viable for a limited period only, during which they must meet and accomplish fusion and fertilization. As a rule, the zygote[25] immediately begins cleavage and embryogenesis—in thorough isolation from other inhabitants of the biosphere (except the mother—as in placental mammals or placental sharks). This period ends with either birth, or the hatching of an egg, when extensive liaisons with biosphere commence. If insulation is necessary in the first stage of development, coupling to the biosphere is as important in the second. Young animals that are either artificially raised without microbes (germ-free youth) or seeded, in a controlled fashion, with either a single or a few microbe species (gnotobiotic animals), revealed through dysfunction, the developmental and nutritional demands of the second stage of ontogenesis (see Chapter 10).

As animal cells are not bounded by cell walls, they can—during embryogenesis as well as later—freely move and change not only their shape, but also their position within the body (as in neural crest, Fig. 7.4, p. 123, or lymphocytes and blood cells in vertebrate organism). Moreover, in contrast to fungal hyphae, they are able to divide in any selected directions and give rise to parenchymal tissues as well as sheets; such sheets can also move (as invaginations in gastrulation), establish plies as well as various kinds of tubes (neural or alimentary tube, blood vessels, etc.).

The growth of an animal body—in contrast to fungi and plants—is in most lineages limited as to the overall shape and internal organization. This allows, often even requires, determining the presumptive body plan early in embryogenesis, so the cells know—and remember for the particular lineage of cells—their position and fate on the virtual body map. Below, we offer three model cases of means that achieve such a layout. Note the decisive roles played by the mother organism; the role of the sperm at these first stages is close to nil.

First, early embryogenesis is a quick process with high demands. Therefore, the egg is usually the largest cell of a given species, carrying reserves of structures, mitochondria, nutrients, RNA and ribosomes, transcription factors, histones, DNA polymerase, etc.—all prepared well in advance of fertilization. Egg maturation is often helped by nurse cells, and is time and energy consuming.

Second, karyogamy is a demanding process that needs a sophisticated restructuration of the newly created diploid nucleus, so zygotic transcription (hence also engagements of paternal genes) may be substantially delayed (see below, *Drosophila*).

Third, even if zygotic transcription took off immediately, a single nucleus would not be able to provide quickly enough for such a big cell.

[25] In some groups, the single cell at the beginning is not a zygote, but a parthenogenetic, gynogenetic egg. For simplicity, we will speak of 'zygote' throughout.

Fourth, the mother organism often sets the heterogeneity of the egg cytoplasm, thus establishing foundations for the presumptive body plan. In sum, early embryogenesis, i.e., processes taking place before organogenesis, is to a high extent matrilineal. The setup and sequence of events, however, may vary even in closely related species.

(i) Model *Caenorhabditis* (roundworm, Nematoda). *Before* fertilization, the egg cell is endowed with a highly heterogeneous cytoplasm—as to composition, structures, and signal molecules. Thus, upon cleavage, the daughter nuclei will find themselves in daughter cells bearing different parts of the original egg cytoplasm, i.e., with different cues to launch particular cell lineages. In extreme cases, deleting (ablating) one of the cells at the 2 (4, 8, 16…)-cell stage will lead to a lack of some embryonal structures of the embryo or larva. The remaining cells do not feel the damage, thus are not able to repair it: they adhere to their automatic subroutines preset by the presumptive map laid down in the egg, without paying attention to the whole. The adult hermaphrodite roundworm contains 759 somatic cells (the numbers of gametes are, of course, copious and variable). It is obvious that such routine development, excluding even wound healing or replacing worn-down cells, can work only in small and short-lived animals.

It is important to mention, however, that in most animal taxa (except vertebrates and insects), tiny *primary larvae* do develop in exactly such a deterministic way. The organism proper will develop only later from an undifferentiated, 'set-aside' mass of larval mesodermal cells, in cooperation with small parts of the larval ectoderm and endoderm (Jägersten 1972, Davidson et al. 1995, Cameron et al. 1998; see also below, Phylotype).

(ii) Model *Drosophila* (fruit fly, Insecta). At the first sight, events are similar to our previous example, but involve much larger amounts of nuclei and cells, so the determination of body axes and cell fates is rather fuzzy. A precise parsing of spaces available requires interactions of embryonic cells and their mutual negotiation of immediate future steps. This allows for variations in egg size as well as cells involved, and also a different distribution of tasks. Such division of labor is different in different species: here, we give only the basics of *Drosophila*.

The anatomy of the egg determines the site of sperm entry and the body axis, but cues are present only *in potentiam*. After fertilization, two coordinated processes run in parallel, karyokinesis and translation of maternal mRNAs accumulated and anchored in cytoplasm. The 'cleavage' is reduced to quick rounds of mitoses and karyokineses (in an order of minutes), resulting in about 5,000 nuclei in a common cytoplasm. They travel to the periphery, with the egg-yolk in the center, and only later they start cellularization. At the same time, translation is initiated of mRNAs anchored asymmetrically in egg cytoplasm. Transcriptional regulators are synthesized by translation of such RNAs on the fore and rear ends, respectively, and diffuse to the center where they meet and inhibit each other, thus parsing the space of the egg (Fig. 7.1, p. 120). A similar process is at work in establishing the dorso-ventral axis. The diffusion of such morphogens represents one of the many generic *physical* processes, important factors in establishing the body plan (Newman et al. 2006; Chapter 5). Their action consists in triggering and regulating transcription patterns in the abovementioned peripheral nuclei, depending on the amount and mutual ratio of regulators present in the egg, or later in the embryo. A cascade of new and new transcription factors, both from egg cytoplasm and (later) nuclei will result in a precise determination of prospective body parts and cell competences therein. Cells in particular segments will receive more and more information, which enables them to further negotiate (and sharpen) boundaries, polarity, differentiation and organogenesis. The resulting blastula, i.e., multicellular cylinder at the brink of gastrulation, contains cells fully determined to build a larva, and even rudiments (so called imaginal discs) of structures that will be built only later in pupa so as to create the adult imago.

The fruit fly embryogenesis is, however, a very specialized process wherein the plan of the *whole* body is established synchronously at the very beginning. In other insects (e.g., beetles), the axis of time also enters the game. Here, development is, from the beginning, accompanied by cell division, and determination; differentiation (even gastrulation) of the fore end is well advanced even before the rear end of the embryo appears. Yet the whole assemble of regulatory factors is homologous and is only applied in different contexts and times. Both processes result in a similar phylotype stage (see below), despite the different origins. This is even more conspicuous in the embryogenesis of the tiny parasitoid wasp *Copidostoma*. Here, the egg gives rise to a 'ply' on which thousands of embryos will 'bud' like mushrooms (Grbić et al. 1996). No body axes are established in advance by the mother; yet the embryogenesis of each such bud will proceed in a manner similar to that described above.

(iii) Model mouse (*Mus musculus*). The result of cleavage is not a gastrula or an embryo, but rather a structure called the blastocyst. Its greater part gives rise to fetal envelopes (allantois, amnion, and chorion; later, also the chimeric organ—placenta); the embryo proper will develop only later from the *inner cell mass.* Blastocysts can also be raised *in vitro*, and only later implanted into the womb. It is thus obvious that the egg cell has no strict cues for development: an individual blastocyst, and even trophoblast cells can start new development without much harm. Blastocysts can also be divided in two (in the case of armadillo even 8 or 16). Even chimeras may occur in litter if two sister blastocysts fuse (this is especially conspicuous when sired by different fathers as may happen, e.g., in mice or dogs). Only *after* the fetal envelopes become established can the embryo proper begin to develop. Moreover, curious type of chimerism exist in mammals. The placental barrier is not absolute, mother cells can cross it to the body of the germ, and vice versa. After several pregnancies, mother organism may present a multiple chimera, containing cells of her descendants. Moreover, she can even pass cells from older children into their younger siblings (Boddy et al. 2015).

The interruption of 'normal' embryogenesis by a peculiar structure (blastocyst) is not a rarity only encountered in amniotes. As mentioned above, many groups (e.g., some sea mollusks, Polychaeta, ascidians, or echinoderms) produce primary larvae.[26] Also, as mentioned above, in addition to differentiated and transitional larval

[26] Don't mistake these with secondary larvae such as caterpillars or tadpoles.

structures, members of these groups contain an amorphous cell mass, which later gives rise to the embryo proper. Here we see an exception to the 'insulation' rule, in that the larva is exposed to the external world even before embryogenesis is complete, but even here both the embryonal primordium and the delayed embryo *are* protected by the larval body. The primary larvae may represent a very ancient—indeed primary—state of animal development, while a phylum-specific embryogenesis is but a later stage in phyletic evolution (see below, Phylotype; Jägersten 1972).

Plants. If fungi can become emancipated from the microbial way of life only when producing fruit bodies, plants came to be emancipated to a much greater extent—even though roots may, from a remote distance, resemble hyphae. Plants cannot move: the only exception in most groups is flagellated spermatozoa, but even this convenience is lost in flowering plants. If we put aside the rare faculty of quick leaf movement (as in *Mimosa*, or in flytraps) as well as the standard slow turning of leaves towards light (tropism), the only way for a plant to reach a target is *growth.* It can be attained by cell division in special areas—meristems, like shoot or root tips, or leaf and flower primordia, or cell prolongation, as in internodia, or in pollen tube. In contrast to fungi, however, plants can both create true tissues, epidermis, vascular tissues (phloem and xylem) and leaf palisades, and also build organs thereof (root, shoot, leaf, flower, etc.).

The last peculiarity to be mentioned here is *metagenesis*, regular alternation of gametophyte (haploid, gamete-producing), and sporophyte (diploid, spore-producing) generation. In some groups, the shapes of both are almost indistinguishable (stoneworts, Charophyta); in ferns, the tiny, free but short-living, gametophytes are followed by the dominant sporophyte (the plant as we know it); in mosses, the haploid gametophyte form prevails, with the sporophyte reduced to a sporangium nursed by the gametophyte. Finally, macroscopic seed plants represent the sporophyte, with tiny gametophytes developing within, and confined to, the flower interior; the flower itself represents a case built by the sporophyte to protect the germ.

Female megaspores (confined to the stamen) give rise to a macrogametophyte, called embryo sac and containing an egg cell. Male microspores germinate into pollen grains (microgametophyte) containing two immobile sperms, and one vegetative cell. The latter has a duty to grow (elongate enormously), to push both sperms along the interior of the style, into the vicinity of the egg. One sperm fuses with the egg to produce a diploid zygote, and later an embryo, of the sporophyte; the second fuses with a diploid central cell to give rise to a triploid endosperm that will nurse the embryo and the new seedling. Both embryo and endosperm, together with tissues originating from the 'mother' sporophyte, will form part of the seed (Fig. 8.1).

Phylotype Again

You are *somewhere* at the side of a mountain ridge, equipped with a set of tools, and the goal is to achieve a *single point* P somewhere across the mountains. Depending on the equipment your have at hand, you can climb and descend the ridge through a pass, or dig a tunnel, or build a balloon, or walk the distance around the range. Nothing will hamper your inventiveness, provided that you can accomplish the task with the given tools. After arriving to P, you are allowed to take off again along many valleys. Keep this parable in mind while continuing below on the 'hourglass' model of development in multicellular organisms (Duboule 1994, Raff 1996; Figs. 8.1 and 8.2).

Quarrels have been under way for almost 200 years as to whether a conserved stage binding all members of a given group exists at all; most animal morphologists have agreed that it is real and bears the name *phylotype* (Raff 1996). For vertebrates, this phylotype stage is called *pharyngula*, which can be distinguished by the presence of embryonal segmentation, pharyngeal pouches, and a neural tube with a neuroporus; in insects, it is the *extended germ-band* stage, with a cephalic lobe, and distinguished gnathic, thoracic, and abdominal segments. Scholars have come to the agreement that the most conservative pattern that defines the phylum is established at this period of development. With the arrival of genomics, it became apparent that there is a similar sort of *mid-development transition* (Domazet-Lošo and Tautz 2010, Kalinka et al. 2010) with gene activation (transcription): the 'molecular phylotype' is characterized by activation of the most conservative genetic networks, which is simultaneously refractory to inputs from outside. Transcriptomics shows a good match of mid-development transition stages with morphologically defined phylotypic stages: it helped in marking such stages, even in groups where morphology was of no help. In this way, Levin et al. (2016) mapped such mid-developmental transitions in ten animal phyla.

In dicot plants, such phylotypic stage is presumably *torpedo* (Quint et al. 2012, Drost et al. 2016). In fungi that produce fruit bodies, such stage can be also pinpointed, but only through molecular cues (Cheng et al. 2015). At the same time, uncertainties emerged as to a strict correlation of transcriptome definition and morphology: for example, in a fungus (*Coprinopsis*), transcriptomics indicated the phylotype stage as a young fruiting body (1–5 cm, still embedded in the protective envelope), whereas morphologically the fruit body is recognizable at a much younger primordium stage (2–6 mm; Cheng et al. 2015). Drost et al. (2017) highlight the fact that the stage of 'phylotype proper' may be flanked by less penetrating but important minima: in animals it is the stage of primary larva that occurs in all groups (phyla) except Arthropoda and Chordata; another such minimum may be in the secondary larva. Similar transitions can be pinpointed in plants in embryogenesis, germination, floral transition, etc. In the future, the modular character of bodies may uncover second-order 'phylotypic' minima at the levels of organ anlagen (Irie and Kuratani 2014).

In animals, where the situation is most conspicuous, the phylotype is the earliest possible final point of insulated development (of course, isolation may last much longer, as, e.g., in birds): from then on, the newly emerging organism (or its parts) starts palpating the surrounding world, and establishing biospheric contacts (entering the web). The notion is most apparent on animals undergoing metamorphosis. As already mentioned, primary larva must establish liaisons with the biospheric web (especially when it is self-feeding), but later the embryo proper will again develop in isolation—from set-aside cells in the body of the larva, whereas larval body will perish. Davidson et al. (1995) and Cameron et al. (1998) stress the fact that the larva develops in a strictly deterministic manner towards a body containing maximally several thousands of cells (as in *Caenorhabditis,* see above)—a first entrenched 'phylotype' stage in organisms that (still) develop through such larvae. The set-aside cells allowed, in evolution, an escape from the cell-number limitation, and an increase in the size, diversity, and ecological achievement of the organism. For illustration (Cameron et al. 1998): The larva of a sea urchin *Strongylocentrotus* at the end of embryogenesis (72 h after fertilization) consists of about 5000 cells and has the longest dimension of about 150 μm. At the onset of metamorphosis, it is about 1 mm long, and contains 150,000 cells, of which 90% are derived from 'set-aside cells', i.e., rudiment with visible adult body plan. Jägersten (1972) gives a much richer list of such larvae, the main differences between them being the appropriations of some further stages of adulthood (so called adultation); the two concepts, thus, are not in conflict. The set-aside cells allowed expansion into new morphospaces, typical for each phylum.

The same may be true for imaginal organ anlagen in secondary larvae, as in holometabolic insects or even amphibians: in plants, they may represent another shallow indentation into the 'hourglass' (Fig. 8.2). Thus, insulation against the biosphere may even be sequential, recurrent in discrete stages. As mentioned above, it can last longer after having passed the phylotype stage. Such is, e.g., the case of hatching young in birds or insects, or birth in viviparous species. Here, however,

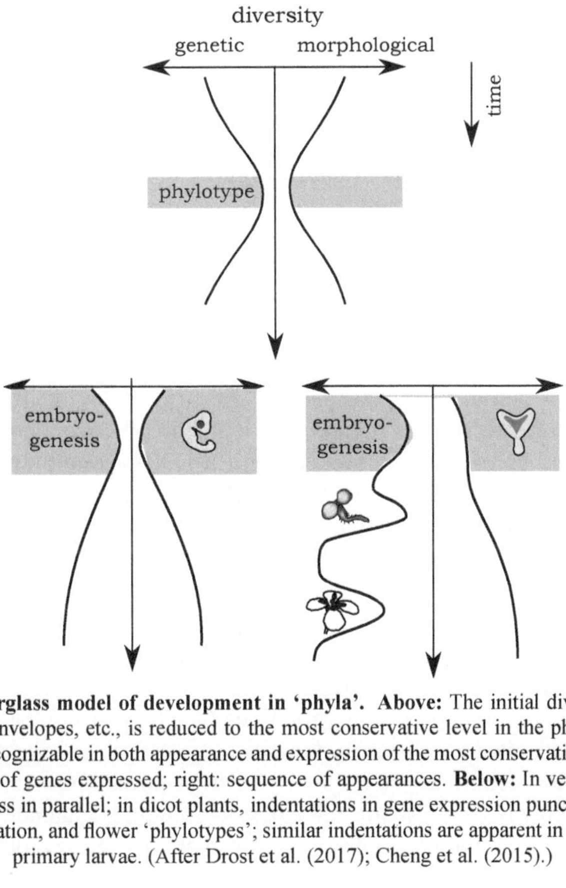

Fig. 8.2: The hourglass model of development in 'phyla'. Above: The initial diversity in cleavage, yolk distribution, envelopes, etc., is reduced to the most conservative level in the phylotype stage, to a common pattern recognizable in both appearance and expression of the most conservative gene complexes. Left line: diversity of genes expressed; right: sequence of appearances. **Below:** In vertebrates, the genes and patterns progress in parallel; in dicot plants, indentations in gene expression punctuate the periods of embryonic, germination, and flower 'phylotypes'; similar indentations are apparent in animal groups with primary larvae. (After Drost et al. (2017); Cheng et al. (2015).)

the growing (albeit insulated) germ is already well equipped with sensors, and can monitor inputs from the surroundings, especially its mother (as in mammals) or even both parents (as in birds). In mammals, some contact with the bacterial world may also exist, as will be discussed in Chapter 10.

Phylotype stages, such as the mid-development transition kit, ultraconservative genetic elements, genetic code, etc., may represent knots of ontogenesis—the Norm, the ring that 'binds them all'. Here, natural selection demonstrates its full power, mercilessly eliminating bearers of whatever mutation in the genetic script and/or sequence of events. Mutations in other parts of the genome may be buffered by redundancy (two alleles, closely related paralogs, DNA methylation, etc.) or epigenetic measures (chaperones, or physiological adjustments). In contrast, 'phylotype' stages are hardwired, genetically and generically; *only* after the 'player piano' phase of these stages—and *only upon them*, the epigenetic jamboree may blossom.

The ultraconservative elements may represent the points of extensive chromatin reorganization that enable cells to gain (or reestablish) the potential to give rise to different specialized cells. The process is relatively common in plants, where specialized cells repeatedly undergo dedifferentiation, to become stem cells for gametogenesis, embryogenesis, or simply replacing lost organs. The situation is more complicated in animals where differentiated states are not that easy to revert, and even stem cells in special organs, hematopoietic, epithelial, etc., resist reverting to more potent stages. Yet there exists many successful procedures as to how to accomplish the task (as we see with Dolly the sheep); however, these have all been achieved empirically, without understanding the underlying mechanisms. The transplantation of a diploid nucleus from a specialized tissue cell into an enucleated oocyte, or the treatment of cells in tissue cultures with cocktails of growth factors, i.e., polymers mimicking extracellular matrix, or combining two or more stem-cell types—all these may serve as examples of such procedures. The same also holds for later stages: as embryos or organoids cultivated *in vitro* (see below). Over the last decade, the literature on the topic has become abundant; below, we offer just a few citations selected more or less haphazardly. For an assembly of mammalian embryo-like structures derived from embryonal stem cell, see, e.g., Harrison et al. (2017). For organoids, or miniature organs, see the short review by Willyard (2015); Seet et al. (2017) report that thymus organoids are able to produce mature T-cells. For a stomach organoid on a dish, see McCracken (2017); for minibrains, see Quadrato et al. (2017), Birey et al. (2017), or Paşca (2018). Rivron et al. (2018) reports on blastocysts grown solely from stem cells.

What is obviously crucial in this respect is the restructuration of chromatin, which involves both a number and a quality of cross-contacts of its various parts, accompanied by changes in the pattern of epigenetic markers on DNA, histones, and chromatin-associated proteins and RNAs, followed by a new pattern of transcriptome, proteome, and epiproteome. The same holds for the external medium, where the necessary precondition is properties of the extracellular matrix scaffold, and an appropriate cocktail of growth factors.

Sophisticated techniques combined with high-throughput statistics allow the study of such changes within single cells. It was reported that gamete nuclei undergo a profound transformation of chromatin while changing into zygotic pronuclei and nucleus, with the structure resembling the 'normal' nuclear status only at the 8-cell stage of the embryo (Wu et al. 2016, Du et al. 2017, Flyamer et al. 2017). At this phase of development, the characteristics include a loosening of higher-order structures in ways that differ between paternal and maternal portions of the genome, and a negotiating of the setup of the new unit. Experiments that successfully transformed cells into just such a primordial state probably (empirically) matched conditions allowing such switch of the nucleus. The task is even more complicated by genomic imprinting, as discussed below.

Transgenerational Epigenetic Inheritance

Epigenetics designates all forms of heredity that is not 'genetic'. The task is 'resetting' development to a 'basic state'—the zygote, and yet to transmit through such reset some of the experience of both (multicellular) parents. Epigenetics thus does not encompass the phenotypic effects caused by mutations of a DNA sequence (point mutations; insertions, deletions, or reversals within the string; or crossing over in gametogenesis, organelle or allele incompatibility, etc.), or experience acquired by learning during the lifetime. The definition is not, however, watertight: is, e.g., gene conversion a genetic or epigenetic effect? Here we will recognize three epigenetic phenomena, but the divide between them is anything but sharp, and mutual influence is frequent.

As previously mentioned (in Chapter 5), in 2009 Jablonka and Raz documented epigenetic inheritance in about 42 species: 12 cases in bacteria, 8 cases in protists, 17 in fungi, 36 in plants, and 28 in animals. Epigenetic inheritance is more abundant in plants than in mammals because plants lack the Weismannian barrier and probably represent the most plastic group of multicellular organisms. For example, a transgenerational epigenetic inheritance that was heritably stable for generations was reported in *Arabidopsis thaliana* (Schmitz et al. 2011) and in *Taraxacum officinale* (dandelion; Verhoeven et al. 2010).

(i) Maternal effects are due to the fact that gene expression may be highly dependent on the intracellular milieu of the egg. Such are the phenotypic effects of egg organization, i.e., uneven distribution of cytoplasmic elements, or gradients of signals, as we have seen above (*Caenorhabditis*, *Drosophila*). As zygotic transcription only starts later in the cleavage process, paternal genes usually can influence neither the structures nor the regulatory networks at early cleavage stages (compare with genomic imprinting below). In some lineages, the site of sperm entry may influence the polarity of the embryo (such as in the brown alga *Fucus*), but usually this site is either prescribed by the egg structure (*Drosophila*; seed plants), or is irrelevant (vertebrates).

Manipulation with egg or zygote cytoplasm mostly leads to drastic breaches in normal development that are often incompatible with survival, or bring about malformations.

(ii) Genomic imprinting is the transfer of maternal or paternal cues via derivatization of nucleotide residues in DNA. Derivatization of DNA by methylation, O-methylation, glycosylation and other groups is common, and helps maintain the pattern of gene expression in specialized cells and their progeny. Probably the most frequent is cytosine methylation in mammals: DNA in eggs, and especially sperms, is highly methylated. Massive demethylation takes place in the zygote and preimplantation phases of development, lasting till that of the inner cell mass stage. Methylation starts again in the implanted embryo; it is interesting that much activity is focused on LINE methylation, especially copies acquired later in evolution (Smith et al. 2014, 2017, Guo et al. 2014, Zhu et al. 2018). As cytosine methylation most often marks areas with inhibited transcription, such a measure makes sense. Differences in methylation are conspicuous between epiblast and extraembryonic ectoderm (Smith et al. 2017). The role of the remaining (i.e., parental) marks, and whether they bear for the progeny information relevant to the life conditions of parents is a topic of extensive study.

In contrast to the situation in animals, Hofmeister et al. (2017) report that in *Arabidopsis* the inheritance of DNA methylation is practically 100%, and spontaneous epialleles are rare.

(iii) Environmentally induced traits. Profound seasonal-, diet- or pheromone-induced changes in appearance are common, e.g., in many social insects (inducing castes in bees, ants, or termites). In insects, variation that lasts across several generations and subsequent seasons can also be artificially induced by food: thus in the butterfly *Nemoria*, the main factor of color differentiation of adults is food available to caterpillars (catkins or leaves; Greene 1989). The sex of the progeny in many reptiles is not determined genetically, but by the temperature of egg incubation. In animals living in groups, the leading (alpha) individual often undergoes remarkable phenotypic changes. Such variations represent simple physiological, developmental, or ethological adaptations.

In this section, however, the topic is the transfer of information concerning the conditions of the lives of the parents into next generations, i.e., instructing the progeny about conditions they are probably going to live in. Under various names (dauermodifications, paramutations, or lingering mutations), for more than a century, this phenomenon has been known to occur in plants (Brink 1973, Jablonka and Lamb 1995, Martienssen 1996). Plants can instruct the next-year seed about the abundance of parasites, so the seedlings have time to prepare for invasion; similarly, species of *Daphnia* instruct their progeny to build a long cap in case of abundant predators (Agrawal et al. 1999). Mountain morphs of plants transferred to lowlands only slowly (over many generations) change their habitus towards the lowland morph; similarly, the dwarf habitus of rice induced by heavy metals will return to norm only after several seasons. In the 1940s, Waddington (1975) coined the term of genetic assimilation for instances when such developmental adaptations became fixed genetically; however, the interpretation for such phenomena came only in the last two decades (Rutherford and Lindquist 1998; see Chapter 5).

In case of fruit fly, *Drosophila melanogaster*, Seong et al. (2011) reported heritable changes of chromatin structure (induced by heat stress); similarly, Siklenka et al. (2015) showed that induced alterations of chromatin modifications were heritable in the absence of any further exposure to the stimulus.

In mammals, there are (nowadays almost iconic) reports of transgenerational epigenetic inheritance in rodents: first regarding A^{vy} allele, which can be epigenetically modified by a sufficient methyl donor diet (methionine, folic acid; Waterland and Jirtle 2003, Cropley et al. 2006, Morgan and Whitelaw 2008), leading to a non Agouti/wild phenotype. Another example involves the behavior of the mother—more specifically, the quality of maternal care, which can influence the pattern of chromatin modifications on enhancers of the glucocorticoid receptor (Meaney 2001, Weaver et al. 2004, 2005, Meaney and Szyf 2005). Further interest in epigenetic inheritance turns toward the study of behavioral aspects of paternal experience, which can be also transferred to the progeny. Thus, Diaz and Ressler (2014) report on the transfer of male olfactory experience in mice.

Obviously, this cornucopia of observations waits for a general theory capable of explaining how epigenetic memory and experience pass through the unicellular bottleneck. This holds not only for sperm, but also for the zygote that must combine the contributions of both parents. Mammals are the most interesting in this respect because the results may have direct consequences within human medicine, from developmental biology to mental variations to cancer.

The task of epigenetics is enormous—in contrast to relatively stable genetic script, it is difficult to catch the dynamics of the ever-floating world of proteins, RNAs, their multiple forms and derivatives and, before all, their crosstalk within complicated networks. As we have seen in Chapters 6 and 7, this is the general and defining feature of any living being. The question is how does life pass the behavioral, mental, social, metabolic, cultural, etc., experience, manners, or idiosyncrasies of both parents to the progeny—*through a single-cell stage*. Most of such experiments are done on mice; for illustration, we give a short review of the difficulties, as seen by Bohacek and Mansuy (2017). They argue, first of all, that the scientific community is still very suspicious when it comes to the transmission of acquired traits. The research is interdisciplinary and different teams use different vocabulary, standards, models and techniques, thus making dialogue between teams extremely difficult. Breeding conditions, group size, animal strains, food, date of pup weaning, stress during the life of parents, details of sexual intercourse, assisted reproduction, *in vitro* maintenance of the blastocyst and number of pups in the clutch—all these "can have profound, lifelong consequences on the offspring", and are not easy to standardize in different labs. Such basic prerequisites are, moreover, followed by high-tech analyses (often on single cells) of the composition and epigenetic markers on proteins and RNAs, demanding statistics as well as—last but not least, interpretation of results, i.e., the positing of causal interconnections between data and phenotypic traits. Below, we offer examples from this vigorously developing field.

In addition to transgenerational inheritance of DNA methylome, epigenetic marks can also be transferred via proteins, especially in histones (Ooi and Henikoff 2007, Hammoud et al. 2009, Brykczynska et al. 2010, Zhang et al. 2016, Inoue et al. 2017, Zenk et al. 2017, Hanna and Kelsey 2017). Even sperm, with its DNA packed mostly by protamins, carries a small number of histones—together with their epigenetic markings (the sins of fathers have many channels to pass forth—see Chapter 5 on organic memory). The egg, with its chromatin and cytoplasmic proteins, may carry even more such information. We can speculate that the initial gamete setting will mark specific alternatives for the massive and vigorous restructuring of histone markings that occur during the preimplantation period.

Both kinds of gametes also carry with them multiple sets of different regulatory RNAs (Vidalis et al. 2016, Jih et al. 2017), and the suggestion exists that these may also influence the development of progeny. Chen et al. (2016) found that males fed on a high-fat diet sired progeny with serious metabolic disorders in male pups. Likewise, fragments of tRNA (tsRNAs, 30–34 nucleotides long) extracted from such sperm also caused a similar effect. Moreover, the effect was not mediated via detectable changes in DNA methylation, thus the authors ascribed RNA fragments a regulatory role. Spermatozoa receive their tsRNA payload from small vesicles (epididymosomes) during maturation in the epididymis (Peng et al. 2012, Sharma et al. 2016, Reilly et al. 2016), i.e., they undergo a kind of direct indoctrination from the male. In similar experiments, Gapp et al. (2014) reported metabolic defects of progeny obtained from traumatized sires—caused by injection of small non-coding RNAs into fertilized eggs. What messages are transferred remains unknown.

Biased Mutations

Epigenetic modifications are not equal to DNA mutation; although they are heritable, they can be quickly replaced, erased or rewritten. Yet, epigenetic modification can, to some extent and in terms of biased mutation, influence genetic information. This phenomenon disputes the neo-Darwinian premise of random mutations independent of environmental perturbances.

Chromatin organization—primarily chromatin occupancy, contributes to differential mutation rates, both in somatic and germinal lines. For example, the single base substitution rate is more suppressed in open, active chromatin, probably because such DNA appears to be repaired more often and more easily. However, such patterns do not concern other mutation types. Along the whole genome, there is a strong heterogeneity in terms of mutations associated with active or silenced chromatin pattern (Makova and Hardison 2015).

Chromatin organization apparently biases the mutation rate in gene regions, and such bias can be adaptive (Chuang and Li 2004). Hernando-Herrarez et al. (2015) show that hypermethylated regions of the genome often correlate with higher mutation rates within these domains. Methylated CpG islands are prone to spontaneous hydrolytic deamination of cytosine into uracil which, in a few steps and with the help of repair enzymes, leads to an A = T base pair in place of the original G = C (Turner 2009). These CpG to TG transitions are more frequent than other mutations and probably led to a decrease of amino acids coded by CpG dinucleotides in some vertebrates. It has been shown that these CpG islands are more susceptible to modifications after exposure to environmental stimuli (Guerrero-Bosagna 2017); therefore, environmental exposure can lead to biased mutations.

9

Evolution by Cooperation

Communication between Different Lineages of Life

Generally, it is impossible to communicate without putting something into the background frame of mutual agreement and assuming that the other is able to access this presupposed knowledge. Otherwise, each speech event would require a complete restatement, with the result that there would be no time to say, or listen to, anything. This is clearly too great an extension for a presupposition as a sentence phenomenon, since the utterance of even the simplest sentence can presuppose all the world in this sense.

—U. Eco 1994b, 228–229

The only utility function of life is DNA survival, [...] *Group welfare is always fortuitous consequence, not a priori drive. This is the meaning of 'the selfish gene'.*

—R. Dawkins 1976, 105, 122

Many times throughout this book, we have come across the motif of understanding, cooperation, or (mutual) manipulation. Such conceptions are not self-evident in original frames of Darwinian paradigm, even if competition (struggle for life) amongst life forms *is* one of centerpieces of the teaching, and can be easily extended even to cooperative behavior. The cultural milieu, however, understood—and still does—Darwinian 'struggle for life' as 'fight for life' (see Markoš et al. 2009). 'Nature red in tooth and claw',[27] social Darwinism in the past, and similar contemporary, excessive interpretations of the teaching (the metaphor of the 'selfish gene', or sociobiology; Wilson 1975), did lot of harm in the past, as they do today.

[27] From a poem of Tennyson (1850), but the poet was not responsible for turning the verse into a slogan.

In brief, the role of communities had always been, in our opinion, underestimated: the individual or its genes always had a priority. The concept of ecological individual as, e.g., a coordinated physiology of a pond or a meadow (Odum 1971) may be an excellent model how to describe behavior of communities, but it was often ridiculed by the opponents from the 'genocentrists' camp as 'mere metaphors'.

From contemporary theories it follows that an individual—'naturally' promoting its evolutionary success at the expense of other inhabitants in the biosphere—must encounter special conditions, when cooperation pays off; otherwise, natural selection always favors defectors and parasites. The game theory offers such models—here we report the 2006 review by the leading personality in the field, M.A. Nowak. He distinguishes (and quotes the authors of) five rules for cooperative strategy payoff:

(1) Genetic relatedness. Kin selection refers to advantage in helping the relatives to propagate genes shared by the group. Hamilton's rule presupposes a relatedness coefficient r, such that $r > c/b$ (c—cost, b—benefit of such interaction), linked to the probability of kindred among cooperators. As relatives share most of their genes, ostensible cooperation turns out to be a mock game, preferring again the selfish genes only.

(2) Probability of next round of game, with memory of previous encounters. Direct reciprocity (repeated prisoner's dilemma) supposes that cooperators know each other and can anticipate partner's moves. Further models elaborated on facts as making errors ('trembling hand'), generosity, duration of the game, etc. Stable cooperation will evolve only if the probability w at the next encounter of players will exceed the c/b ratio: $w > c/b$.

(3) Social status. Indirect reciprocity is a game of reputation typical for humans only: generosity to refugees, charity and similar cognitive demands will increase the status of the donor, and others tend to cooperate with a person with such a high reputation.

(4) Network reciprocity concerns a well-mixed population with many neighbors of different desires. Evaluation depends on the number of neighbors, the fraction of defectors among them, and the ability of the group to punish them. Depending on such interlinking, cooperators may win part, or even all games (as when thoroughly raising their young).

(5) Finally, group selection concerns organized groups who enter the game as such (i.e., not as individuals): the game within the group proceeds, as elsewhere, along some of variants listed above, and groups as a higher-order individual enter a similar game with other groups—social animals serve as good examples.

Explicatory power of such models is high, yet notice that such theories take into account only short periods and limited number of players. We, however, have in mind the biosphere functioning over ages, and with a great number of ever-changing partakers. Moreover, the very character of symbiosis may change over ontogenesis: e.g., harmless endophytic fungi living in a tree may turn into killers of the same tree if it becomes infected by parasites or otherwise damaged. What also deserves a comment is so often criticized teleology, '*a priori* drive' towards new partnerships.

The umwelt of living beings is much greater (as to capacity) than their appearances realized in their lifetime (see Chapter 4); we all are born endowed with this memory and experience of long ages. If possible, individuals prefer simplex solutions without much hesitation (on simplexity see Chapter 7), while the capacity of alternative actions (strategies) may lie idle even for many generations. Models of the game theory, on the other hand, take as their axioms extremely naive individuals. The stance is appropriate, in order for the model to uncover basic rules of the game. Let us take as *our* axioms extremely experienced partners who know many survival tricks on the playground of the 'commons', and usually avoid its 'tragedy' (Hardin 1968). Only when confronted with new tasks, they start searching across their stores of communicative tools. Of course, 'trembling hands', catastrophes or even extinctions, will often take place too.

In Chapter 4, we discussed symbiotic emergence of three types of cells, of cell organelles as mitochondria and plastids, and came to a conclusion that the longer the time elapsed from the beginning, the lesser the probability of 'absolute' fusion into one being, as it used to be common—as we believe—in times of the LUCA. The basic lineages of life diverge and mutual understanding gets less probable in time—as in human cultures and languages. The last major symbiogenetic event is probably acquisition of secondary and tertiary plastids—by engulfment of eukaryotes (not prokaryotes any more) by different lineages of eukaryotes. Such a neat scenario is somewhat thwarted only by recent primary chloroplasts of *Paulinella*, but even here the chloroplasts can perhaps be considered as vertically transmitted bacteroids (see below). More interesting are the tight symbiotic events lasting for hundreds of millions of years. The participants of such tight cooperation remain discretely 'self', mostly accomplishing their propagation, or parts of their life cycles, autonomously.

If one accepts the symbiotic perspectives, he/she will discover manifestations of symbiosis everywhere. Here we restrict the perspective somewhat, discussing symbioses embracing only two or a small amount of participants. Symbiogenetic events leading to organelles were dealt with in Chapter 4, complicated ecological webs of holobionts will be discussed in the next chapter. What we leave out of our focus is 'ecology proper', i.e., communities consisting of many participants, like meadows, forests, water bodies, or sediments. Roughly, then, what follows can be divided into three categories, as to the kind of mutual understanding:

(i) Intimate endocytobioses, consisting of bacteria (or bacteroids), archaea, or even eukaryotes living in a host cell. The host is usually eukaryotic—but a nested symbiosis of bacterium-in-a-bacterium, a tandem itself living in the eukaryotic cytoplasm, was also described (McCutcheon and von Dohlen 2011). Some protists, like Ciliata or Hypermastigina, are masters in housing a full symbiotic 'zoo' in their cytoplasm. Bacteriomes (sometimes also mycetomes) are common in special organs of insects, Pogonophora (*Riftia*), molluscs (Bivalvia), cnidarians, plants (nitrogen-fixing bacteriomes in root nodules or on thalli), and many others. With a single unimportant exception (algae living in the *Ambystoma* salamander embryos), such special intracellular symbionts are not known in vertebrates—perhaps because they so thoroughly see after sterility of their embryonal development. Intracellular symbionts in later stages of vertebrate ontogenesis mostly behave like pathogens, e.g., *Salmonella*, *Plasmodium*, *Leishmania*, or *Trypanosoma cruzi*.

As to their transmission into the next generation, bacteroids may become scavenged from the external pool of free-living bacteria or archaea. When intracellular, they may lose their cell wall, and grow in size while getting unshapely; they, technically speaking, become an organelle. Their eukaryotic partner will usually also undergo substantial ultrastructural rebuilding (see nitrogen-fixing nodules below). If there is no such free-living pool, bacteroids must safeguard their transmission either by producing propagules, or, as in animals, infecting the progeny of their host at some stage of its development (gametes, embryos, larvae). In protists, of course, they may simply synchronize their cell cycle with that of the host; even then, however, they must overcome the bottleneck of occasional host's sexual reproduction.

With some hesitation, we rank into this group also a bizarre case of 'symbiosis' —kleptoorganelles (Nowack and Melkonian 2010, Rauch et al. 2017) known in protists, cnidarians (*Hydra*, corrals), and even molluscs (*Aplysia*). For example, the dinoflagellate *Amylax* hosts plastids from its cryptophyte prey, and when with plastids, it is capable of phototrophic growth, independent of the presence of any prey (Park et al. 2013). In such cases, ingested algae will not get digested completely, and their chloroplasts survive in the cytoplasm of the predator cells for a longer time (even during their whole lifetime, as in *Aplysia*), serving their primary purpose to an extent that their host stops feeding and, technically speaking, turns into a 'plant'. Even more sophisticated is the kleptosystem in *Myrionecta* where also nuclei of the ingested alga will survive for some time (thus turning the host into animal-plant chimera), and supply the chloroplast with necessary gene products (Johnson et al. 2007, Nowack and Melkonian 2010).

Parasitic plants (*Cuscuta, Striga, Viscum*, etc.) develop special organs—haustoria—that 'plug in' the symplast of their hosts, thus breaching the closures of both partners. Extensive, yet controlled exchange of nutrients, signals, and gene products takes place through the connection, so, technically speaking, we again encounter a chimera. Bacterial parasites connected through a cytoplasmic bridge to their hosts (e.g., *Agrobacterium*, see Fig. 3.1) also belong to this category of symbiosis.

(ii) There also exists very intimate cohabitations of partners who, however, remain all the time 'for self', thoroughly maintaining their cell closures. Such is the case of lichens, of mycorrhiza and of endophytic fungi. In protists, an intimate symbiosis of a unicellular alga with nitrogen-fixing non-photosynthesizing cyanobacterium may serve as an example (Thompson et al. 2012). Bacterial ecosystems like mats, sediments, etc., also can harbor groups of organisms intimately connected through their way of living.

(iii) Tentatively, we introduce to our list some cases of mimicry and imitation. Such phenomena can be defined as extreme mutual morphogenetic coupling—based again on understanding some external cues—of organisms belonging to different lineages.

On Lichens, Squids, and Coral Bleaching

Here we present three 'typical' examples of symbiosis as known to general public: fungus-plant partnership, animal-bacterium symbiosis, and intracellular symbiosis.

(1) Lichens have been mascots of symbiotic interactions since the 19th century. This kind of symbiosis of a fungus with an alga or cyanobacterium has become a textbook example, and need not be discussed here in detail. The intimacy of cohabitation earned the creatures even a distinguished position in the system of organisms. Even if intimate symbioses abound everywhere, probably only lichens invented a special form of hand-in-hand propagation of both partners. It came as a recent surprise that lichens can contain in their cortex even a third partner—a yeast, living in a stable symbiosis with the 'dominant' fungus and alga (Spribille et al. 2016). In the light of such findings, less known symbiotic assemblages may bring even more surprises. Both fungi and plants are apparently prone to a mutual symbiosis: Hom and Murray (2014) describe relatively easy establishment (in the laboratory) of mutualist cohabitation of yeasts or even hyphal fungus, with a unicellular alga. It must be more sophisticated in the case of, say, mycorrhiza, lasting for some 400 millions of years (see below).

(2) The recent mascot for demonstrating symbiosis is the tiny squid *Euprymna scolopes*, hosting marine light-emitting bacterium *Vibrio fischeri* in its special 'light organ'. The pelagic, free-living bacterium infects the prospective light organ of juvenile squid, thus inducing far-reaching metamorphosis of both cephalopod tissue and bacteria, with circadian fluctuation of many functions (Wier et al. 2010). In fact, such infection is necessary for proper development of the organ (McFall-Ngai et al. 2012, Kremer et al. 2013).

(3) The third example penetrating recently into the conscience of general public is the worldwide ecological tragedy of coral reefs—so called coral bleaching. Tiny cnidarians building the reefs live in an obligatory intracellular symbiosis with algae or cyanobacteria. This partnership has a tendency to fail recently for unknown reasons (water pollution, acidification, or global warming being frequently accused): the green symbionts disappear and the animals, though able to catch the prey, die anyway.

Protists

Many protistan lineages may serve as an example of multiple, parallel and pervasive evolutionary experiments with endocytobiosis, i.e., intracellular symbiosis with bacteria or other protists that, at least theoretically, can also live freely, outside the host. Symbiogenesis, i.e., the process of transformation of bacteria and/or algae into semiautonomous organelles, was discussed in Chapter 3 and will not be treated here. We draw, however, reader's attention to numerous bizarre transformation of organelles because it may be connected with symbiotic interactions too. For example, in many lineages living in anaerobic habitats (such as Parabasalia, some ciliates, fungi, and even in animals—Loricifera from deep anaerobic Mediterranean waters), mitochondria underwent transformation into hydrogenosomes, hydrogen-producing non-respiratory organelles. The group Kinetoplastida is characterized by the kinetoplast, a complicated mitochondrial structure with extensive editing of RNA transcripts; in the same order, a special organelle—glycosome—houses the glycolytic pathway that is, in other eukaryotes, located in the cytoplasm. In Chapter 3, we already mentioned occelloids in dinoflagellates originating in a transformed mitochondrion and chloroplast (Gavelis et al. 2015). The list could be longer; here, however, we just point towards inventiveness in protists as to their intracellular symbioses. Such teams get established to break into a new niche, or create a new one; often, it is not easy to distinguish the form of symbiosis (mutualism, enslavement, pathogenicity, etc.; Nowack and Melkonian 2010).

Essentially, three big areas of intracellular symbiotic interactions can be distinguished in protists. First, methanogens or sulphobacteria help removing hydrogen—a waste product of hydrogenosome metabolism. To maintain glycolytic flow with protons as final electron acceptors, hydrogen concentration must be kept extremely low. The task is fulfilled by above mentioned prokaryotes present in the close environment, or as intracellular symbionts, and showing extremely high affinity towards hydrogen, as electron donor for their respiration.

The second area is processing of nutrients that are hard to metabolize by eukaryotic metabolic pathways—like cellulose or lignin. Very often, the symbiotic assemblage also contains bacteria scavenging nitrogen-containing wastes, or even fixing molecular nitrogen.

Finally, as already shown above, the third group covers protists hosting photosynthesizers, prokaryotic or eukaryotic, or even kleptochloroplasts released from algal food.

In general, bacterial symbionts grow in their hosts to much bigger size when compared with their free-living kin: they lose the cell wall and become intimately interdigitated with membrane systems of the surrounding cytoplasm and organelles. The advantage for bacterial symbionts often lies in the fact that the protist is capable of efficient translocation: it carries them into zones with optimal concentration of nutrients, towards optimal light intensity, or it can escape the pressure of predators. The host can also provide many nutrients, e.g., providing high partial pressure of CO_2 to photosynthesizing symbionts, much higher than in the ambient water column. We give some examples below.

Cyanobacteria enter endosymbiotic interactions in two ways. Better known is the role of a photosynthesizer in various groups, with the bacterium attaining different status of autonomy—up to a bona fide chloroplast as in *Paulinella*. The second alternative is stripping off photosynthetic abilities, and retaining only the nitrogen-fixing pathways; such reduced nitrogen-fixing bacterium may live in a close association with other organisms (Adams and Duggan 2012), or enter endosymbiotic relations, again up to the level of a protoorganelle. Such is, for example, the status of a cyanobacterial symbiont in diatoms (family Rhopalodiaceae; Bothe et al. 2010, Nakayama and Inagaki 2011).

Termites, cockroaches, and other wood-consuming insects, and herbivores in general must rely on their gut symbionts to decompose cellulose and lignin, and often also to recycle nitrogen compounds, or even fix molecular nitrogen into such compounds. Many bacteria, fungi, and protists are free-living in the innards of such animals, live closely attached to the protistan surface (in some cases, attached spirochaetes even confer motility to the host), or housed as endosymbionts in special organs, as in insect bacteriomes/mycetomes (see below). In termites, it is cellulolytic gut flagellates (Parabasalia, Oxymonadida) with nitrogen-fixing or -recycling Bacteroidales or cyanobacterial endosymbionts (review Ohkuma 2008). Methanogens frequently join the team and inhabit flagellates of the Parabasalia and Oxymonadida (Embley and Finlay 1993, Ohkuma 2008), using the hydrogen produced by the hydrogenosomes to reduce CO_2 (Fenchel and Finlay 1991). Another example is the ciliate *Cyclidium porcatum* containing a complex of hydrogenosomes, extensively interdigitated with a bacterium and a methanogenic archaeon in its cytoplasm (Clarke et al. 1993).

Endosymbiotic eukarya are usually photosynthesizers—as, e.g., the ciliate *Paramecium* hosting the alga *Chlorella*. The huge cells of foraminifers host in their vacuole a vast array of dinoflagellates, diatoms, chlorophytes, rhodophytes, chrysophytes and haptophytes (Richardson 2001, Holzmann et al. 2006). Symbionts are propagated in synchrony with their host, but must be replenished (reacquired from the environment) after each gametogamy (Röttger et al. 1998).

Intracellular Symbiosis in Multicellular Organisms

Here we bring some examples of life cycles of intracellular symbionts, bacterial or eukaryotic, in multicellular eukaryotes. The pattern is in many aspects similar to that described in protists; the difference is that the symbionts mostly live in special cells or tissues, not in all cells of the host (bacteriomes or mycetomes, nitrogen-fixing modules, etc.). Some such symbionts also represent commensals, invaders, or pathogens, as, e.g., bacterium *Wolbachia* in insects or nematodes, or eukaryotic malaria parasites in vertebrates. As the hosting organ is mostly not the gonad or the germline, a logistic problem must be overcome on how to transmit its inhabitants to the host's progeny, especially if the symbiont in question is not available in the environment in form of free-living cells or spores (Bright and Burghelesi 2010). Thus, in animals living in obligate symbiosis, ovaria or the eggs must become infected, usually during a specified time-window. Plants and their symbionts usually grow towards each other while using special communication systems; pathogens are spread by insect vectors (as mosquitoes in case of malaria parasites). In all cases, partners recognize each other with high sensitivity. (This is, of course, true also in other types of symbiosis: for example, the light organ of *E. scolopes* becomes receptive to *Vibrio* when exposed to only several bacteria.) Here we touch eukaryotic parasites only marginally, and focus our attention to cases of closer mutual bonds (even if not always mutualistic).

Insect bacteriomes/mycetomes can be found in most insects: they are especially common in the order Blattaria (cockroaches), Homoptera (aphids, mealybugs, etc.), and Coleoptera (beetles, especially family Curculionidae); below, we discuss only bacteria, but yeasts and other fungi may inhabit the organ too (as in grasshoppers). As a rule, the symbionts cannot multiply outside their hosts (or vectors), and must reinfect the new generation of hosts by hitchhiking the germline, or a vector (bloodsucking or plant sap-sucking insects or ticks). It follows that attempts of cultivation of symbionts mostly failed (but see Takeshita and Kikuchi 2017, who succeeded with a bug symbiont).

Roughly, three kinds of endosymbiosis can be distinguished in insects (Ishikawa 2003); below, we briefly characterize them, later to pay greater attention only to the *first type*—obligatory inhabitants of special cells, comprising well defined, vertically transmitted, 'domesticated' endosymbionts so that each insect lineage has its own specific partner.

The *second type* of symbiosis is represented by 'guest microbes': various bacterial symbionts of unknown status or importance, living in many types of body cells, and arousing a lot of embarrassment in researchers. Probably, they are mere opportunists that discovered ways of how to circumvent host's immune surveillance (whereas pathogens are kept at bay; Douglas 2014, 2015). Such is probably the case, for example, of the bacterial genus *Sodalis* living in tsetse flies (Aksoy 2003).

As a *third group,* we can consider reproductive parasites as *Wolbachia*, manipulating their hosts towards higher efficiency of symbiont transmission (by cytoplasmic incompatibility of gametes, killing males, stimulating parthenogenesis, etc.). Only symbionts of the third type are distributed in all tissues (except sperms), which can be directly transmitted vertically in the eggs, as bona fide organelles. In contrast to organelles, however, occasionally they can also spread horizontally, infecting new lineage of hosts (see the case study on *Drosophila simulans*; Weeks et al. 2007). *Wolbachia* is a member of Rickettsiales, bacteria specialized to such intracellular way of living, infecting millions of different insect species (Douglas 2010). Though interesting as possible motors of evolution of their hosts (see Chapter 3 in connection with mitochondrial origins), and possibly as means of pest control, we will not discuss them here.

Vertical transmission of obligate intracellular symbionts (i.e., the first group listed above) requires bacteria to be released from their resident tissue in the body, and to infect the progeny at some stage of its development. For example, in aphids reproducing mostly as viviparous parthenogenetic females, embryos get infected by symbiotic *Buchnera* from periovarial bacteriocytes, through an opening in the embryo (Braendle et al. 2003); the newly acquired bacteria find their way to prospective bacteriocytes and become established there, and only there. If aphids enter sexual cycle, eggs are contaminated directly.

Tsetse flies (also viviparous) deliver the inoculum of *Wigglesworthia* with 'milk' produced as food for the ovoviviparous larva (Aksoy 2003, Pais et al. 2008). Bugs may produce 'symbionts capsules', smearings, or bacteria-contaminated jelly (Hosokawa et al. 2006, Engel and Moran 2013) surrounding the eggs; newly hatched larvae take such material as their first food, and as an inoculum of the symbionts.

There exists a high degree of complementarity and syntrophy between both partners, so the symbionts can rely on the host as to delivery of most nutrients, while delivering essential building block or vitamins. Thus, in aphids consuming sugar saps from plants, efficient recycling of nitrogen wastes is a prime imperative—and *Buchnera* secures the task. In blood-sucking insects (Braendle et al. 2003, Douglas 2010, 2015), it is the supply of vitamins that is scarce in the prey's bloodstream (Aksoy 2003). The genome of obligatory symbionts may be drastically reduced: for example, *Buchnera*, a relative of *E. coli*, has a 7-fold shorter genome (Nikoh et al. 2011) and is, moreover, full of repetitions; the big bacterocyte harbors more than 100 copies of it. For more examples of insect symbionts, see, e.g., three volumes devoted to insect symbiosis (Bourtzis and Miller 2003).

The last example in connection with insect came as surprise. The mealybug *Planociccus* houses in its bacterocytes a nested system of bacteroid *Tremblaya* containing, in its own cytoplasm, another bacterium, *Moranella* (McCutcheon and von Dohlen 2011). All three organisms cooperate in complementing their metabolic pathways.

A peculiar intracellular symbiosis developed in tubeworms (Pogonophora, genus *Riftia*) who inhabit oceanic bottom oases around black smokers. The smoker emits large amounts of sulfane (SH_2), whereas the ocean around contains oxygen. The worm has no alimentary tract, its body is full of bacteriome cells containing sulphobacteria instead, and the blood circulation brings both gases to them. Bacteria oxidize sulfane and energy gained from the process powers the needs of both bacteria and their host (Nussbaumer et al. 2006). Many mollusks (Bivalvia) living at the interface of oceanic waters and hydrothermal vents adopted a similar partnership with sulfobacteria (Laming et al. 2015). Similar symbiotic bonds between metazoans and hydrogen-oxidating bacteria are also assumed in such habitats (Petersen et al. 2011).

The last group of animals to be mentioned here is corals who live in tight endosymbiosis with algae or cyanobacteria scavenged by endocytosis from the environment. As said above, the recent event of coral bleaching brought the whole ecosystem of coral reefs at the verge of extinction.

Nitrogen-fixing nodules resulting from symbiosis of legumes with *Rhizobium* appear upon mutual recognition of free-living soil bacteria with the receptive root (Miller and Oldroyd 2012, van Zeijl et al. 2015). Upon infecting root cortex cells, both plant cells and bacteria undergo deep transformation: root cells grow in size and become polyploid, and bacteria lose their cell walls, grow in size and ploidy, and start expressing several hundreds of genes necessary for nitrogen fixation. The plant grows the nodule, accumulates energy resources, and keeps oxygen concentration in the nodule close to nil (Burghelesi 2016). The whole nodule, including specialized bacteria, dies after several days, but it turned out that it also contains untransformed bacteria that feed saprophytically on decaying tissue (Timmers et al. 2000). Such may be the reward for the bacterial lineage for keeping the nitrogen-fixing machinery while living free in soil (oxygen concentration would inhibit its operation in soil), for entering the process, and for sacrificing its irreversibly modified nitrogen-fixing members. In the next section on mycorrhiza, we return to rhizobial symbionts, to point at similarities of both symbiotic systems.

Mycorrhiza

Tight cooperation of plant roots and/or rhizoids with fungi has lasted from the first appearance of land plants in the Ordovician, more than 400 millions of years ago (Kenrick and Strullu-Derrien 2014). The same may hold for cooperation of roots with bacteria (see previous section). The principal goal of such alliances is a tradeoff of nutrients—products of photosynthesis (e.g., fatty acids; Luginbuehl et al. 2017) for inorganic salts (especially phosphate and ammonia) mined from the rock or from litter. Such processes are insured by reciprocal rewards between organisms involved (Kiers et al. 2011). Only a small group of plants lives without mycorrhiza: usually, it is only ephemeral plants that complete their life cycle in few weeks, and live on newly-exposed posts as landslides or burnt places (e.g., *Arabidopsis*).

There exists several well-established protocols of such cohabitation. Probably the most ancient, and most widespread, is arbuscular mycorrhiza (review Genre 2012), present in 80–90% of land plants (Fig. 3.1, p. 31). All fungi involved in this type of mycorrhiza belong to the phylum Glomerulomycota, obligatory symbionts with an idiosyncratic life cycle. They are asexual with a huge single-cell, syncytial propagule ('spore', 500 μm in diameter) containing thousands of nuclei, and originating by merger of hyphae of different 'individuals'. Hyphae themselves are aseptate, so formally they present single-cell multinuclear chimeric plasmodia with readily streaming cytoplasm. Hyphae of different individuals may merge upon encounter, thus creating a new 'individual' with an augmented nuclear heterogeneity.

The germinating spore radiates highly branched, transient germination hyphae that must find a suitable plant within several days. It is not surprising that the search is not haphazard: both plant and fungus sense each other's presence and proceed towards the encounter actively (on the role of diffusible signals in mycorrhiza and rhizobial symbioses, see Miller and Oldroyd 2012). As an attractant, the root-tip secretes a pheromone strigolacton (in plant body functioning as one of the hormones), whereas hyphae emit a plethora of chitine derivatives called Myc factors (lipo-chitooligosaccharides; Rasmussen et al. 2016). Establishment of the contact must also inhibit to some extent the plant immunity aimed against fungal pathogens (Zipfel and Oldroyd 2017). We can anticipate discovering a broader palette of such morphogens in near future.

Upon encounter, both organisms cooperate in building a 'landing' interface called hyphopodium on an epidermal cell of young root. From now on, both partners switch their developmental pathways: the fungus sends arbuscular hyphae into plant cell interior; in turn, the plant cell undergoes a deep restructuring of its cytoplasm and builds special kinds of cell wall—perifungal membrane surrounding hyphal offsets. Of course, the process is accompanied by extensive mutual activation of signaling pathways and, by deep changes in the pattern of expressed genes, and by reconfiguration of the cytoskeleton (Rasmussen et al. 2016).

Fully grown arbuscule will fill almost the whole volume of the plant cell, yet both organisms remain clenched to each other without direct cytoplasmic contact. The structure lasts for several days; in the meantime, the fungus sends away branches, via extracellular space as well as through the symplast anastomoses of plant cells, towards the advancing root tip that produces cells suitable for establishment of new contacts.

In plants hosting arbuscular mycorrhiza, the symbiosis of root with nitrogen-fixing bacteria (gramnegative *Rhizobium*, see above) shows a very similar pattern of establishing contact (Denison and Kiers 2011). Mutual recognition precedes infection: the bacterium secretes signal—in this case called Nod-factor, based, as in case of mycorrhiza, on lipo-chitooligosaccharides; the signal is, again, sensed by special receptors and responded by plant pheromones (Markmann et al. 2012, van Zeijl et al. 2015). Upon establishing contact, special morphological unit—root nodule—is built, as described above. In contrast to mycorrhiza, nitrogen fixation can proceed only in an oxygen-free environment—this is safeguarded by the plant tissue. It is worth mentioning that a close relative to *Rhizobium* is the plant pathogen *Agrobacterium*. Nodular symbiosis may have started in evolution by bacteria hijacking components of coupling tools (necessary to erect hypopodia) from the fungus. Whereas chitin is a component of cell wall in fungi, and producing lipo-chitooligosaccharides may not be a problem, bacteria produce such signal molecule without bonds to general metabolism.

From other recognized types of mycorrhiza, we mention only the ectomycorrhiza because it is typical for temperate and boreal trees. In this case, fungal symbionts belong to Basidiomyceta, i.e., producers of common mushrooms encountered in the forest. Fungal hyphae do not penetrate the root cells as in case of arbuscules, but form a thick meshwork around them (Felten et al. 2012, Chen et al. 2016). Ectomycorrhizal fungi are better in gathering nutrients from organic litter—this may represent advantage in forests with a massive leaf fallout. Plants with ectomycorrhiza also bind nitrogen-fixing symbionts—this time they build nodules with actinomycetales (*Frankia*; Franche and Bogusz 2012).

Symbiotic nitrogen fixation can be accomplished also with cyanobacteria. Liverworts and hornworts produce, on their fronds, slime cavities called hormogonia containing bacteria; in this case, no signalling and attraction is needed, as infection by the bacterium is constitutive (Adams and Duggan 2012).

Bacterial Consortia

Bacteria enter all possible symbiotic interactions. Above, we discussed endocytobiosis and extracellular cohabitations with eukaryotes; holobionts will be the issue of the next chapter. Usually, bacteria do not live as unicellular individuals (save dispersion stages). They adhere to substrates (organic or inorganic) and build extremely complicated ecosystems—often consisting of hundreds of different taxa intimately interconnected by extracellular matrix, diffusing signals, metabolite and even gene transfers. Examples are mats, plaques, stromatolites, sediments, soil, floating fluffy assemblies, films or foams on the water surface, or holobionts (Shapiro 1997, O'Toole et al. 2000, Webb et al. 2003, Keller and Surette 2006). Such structures bear comparison with tissues of multicellular organisms.

Essentially, two multicellular bacterial forms do exist: multispecies consortia (biofilms), and monospecific colonies. For initiation, both require a substrate: solid (e.g., stromatolites), semi-solid (organic surfaces, sediments), or even air-water interface; subsequently, they transform, reinforce, and shape such an assembly by co-construction of extracellular matrix, microcolonies, channels, networks, etc.

Colonies arise when a single clone is grown axenically ('germ-free'). For simplicity, we leave out here monospecific bacterial bodies that build a thallus (like in *Streptomyces*), aggregate into predatory swarms (myxobacteria), or grow as filaments (cyanobacteria): such bodies are able to rise even from the background of bacterial consortia present in the environment (for development of such forms, see Brun and Shumkets 2000). We focus here on colonies *sensu stricto* growing on a sterile substrate. Growth of an axenic colony, mutual interactions of colonies, and negotiation of the morphospace are results of intricate regulatory and metabolic networks (Shapiro 1998, Ben-Jacob and Levine 2005, Cho et al. 2007, Rieger et al. 2008, Čepl et al. 2010, 2016).

In Chapters 6 and 8, we argue that to build a multicellular clonal body, it is indispensable to insulate the early embryo from interfering effects coming from the surrounding biosphere. In contrast, from some stages on, establishment of biospheric liaisons is highly wanted: organisms devoid of such contacts will most often develop as freaks of a sort. Bacteria and Archaea (as well as most protists) are—in most circumstances—not able to arrange for such isolation; mostly, they are not able to even grow under artificial axenic conditions, i.e., in a monoculture, or in gnotobiotic co-cultures. In fact, most bacteria are not even able to construct a colony in the rare occasion when left in isolation.

The 'tree of life', as in Fig. 4.1, contains whole branches (even of the highest importance from evolutionary point of view) constructed solely on the basis of DNA extracted *en mass* from some biotope, and reconstructed (by software) into virtual genomes: the appearance, metabolism, density and roles played in the ecosystem by such ghosts remain mainly unknown. More valuable for untangling such relationships are settings making such cultivations and co-cultivations possible—at least for a handful. As said previously, in nature, chances for starting a colony are rare, as suitable sterile substrate occurs only sporadically (as cracked fruits, surface of marmalade, milk, etc.); it follows that a colony is but an occasional phenomenon, with bacteria spending most of their lives in multispecies consortia. In laboratory, colonies are grown on semi-solid support (like agar) clonally from single cells, or even of cell mass when planted to small area as dense suspension (as much as 10^7 cells). Such a colony is a genuine multicellular body with development akin to more stringent developmental programs grown in multicellular eukaryotes (Ben-Jacob et al. 2004, Ben Jacob and Levine 2005). As not bound to procreation, a bacterial body is, to a great extent, freed from the pressure of natural selection, and is thus highly flexible in reactions to changes in temperature, composition, salinity, or pH of the medium, or hardness of the substrate (Rieger et al. 2008, Pátková et al. 2012, Čepl et al. 2010, 2016). As such, it offers three advantages for research: first, as a clonal multicellular organism; and second, as an extremely reduced microbial mat. A third perspective is ecological, studying interaction between colonies (Kerr et al. 2002, Nahum et al. 2011, Pátková et al. 2012). This alternative—as a possible model for interactions in mats—will be dealt with below.

Colonies can effectively communicate with other inhabitants of their environment (i.e., foreign from the point of the colony), be it conspecific or foreign colonies, a simple layer of bacteria in the vicinity, or in chimeric settings. They can also assess the area of free substrate in the vicinity, to steer its growth towards appropriate size and shape (Rieger et al. 2008, Čepl et al. 2010), or compete with siblings (Be'er et al. 2009, 2010). Airborne volatiles were shown to play a role of interspecific information carriers (Kim et al. 2013); signal(s) diffuse through the agar substrate also (Čepl et al. 2010, 2014).

Thus, colonies are shaped by both autonomous patterning and by signals generated by co-habitants of the morphospace (e.g., a Petri dish), mediating both internal shaping of the body and communication between bodies sharing the same living space. Chimeric colonies, or ternary constellations leading to rock-paper-scissors interactions, are also easy to arrange (Kerr et al. 2002, Nahum et al. 2011, Pátková et al. 2012).

The result of development is affected by the overall distribution of neighbors in the dish. Ben Jacob et al. (2004) propose genuine semantical communication between bacteria comprising the colony. To disentangle such complicated interactions, development under germ-free (colony as an extremely simplified biofilm) or gnotobiotic conditions (involving two or at most a small number of interacting species) is often of a great help.

It came as a surprise that bacteria growing even as a monoculture generate diversity in predictable ways, to fulfill different metabolic or other survival tasks. The phenomenon may belong to 'insurance effects'—differentiation of specialized cells prepared to solve quickly rare, but foreseeable, conditions. Such are dormant persistent cells (Harms et al. 2016), formed to ensure dissemination (Boles et al. 2004), or division of metabolic labor (Rosenzweig and Adams 1994), and/or 'social' liasons within a community. Horizontal gene transfer is also involved in harmonizing the community (Shapiro et al. 2012). All such ways optimize cooperation tasks (and also competition and other forms of living together) within the 'public-goods' arena (Boles et al. 2004, West 2007, West et al. 2007, Brockhurst et al. 2008, Liu et al. 2017, Marijuán et al. 2018). Some authors do not even hesitate to endow bacteria with social intelligence based in linguistic communication (Ben-Jacob et al. 2004, Joint et al. 2007).

Mimetic Rings

What follows may be taken as a science fiction by many readers, but it gives examples that the scope of symbiotic interactions abound at all levels of biosphere. The topic is mimicry, imitation of the appearance of living or non-living entities (the model) by unrelated living beings (the mime; Komárek 2003). Often they form mimetic 'rings', i.e., several mimes resembling a model that is dangerous, poisonous, unpalatable, etc. (butterflies, dipterans, snakes, etc.), or cryptic (resembling leaves, twigs, droppings, or some debris). We leave aside such cryptic and aposemantic mimicry, and will pay attention to animals living in close cohabitation with social insects (ants, termites, wasps, or bees). Such *inquilines* spend their whole life cycle in the nest. Insects (and perhaps also mites) are by far the prevailing inquilines—especially beetles, and rove beetles (Staphylindae) among them (tens of cases described; Parker 2016). Such social parasites must ensure their invisibility for nest owners, and there exists a plethora of solutions. For example, the appearance may correspond to free-living relatives of the beetle, but the organism is matching its host by odor (as in the case of phoretic[28] members of the genus *Thorictus*; Lenoir

[28] Phoretic: being attached to the host and carried around for the most part of its life.

et al. 2013). They may get a limuloid appearance (as of a horseshoe crab), making them uneasy to grasp and kill. They may develop an active defense by excreting repellents, tranquilizers, or even providing drugs pleasing the hosts. However, the most interesting is the myrmecoid (or termitoid) overall appearance of many species, i.e., visual resemblance to their hosts. Such resemblance is often very sophisticated and surprising from the point of the human observer. Yet it is easy—to a human observer again—to reveal the camouflage: some aspects of host morphology often come much exaggerated, whereas the others are sloppy. What is also conspicuous is that the hosts are often blind, it is dark in the hill anyway, and even superficial palpation would immediately reveal the 'incompleteness' of the inquiline. See, e.g., the monstrous physogastric termitophilous rove beetle *Coatonachthodes ovambolandicus*. Physogastric abdomen has a shape of termite worker with three pairs of exites resembling appendages. It resembles a termite worker, but only from above. For termites in the dark termite hill, it must obviously evince other aspects of being termite-like (Kistner 1968, 1990). Perhaps the hosts pay extra attention to exactly those traits that became accentuated in the social parasites. Quantitative analysis of such imperfect, yet effective, mimicry was performed in hoverflies (Diptera, Syrphidae; Penney et al. 2012).

Many groups had undertaken the trajectory in parallel. After all, the reward is appealing: the inquiline finds itself in a nest full of food, and is protected from outside, often even coddled. It is not surprising, then, that groups somehow 'predetermined' for communication, as in aleocharine group of rove beetles, contain many lineages evolving convergently for some 100 millions years, as cohabitants of army ant lineages (Maruyama and Parker 2017).

Now comes our speculation: inquilines are exposed to regulatory factors and cues common for the host community—pheromones, food and wastes, and overall 'atmosphere'. Consequently, their original developmental program may be shifted somewhat towards host-like appearance, and they acquire some (never all) characteristics of a 'myrmecoid' body plan 'hanging in the air'—petiolate abdomen, geniculate antennae, long legs, ant-like locomotion, texture of the cuticle, smell, etc. and are complemented with novelties from the side of the inquiline—as glands providing pheromone manipulation mentioned above. It is well known that environment can, to a great extent, manipulate gene expression even in humans (Gibson 2008); epigenetic shifts may be even more drastic, and if long-lasting, they can be followed by genetic assimilation.

Developmental body plans are similar in all insect (even arthropod) lineages, canalized along chreods towards lineage- and species-specific appearance (phenotype). In a new environment, factors in the host's nest may deflect the development along other, 'myrmecoid', chreods. The inquiline starts to imitate its host (by scanning the memory and experience of its umwelt, see Chapter 3). After some generations, genetic assimilation will fix the new developmental pathway. To push the scifi even further, we mention the speculation by Kleisner and Markoš (2005, 2009): they suggested a term 'semetic ring' for such semblance associations, stressing the fact that something similar to fashion may also play a role. The mime (in order to adapt to the environment) would 'guess' the common way of appearance and will actively experiment with its own. The inquilines barter host's protection for disseminating its appearance.

The idea of imitation by grasping the common 'air' of the place or epoch may not be restricted to intimate associations, as in the case of myrmecophiles or termitophiles. Mimetic rings of flies, butterflies, or snakes, may also fall into the category. Very conspicuous is the parallel evolution of traits in unrelated lineages sharing the same geographical area. Thus, Went (1971) gives examples of such phenomena in dicotyledonous plants; for example, mass appearance of divaricate shrubs in New Zealand and Australia (51 species in 23 families, not exotic at all, belonging to globally dispersed ones), or massive red autumn leaf coloration, restricted to Japan and Northeastern part of North America. At all such occasions, a common factor may be present in the environment, pushing inhabitants of the community to accentuate some—often marginal from the point of human observer—trait. It may be marginal also from the point of survival, yet the members of the ring take much 'care' to develop the trait, to resemble their compatriots in the ecosystem. Selection pressure, reflected dictate of some model, or even pressure of 'fashion'—all such factors may drive ecological speciation (Chamberlain et al. 2009). Interspecies hybridization was also described for the spread of mimicry, but so far it was described only at the level of a butterfly genus (*Heliconius* genome consortium 2012). Described phenomena may also represent reactions to a 'cocktail' of environmental tools, raising—in some members of the community—a new, reflected usage of old ontogenetic tools, "that so many cases of parallel development of shrub shapes, and leaf form are not climatically, but geographically circumscribed" (Went 1971). For interspecific polymorphism, the term 'supergene' is coined, switching development to this or that form, with no intermediates—as in the case of butterfly *Papilio polytes* (Kunte et al. 2014), or in *Heliconius*, where "cis-regilatory evolution of a single transcription factor can repeatedly drive the convergent evolution of complex color patterns in distantly related species, thus blurring the distinction between convergence and homology." (Reed et al. 2011)

10
I, Holobiont

> *Our task now is to resynthesize biology; put the organism back into its environment; connect it again to its evolutionary past; and "let us feel that complex flow that is organism, evolution, and environment united." The time has come for biology to enter the nonlinear world.*
>
> —C.R. Woese 2004, 179–80

Lately, the study of symbiotic interactions between a eukaryotic host and its microbiota, i.e., a holobiont, has become a primary topic of biological science. It turns out that such interactions are essential for our physical and mental wellbeing as well as for evolution as such. The term holobiont denotes a 'whole' living entity, which involves not only a host but also all its symbionts. It usually represents a single macroscopic host that is in mutually dependent, synergistic relationships with viruses, bacteria, archea, fungi or other eukaryotes. These symbiotic relationships include parasitism, commensalism and mutualism. Some consider the holobiont to be the genuine biological individuum (Gilbert et al. 2012, Booth 2014). Indeed, few relationships in living nature are permanently neutral *to each other*, perhaps none, because evolution proceeds through relations of mutual understanding *of the Norm*. The term superorganism has also been used to describe these sorts of mutual relationships; however, the term is not accurate: superorganism originally meant a higher level of organization composed of individuals of the same kind as eusocial insects, and it also points toward the global ecosystems like Gaia. Holobiont, on the other hand, covers several domains of life and a huge variability of genomes. Thus, a superorganism, such as a termite colony, is also full of holobionts (Zilber-Rosenberg and Rosenberg 2008, Rosenberg and Zilber-Rosenberg 2013).

The vast complexity of symbiotic relationships and its influence on the health and development of any specific host is exhaustively gathered in Gilbert and Epel's book (2015); some examples of symbiosis are also mentioned in the previous chapter. We try not to repeat what has been already said. Gilbert and Epel mention iconic examples of mycorrhiza, the previously discussed biosphere of permeating and interconnecting symbiotic relationship between plants and fungi or plants and bacteria (rhizobium and endophytes). Further, they explain the problem of coral bleaching (as a result of disturbances in the relationship between the unicellular algae and the coral, most probably caused by elevated sea temperatures), as well as holobionts par excellence—such as termites *Mastotermes darwinensis*, which can only digest lignin thanks to *Myxotricha paradoxa*, a protozoan which itself is a holobiont living together with four other bacterial species. They also offer many well-known examples of developmental symbiosis: many species gain gut symbionts from feces (in groups of insect or vertebrates) or capsules; without such horizontal transfer, the individual development is compromised. Iconic examples of developmental symbiosis include the relation of *Euprymna scolopes* with bacteria *Vibrio fischeri*, as well as the vertical transfer (through egg or sperm) in many species of the bacterium *Wolbachia*, which influences the sex ratio through killing male embryos but is also important for normal development. A similar scenario can be found in the case of *Drosophila* species, which lives in tight relationship with *Spiroplasma poulsonii*, a gram-positive symbiotic bacterium whose presence is usually fatal for male embryos, but not female. Male embryos are killed through overexpression of the Spaid protein, which causes massive apoptosis and neural defects, and thus distorts the *Drosophila* sex ratio (Harumuto and Lemaitre 2018).

Nowadays, the most frequently studied holobiotic niche is the human gut. Many of the lectures and papers devoted to this topic begin with the fact that about 10^{14} bacteria reside in the human body, with the majority in the alimentary tract (over 1000 species, dominated by two strict anaerobic bacterial phyla—Bacteroidetes and Firmicutes). The widely cited ratio of bacterial to human cells 10:1 has recently been questioned (Sender et al. 2016) and now tends towards a more realistic 1:1. The estimation of the number of species has settled to the range of 500–1000 (Gilbert et al. 2018). Such studies are metagenomic, and sequence the whole range of symbiotic genomes that, together with the host, form the hologenome. (There is a difference between the metagenome, which represents a specific environmental sample, and a hologenome, which covers the whole symbiotic unit of various organisms and their respective genomes.) Sequencing has become an effective and quite necessary method to identify our inhabitants—as only about 7% of known bacterial species can be cultivated on a Petri dish (although a study by Browne et al. 2016 shows that the number of culturable bacteria could be higher; they managed to cultivate up to 137 different species). It turns out that there are roughly 2 million holobiont genes, i.e., roughly 100 times the number of human genes (fungi, viruses and phages are not taken into consideration in the estimate; Gilbert et al. 2018). The terms holobiont and hologenome were independently introduced by many authors such as L. Margulis, D. Mindell, R. Jefferson, E. Rosenberg and I. Zilber-Rosenberg (for a review of the history of this term, see Rosenberg and Zilber-Rosenberg 2013), though to be clear, the terms are not convertible (Bordenstein and Theis 2015).

Human Holobiont

Viruses, bacteria, fungi and archea can all be found on our skin, in the urogenital tract, in the intestines, in the lungs and in the mouth, i.e., in places which are exposed to the external environment (they are not inside cells). The human microbiome project has gathered data about various microbial communities associated with different sites of the body (Human Microbiome Project Consortium 2012).

Here, we focus only on the gut: it is the most studied and perhaps the most diverse holobiotic site. Human gastrointestinal tract involves stomach, small intestine (duodenum, jejunum, ileum) and large intestine (caecum, colon and rectum). The phyla dominating the human stomach are Proteobacteria, Firmicutes, Bacteroidetes and Actinobacteria (Maldonado-Contreras et al. 2011). The dominating genera are *Helicobacter, Prevotella* and *Streptococcus*. The small intestine is a less studied site, but researchers have identified the dominating facultative and obligative anaerobes, such as *Streptococcus* sp., enterobacteria, *Clostridium* sp., Bacteroidetes. The large intestine consists of the caecum, colon, rectum and anal canal. According to stool samples, our large intestine contains phyla such as Bacteroidetes (up to 40%), Firmicutes (up to 50%), Proteobacteria, Fusobacteria, Verrucomicrobia, Actinobacteria, Ascomycota and Euryarchaeota[29] (Riedel et al. 2014). Only about 40 species of bacteria make up 90% of the intestinal flora. There were 16 identified species of fungi (mainly Ascomycota, Basidiomycota), and some species of archea such as anaerobic *Methanobrevibacter smithii a Methanosphaera stadtmanae* (Riedel et al. 2014).

Our intestines contain 10^{13} different types of viruses including bacteriophages, archaeal, plant and mammalian viruses (Reyes et al. 2010). Moreover, about 8% of the human genome is of viral origin (Horie et al. 2010). Viruses play many important roles in mammals, although these are still less understood than those of bacteria. Adenovirus reduces cancerous tumors in hamsters (de la Maza and Carter 1981); both the hepatitis G virus and the cytomegalovirus slow the proliferation of HIV in humans (Timmons et al. 2013). Hepatitis A virus inhibits hepatitis C virus (Deterding et al. 2006). Herpesviruses can suppress bacterial pathogens (Yager et al. 2009); for example, murine gammaherpesvirus 68 and murine cytomegalovirus both help build resistance to bacterial infections of *Listeria monocytogenes* and *Yersinia pestis* (Barton et al. 2007).

Kernbauer et al. (2014) showed that an enteric murine norovirus plays a role similar to that of beneficial bacteria, i.e., protects the intestine from damage and pathogens. Infecting mouse intestines with this common, non-pathogenic virus repairs morphological defects found in both germ-free and antibiotics treated mice intestine. The infection helps restore the number of $CD4^+$ and $CD8^+$ T cells, increase the levels of antibodies and the number of innate lymphoid cells, and restore the thickness of the villi in the intestine and the number of granules in the Paneth cells.

As mentioned in previous chapters, the defining mammalian trait, the placenta, is partly of viral origin: viral syncytin, a gene of endogenous retrovirus HERV-W[2],

[29] These are formed by genera *Candida* (less than 1% of total), *Methanobrevibacter* (less than 1%), *Anaerostipes* (less than 1%), *Clostridium* (1–2%), *Eubacterium* (1–5%), *Ruminococcus* (1–8%), *Roseburia* (1–8%), *Dorea* (0–2%), *Blautia* (0–2%), *Faecalibacterium* (5–15%), *Streptococcus* (less than 1%), *Lactococcus* (less than 1%), *Lactobacillus* (5–15%), *Bacteroides* (5–35%), *Parabacteroides* (1–3%), *Prevotella* (1–5%), *Porphyromonas* (1–2%), *Alistipes* (2–8%), *Escherichia* (less than 1%), *Fusobacterium* (less than 2%), *Akkermansia* (1–3%), *Bifidobacterium* (1–10%) and *Collinsella* (1–8%): the list was taken from a review Riedel et al. (2014).

enables the cells to clump and the virus to spread easily from cell to cell. *In utero*, the syncytin enables cells fusing together to create syncytiotrophoblast. Hence, the integration of the viral gene into our genome enables us to form such structures as placental syncytia (Mi et al. 2000).

These relationships can get even more complicated, as some bacteria can use viruses to kill competing bacterial species while invading new territories (Gama et al. 2013) and sometimes even gain either antibiotic resistant genes upon destruction of such competitors (Haaber et al. 2016), or 'domesticated' phages which produce toxins in order to kill their competitors.

Variation in Microbiome Within Individuals

The final decision as to whether the microbiome composition is environment dependent or rather gene dependent (Goodrich et al. 2014) is still debatable, but it is strongly inclined towards the environment (Rothschild et al. 2018). Couples from the same household share more similar microbiota than individuals from different households (mostly skin microbiota) and dog owners also share the skin microbiota with their pets (rather than with other dogs, Song et al. 2013). Physically interacting individuals, usually couples, share more similar microbiota than people who share the same living environment but do not interact physically (Lax et al. 2014). The fact is that there is an enormous variability of microbiome among people, although some phyla are more common than others; yet what influences this variability may produce is still unknown. Generally, the most changeable site is skin, followed by gut and oral cavity. In some persons, the microbiome remains quite stable, whereas in others it is much more changeable (Caporaso et al. 2011). In one study, David et al. (2014a) studied stool and saliva samples of two men over the period of a year. They confirm that about 5% of gut and saliva microbiota are stable and belong to the core microbiome. Yet both individuals had a very personalized microbiome (which can have some forensic applications) and the differences between these two individuals were greater than the annual fluctuations in either of them. Interestingly, both subjects responded quite differently to gut microbiome perturbation (due to diarrhea, etc.). The microbiome of one of the individuals returned to its original state, whereas the other subject's microbiome remained changed, although with a phylogenetically related species of bacteria. Thus, the functions were preserved even though the compositional stability was perturbed (David et al. 2014a). This shows that there is an apparent metabolic conservation within bacterial species.

However, the microbiome as a whole is a constantly changing and fluctuating community: every day we consume about 10^{10} microorganisms; it can be quite difficult to distinguish a permanent resident from a nomad. In addition, sequencing methods cannot determine whether or not the bacterium is dead, and what is its role in the overall body economy.

Colonization Determinants

The colonization of a newborn individual is mostly determined by the maternal organism. Many organisms, if not infected by internal symbionts through egg or sperm, as in the case of *Wolbachia*, must go through a specific phase of inoculation with the beneficial bacteria, usually from the maternal, but sometimes from fellow, organisms.

The first and primary source of microbiota for humans is the mother's vagina and feces, later also her skin and breast milk. Already during pregnancy, the hormones of a pregnant woman change the microbial composition of her intestines and vagina, which in turn affects her physiology. These changes are especially significant during the third trimester, when the abundance of Proteobacteria and Actinobacteria increases (Koren et al. 2012). If these intestinal bacteria are transferred to a non-pregnant, gnotobiotic mouse, it changes its physiology: the mouse became insulin resistant and gained weight quickly. In the third trimester, the structure and composition of the bacterial community of a pregnant woman resembles the metabolic syndrome associated with weight gain and dysbiosis, and often causes type 2 diabetes; however, such effects are in place during pregnancy because they promote fat storage and fetal growth.

There are also speculations as to whether some bacteria can pass through the amniotic barrier (Funkhouser and Bordenstein 2013; for review, see Willyard 2018), challenging the 'sterile womb paradigm'. Distinct types of microbiota have been found in placenta (Stout et al. 2013, Aagaard et al. 2014), amniotic fluid (Collado et al. 2016), meconium—the first stool of a newborn (Jimenez et al. 2008), and in umbilical cord blood (Jimenez et al. 2005).

In meconium, the microbial community includes the family of Enterobacteriaceae as the most abundant taxon (up to 60% represented by the *Escherichia* genus). In some individuals, samples of Firmicutes has also been found, particularly in families Leuconostocaceae, Enterococcaceae and Streptococcaceae (with the most abundant genera of *Leuconostoc, Enterococcus, Lactococcus, Staphylococcus* or *Streptococcus*). The abundance of Bacteroidetes is very low (Gosalbes et al. 2012). Two sources of such microbiome are possible: it can come from swallowed amniotic fluid or it can get into the infant's gut easily after amniotic rupture. However, Gosalbes et al. (2012) deny the latter option because such microbial composition would then be more similar to skin, fecal or vaginal microbiota, hence different from their own results. After studying the composition of human meconium, Jimenez et al. (2008) inoculated pregnant mice with genetically marked *Enterococcus faecium* isolated from the human breast milk, and later found this strain of bacteria in meconium of offspring born by cesarean section. This implies that neither skin, fecal nor vaginal sites of the mother are a source of meconial microbiota. Genera *Enterobacter, Enterococcus, Lactobacillis, Photorhabdus* and *Tannerella,* collected from meconium, are supposed to have a causal role in premature birth (Ardissone et al. 2014). Umbilical cord blood contained species such as *Enterococcus faecium, Propionibacterium acnes, Staphylococcus epidermidis,* and *Streptococcus sanguinis* (Jimenez et al. 2005).

Additionally, Stout et al. (2013) showed that 27% of all placentas, both in-term and preterm pregnancies, contain intracellular bacteria. Aagaard et al. (2014) examined the placentas of 320 women and found that many of them contained the bacterial DNA of nonpathogenic commensal microbiota from the Firmicutes, Tenericutes, Proteobacteria, Bacteroidetes, and Fusobacteria phyla. These usually reside in oral microbiota, presumably transported from the gut lumen to the bloodstream with the help of dendritic cells (Rescigno et al. 2001, Collantes-Fernandez et al. 2012). Collado et al. (2016) revealed that both amniotic fluid and placenta contain low diversity microbiota dominated by Proteobacteria; they believe that fetal gut colonization already begins prenatally. However, there are also studies that deny the existence of placental microbiota, suggesting that the false positive evidence is a result of contamination (Lauder et al. 2016), and is generally weak due to insufficient detection methods which identify low concentration of microbial populations, or due to the lack of controls for contamination (Perez-Muñoz et al. 2017).

Nevertheless, bacteria present in mother's gut can (at least in the mice model) increase the innate immunity of the pups *in utero*, thus preparing them (through the antibody-mediated transfer) for the load of microbes they will encounter during birth (Gomez de Agüero et al. 2016): the immune system recognizes these symbiotic bacteria as its own. A recent study by Donaldson et al. (2018) shows how communication between commensal bacteria and host immune system looks. *Bacteroides fragilis* can change its surface ways that allow it to invite binding of the host immunoglobulin A (in mice model), which facilitates bacterial association with gut mucosa, thus leading to selective colonization of the gut. Whereas in adults the core microbiome communities are generally stable, in infants, especially during the first three years of life, the development of microbiota composition is quite dramatic (Caporaso et al. 2011). During the first year of life, the gastrointestinal tract is colonized by 10^{13}–10^{14} microbes from between 500 and 1000 different species (Weng and Walker 2013), the phylogenetic diversity is the lowest at birth and slowly increases with age (Koenig et al. 2011). The uniqueness of the holobiotic trace of every individual is already in progress during these first stages of life (Raveh-Sadka et al. 2015), but is stabilized into a lifelong microbial signature only after this sensitive and dramatic period (Caporaso et al. 2011). Interpersonal differences in gut microbiome are thus greater among infants than among adults (Kurokawa et al. 2007); this is probably also due to historical contingencies, differing modes of delivery, nutrition, and other environmental aspects (Azad et al. 2015, Stokholm et al. 2016, Bokulich et al. 2016). Also, the first colonizers are a decisive factor: who comes first sets the rules for who comes later.

The gut microbiome of children who are born by vaginal delivery differs from those born by cesarean section (Penders et al. 2006, Dominguez-Bello et al. 2010, Bäckhed et al. 2015). The microbiome of vaginally born is mostly occupied by *Bacteriodes*, whereas children born by cesarean lack *Bacteroides* and *Bifidobacterium*—sometimes until the age of 18 months. Vaginally born children inherit the microbiome from their mother's vagina (*Lactobacillus, Prevotella, Atopobium*, etc.) whereas the microbiome of children born by cesarean section resembles the skin microbiome of medical workers or the hospital environment, and is dominated by *Staphylococcus, Corynebacterium* or *Propionibacterium* (Dominguez-Bello et al. 2010). Premature infants are even more vulnerable to microbial dysbiosis because they are separated from the natural environment, placed in sterile incubators and given courses of strong antibiotic treatment. Such babies thus lack microbes such as *Bifidobacterium* or *Lactobacillus*, suffer from lower microbial diversity in their intestines, and have a greater amount of possible pathogens—often leading to necrotizing enterocolitis.

This disease is not caused by a particular microbial agent, but rather by the dysbiosis of the whole microbiome. The more medical interventions a child experiences during the first six months of its life, the more medical issues that child will have later in life. An early antibiotics intervention, which alters the microbiome composition, may be associated with later development of obesity, Crohn disease and asthma (Koenig et al. 2011, Cho et al. 2012, Trasande et al. 2013, Cox et al. 2014, Arrieta et al. 2015, Yassour et al. 2016).

Breast milk contains numerous proteins like immunoglobulin A, cytokines, active proteins and enzymes such as lysozyme or lactoferrin (Lönnerdal 2013). These modulate the immune system of the infant, help with nutrient absorption, and play a role in defending the infant against pathogens. Lactoferrin, for example, can suppress the growth of bacteria or even kill them; it can likewise prevent infections in preterm babies, and is important especially during the first days of life (Villavicencio et al. 2016). Breast milk is also a major source of bacteria (*Staphylococcus, Streptococcus, Micrococcus, Lactobacillus, Enterococcus, Bifidobacteria,* and many intestinal anaerobes; Martín et al. 2004). Even viruses can be transmitted by breast milk, although known cases concern pathogens only (Lawrence and Lawrence 2004). An infant consumes 10^5–10^7 commensal bacteria daily (Heikkilä and Saris 2003).

Additionally, breast milk also has a prebiotic effect: some oligosaccharides found in breast milk, the second most abundant after lactose, are not for the newborn to digest, but provide food for bacterial symbionts such as *Bifidobacterium* (which is a dominant family in breast-fed babies; Roger et al. 2010). The genome of *Bifidobacterium longum* subspecies *infantis* contains a region within genes whose products are involved in the metabolism of these complex sugars. In phylogenetically related species of *Bifidobacterium* that do not reside in the human intestine, these genes are absent; this indicates a coevolution between *B. longum infantis* and the host (Sela et al. 2011).

Functions of Microbiome

Germ-free mice are commonly used to study the complex role of human microbiota, a fluctuating community impossible to capture as a whole. These mice are kept in sterile bubbles, born by caesarean section and completely without symbionts. Such a germ-free animal is, of course, an artificial manipulation and becomes gnotobiotic only when inoculated with a specific community of bacteria. The effects of such inoculation are then observed (Luczynski et al. 2016). Similar experiments are also made on piglets, which serve as an animal model more resembling the human model.

Human intestines serve as physical, chemical and immunological barriers: the first barrier is epithelial cells (ECs) such as enterocytes, goblet cells, Paneth cells, enteroendocrine cells and antigen-sampling M cells. The surfaces of ECs are covered by various glycoproteins and glycolipids, which serve as attachment substrates for commensal bacteria or pathogens. ECs also produce glycosylated mucins, which are the main protective barrier of the intestines (Goto et al. 2015). Bacteria cover this mucosal barrier, and form a barrier of their own, especially for potential pathogens from the environment, as well as produce antipathogenic substances. For instance, pathogenic bacteria *Staphylococcus lugdunensis*, living in the nose, produces the antibiotic lugdunin, which prevents the colonization of this niche by *Staphylococcus aureus*, an opportunistic pathogen causing blood and heart infections (Zipperer et al. 2016). In a similar manner, *Bacteroides fragilis* protects the host from colitis caused by *Helicobacter hepaticus* through the production of polysaccharide A, which blocks the inflammatory response by suppression of IL-17 evoked by *H. hepaticus* (Mazmanian et al. 2008). Furthermore, the non-resistant strain of *Enterococcus*, which is a part of the human microbiome, can produce a pheromone cOB1, which kills antibiotic resistant bacteria *Enterococcus faecalis* V583 (Gilmore et al. 2015).

A healthy microbiome (mainly Bacteroides fragilis, Bacilus subtilis and Bifidobacteria); also induces the formation of mammalian gut-associated lymphoid tissue; it has been shown that germ-free mice cannot form these structures completely (Bouskra et al. 2008). A bacterium called *Bacteroides thetaiotaomicron*, which produces a protein colipase, helps forming blood vessels (as well as hydrolyzes fats). In germ-free mice, the vessel network is much poorer, but it can recover within 10 days after a fecal transplant (even in the case of adult mice). *Bacteroides thetaiotaomicron* induces transcription of angiogenin-4 gene in the Paneth cell in intestinal crypts, which is important in inducing blood vessel formation (Hooper et al. 2013). This bacterium also helps induce epithelial fucosylation, i.e., glycosylation of the epithelial cell surface with fucose residues, by inducing IL-22 production (Goto et al. 2014). Fucosylation is an important factor for gut homeostasis because fucose residues serve as attachments receptors and nutrient source for commensal bacteria (Goto et al. 2015). Such effects are also observable within other microbial sites as microorganisms contribute to a normal development of olfactory epithelium: germ-free mice had a thinner cilia layer and reduced cell proliferation in olfactory epithelium (François et al. 2016). Microbial colonization of the gut influences the development of B-lymphocytes (Wesemann et al. 2013). In general, the development of our immune system would not be possible without gut bacteria (Lee and Mazmanian 2010).

The main function of gut microbiota is to digest food components such as fiber or starch. In so doing, they produce short-chain fatty acids such as propionate, acetate or butyrate, which are the main products of complex sugar fermentation (especially Bacteroidetes or Verrucomicrobia; Koenig et al. 2011, den Besten et al. 2013). Short-chain fatty acids account for about 10 per cent of dietary intake in humans. Bacteria are also an important source of amino acids (tryptophan), bile acids, choline metabolites, lipids or vitamins such as K (Firmicutes). Up to one third of the metabolites of mammalian bodies are products of the gut microbiome (McFall-Ngai et al. 2013). In addition, they help with detoxification of harmful substances and with digestion and biotransformation of drugs—influencing the drug's effectiveness in accordance with the microbiome composition. Some drugs can even be toxic for people with certain microbiome composition. Gut microbiota can also influence the reactions to cancer immunotherapeutic treatment (Gopalakrishnan et al. 2018).

The human gut is innervated and contains about 500 millions neurons (including motor neurons, intrinsic primary afferent neurons or sensory neurons, and interneurons; Nezami and Srinivasan 2013, Luczynski et al. 2016). Additionally to neurons, it contains enteric glia. They are part of the enteric nervous system (ENS), which is collected into two types of plexuses: submucosal and myenteric plexus.

Myenteric plexus is able to control the peristaltics, i.e., the relaxation and contraction of intestinal wall, whereas submucosal plexus regulates enzyme secretion, absorption, blood flow, and chemical conditions in the gut (Nezami and Srinivasan 2013). These processes are regulated with similar neurotransmitters such as acetylcholine, dopamine, serotonine, norepinephrine, bile acid, etc. The enteric nervous system can communicate with the central nervous system via parasympathetic (vagus nerve) and sympathetic (prevertebral ganglia) nervous systems, thus forming a brain-gut-microbiome axis. The brain-gut-microbiome axis is bidirectional, which means that environmental stress processed by our brain can regulate digestive tract activity and influence the composition of gut microbiota and its metabolic activity, and also, that metabolites produced by bacteria can influence the brain activity.

Microbial symbionts produce many metabolites that modulate both behavior and health. *Lactobacillus* and *Bifidobacterium* produce the neurotransmitter GABA and through the vagus nerve (gut-brain axis) regulate the GABA receptors and thus emotional behavior (Barrett et al. 2012). Treatment with *Lactobacillus rhamnosus* reduces stress-induced anxiety in mice and leads to higher expression of GABA receptors in the brain (Bravo et al. 2011). Symbiotic bacteria also produce dopamine, serotonine, adrenaline, nonadrenaline or acetylcholine (Stephenson and Rowatt 1947, Yano et al. 2015). Germ-free mice produce less dopamine and adrenaline (Asano et al. 2012). Studies also show that *Bacteroides fragilis* is important for normal social behavior in mice (Hsiao et al. 2013). Germ-free mice have been reported as asocial; they often have repetitive, anxiety-like behavior as well as a 'leaky gut'—which is also found in autistic children. After inoculation of germ-free mice with *Bacteroides fragilis*, both behavior and the integrity of the intestinal wall improved toward the normal spectrum. Levels of metabolites, usually associated with anxious behavior, such as indolpyruvate or ethylphenylsulfate, also decreased (Hsiao et al. 2013). Autism spectrum disorder is thus thought to have both genetic and environmental predispositions. Concerning the environment, microbial dysbiosis or reduced variability of microbiome is often considered a primary suspect (Adams et al. 2011, Hsiao et al. 2013, Kang et al. 2013, Son et al. 2015). As mentioned above, the microbial symbionts in our gut produce many important metabolites that modulate our social and emotional behavior, and general mental health. Similar effects have been reported in cases of oral microbiota influencing the onset of diseases such as Parkinson or Alzheimer (Shoemark and Allen 2015, Pereira et al. 2017). Also, Qiao et al. (2018) reported altered oral microbiota in autistic children, with a higher abundance of Proteobacteria and lower abundance of Actinobacteria and Bacteroidetes, together with reduced genera such as *Prevotella, Selenomonas, Actinomyse* and more, as well as increased amounts of *Haemophilus, Streptococcus, Corynebacterium* and other possible pathogens which can affect brain development (Qiao et al. 2018). Similar dysbiotic signatures were also shown in the gut of autistic children (Kang et al. 2013).

Psychiatric problems such as anxiety and depression often coexist with diseases such as irritable bowel syndrome, which shows a link to microbial dysbiosis. Germ-free mice often produce less dopamine, adrenaline or serotonin. The microbiome profile of patients with this syndrome has an increased ratio of Firmicutes to Bacteroidetes, reduced beneficial species such as *Bifidobaterium* or *Faecalibacterium* and increased number of *Ruminococcus, Dorea* and *Clostridum* species which are associated with infection (Rajilić-Stojanović et al. 2011). The syndrome can also be caused by antibiotic treatment early in life (O'Mahony et al. 2014).

The offsprings of mice who were fed a high fat diet during gestation also display less social behavior than offsprings of mothers on a normal diet. Compared with control populations, socially deprived offspring also shows a reduced bacterial microbiome, their brains contain less oxytocin, and the brain signaling in regions connected to reward stimuli are weaker. These abnormal effects were reversed, when mothers on high-fat diet were inoculated with *Lactobacillus reuteri* (Buffington et al. 2016).

Microbial dysbiosis has also been linked with the prevalence of obesity in human population. This is readily demonstrated through microbiome transfer from obese human individuals to germ-free mice, which usually leads to the development of an obese phenotype. Moreover, a microbial transplant from a lean individual can reduce weight gain in obese recipients (together with the right diet; Ridaura et al. 2013). In obese human individuals, the microbiome profile often shifts towards a higher abundance of Firmicutes and an almost 50 per cent reduction of Bacteroidetes. Bacteroidetes are considered protective against increased adiposity, but may well involve far more complex interactions (Ridaura et al. 2013). The bacterial richness in the human microbiome has been connected to higher overall levels of body fat in the Danish population, wherein individuals with lower microbial diversity usually gained more body weight and fat than people with richer microbiome (Le Chatelier et al. 2013). For a healthy microbiome, diet is obviously an important factor. A plant-rich diet promotes a healthy microbiome and metabolite production that furthers the abundance of Bacteroidetes, which produce short-chain fatty acids, which eventually inhibit the storage of fat.

Perry et al. (2016) also observed that, when exposed to a high-fat diet, gut microbiota promotes insulin secretion through acetate signaling to the brain. Rats on a high-fat diet had higher production and turnover of acetate compared with rats on a normal diet. Acetate can mediate insulin secretion through the parasympathetic nervous system, and is also involved in the higher levels of ghrelin. This 'hunger hormone' stimulates a constant drive to eat, which is another contributor to obesity in these rats. These days, there is about 60 per cent accuracy in guessing whether a human stool sample comes from lean or obese individual.

Naturally, diet has a major and rapid impact on microbiome composition: a lack of dietary fibre can lead to a decrease in microbial diversity (Sonnenburg et al. 2016), and a meat diet leads to an abundance of bile-tolerant microorganisms such as *Alistipes, Bilophila* and *Bacteroides*, as well as to decreased levels of Firmicutes (David et al. 2014b). Suez et al. (2014) demonstrated that the consumption of non-caloric artificial sweeteners changes the composition of gut microbiota which can lead to glucose intolerance.

Human dietary habits involve another example of human-bacteria coevolution (the coevolution of *Bifidobacterium* and human breast milk was mentioned above), this time in the case of marine bacterium *Zobellia galactanivorans* living on the alga *Porphyra*, which is used for sushi preparation and therefore widely consumed by Japanase population (they consume more than 4 kilos per year). A gene that serves the digestion of this alga, more specifically an enzyme porphyrase, gained from *Zobellia* through horizontal gene transfer, is present in *Bacteroides plebeius*, a bacterium residing only in intestinal tract of the Japanase population (Hehemann et al. 2010).

Insulin sensitivity is also linked with changed microbial communities and microbial dysbiosis (Pedersen et al. 2016), as is cardiovascular disease (Hansen et al. 2015), type 2 diabetes (Qin et al. 2012, Karlsson et al. 2013), irritable bowel syndrome (Jeffery et al. 2012), inflammatory bowel disease such as Crohn disease or ulcerative colitis (Peng et al. 2014, Sheehan et al. 2015). As many papers mention, sometimes it is hard to tell whether such microbial differences are a biological cause or a consequence of a specific disease.

The microbiota can also influence gene expression in multiple tissues by regulating the histone proteins through acetylation or methylation (Krautkramer et al. 2016). When mice are fed with a high-fat and high-sugar diet, bacteria do not get required nutrients and do not influence DNA conformation. However, when supplemented with short-chain fatty acids, these effects were restored as they were with normal microbiotic colonization.

Fecal Transplants

The increased interest in microbial communities has brought back an old treatment practice: fecal transplants. These have turned out to be a very efficient means to cure *Clostridium difficile* infection. Unlike antibiotics such as vancomycin, causing disrupting effect to the microbiota, the fecal transplants treat the infection of *C. difficile* within a day—and with 94 per cent efficiency (Smits et al. 2013, Kassam et al. 2013, van Nood et al. 2013). In the 4th century, a doctor named Ge Hong began such treatments in China (the yellow soup); in the 17th century, veterinarians began using them to cure cows of intestinal problems (de Vrieze et al. 2013). In the elderly, there is a reduced variability in the microbiome, and facultative anaerobes increase, whereas the abundance of *Bifidobacterium* decreases, i.e., in general, the number of Firmicutes overrules the number of Bacteroidetes. More than half of the Western population over 60 years contain *Helicobacter pylori* in their microbiome. Such bacterium can overrule the microbiotic composition, as when the person is treated with antibiotics, for example. Fecal transplants can be a very efficient treatment for such cases.

Holobiont as a Unit of Selection

The holobiotic view has strongly infiltrated the consensus of mainstream science; in 2017–18, the number of papers concerned only with microbiota in the journal consortia of Nature and Science has exceeded 2000. It cannot be overemphasized: there exists *no* eukaryote without symbionts, and the first holobiont may well have been a eukaryotic organism with endocytic bacteria. There are many examples of how the symbiosis influences organismal development (see this and previous chapters), and that these co-evolving relationships are both function- and host-specific. Evolution can rightly be understood as proceeding through symbiotic events (as discussed in earlier chapters).

The hologenome theory of evolution emerged with a paper by Zilber-Rosenberg and Rosenberg (2008) wherein they used four generalizations to define the theory: "(1) All animals and plants establish symbiotic relationships with microorganisms; (2) symbiotic microorganisms are transmitted between generations; (3) the association between host and symbionts affects the fitness of the holobiont within its environment; and (4) variation in the hologenome can be brought about by changes in either the host or the microbiota genomes; under environmental stress, the symbiotic microbial community can change rapidly."

Since then, there has been an ongoing discussion as to whether hologenomes or holobionts should be regarded as single units of selection co-evolving through many generations, and whether the concept of holobiont is helpful or hollow (Bordenstein and Theis 2015, Moran and Sloan 2015, Theis et al. 2016, Douglas and Werren 2016). We reflect on this discussion in the last part of this chapter.

The most widely known examples of symbiotic co-evolution are the mitochondria and chloroplasts of eukaryotic cells. In their book *Major transitions in evolution*, Maynard Smith and Szathmáry (1995) argue that reducing conflict and competition on one biological level can lead to biological transition on another—e.g., the evolution of chromosomes, multicellularity and eusociality. Bordenstein and Theis (2015) later gathered ten principles that compile the defining information of the holobionts' symbiotic relationships. They define holobionts and hologenomes as units of biological organization: eukaryotes have never been individuals (Gilbert 2014), and cannot be grasped as autonomous individuals, but rather as units composed of various microbial symbionts. All genomes of these symbionts form the hologenome of a specific holobiont, and there is variation across all these genomes including nucleus, organelles and the whole microbiome (with variation provided by beneficial, deleterious or neutral mutations, recombination, gene loss and duplication, but also horizontal gene transfer and other processes). The evolution of a holobiont can be also grasped as a change of allele frequencies. Thus, microbial symbionts provide a new source of variability for the host organism. Several authors also argue that, as the microbiome is inherited environmentally and not genetically, this line of thinking thus involves a Lamarckian dimension of heredity (Zilber-Rosenberg and Rosenberg 2008, Bordenstein and Theis 2015). As in epigenetics, this brings a controversy, even though Lamarck did not know microbes. In fact, the term serves rather poorly as a vague and general term for non-genetic inheritance, and we think should rather be avoided (Švorcová and Kleisner 2018).

The concept of the holobiont also encompasses the concept of multilevel selection theory (Kerr and Godfrey-Smith 2002, Okasha 2013), which argues that units of selection can function simultaneously at and across various levels of biological organization. Moreover, the same behavior that results in a real loss of possibility on one scale of life can also generate successful opportunism on another (Ostdiek 2016). Borderstein and Theis (2015) argue that natural selection can drive the loss of deleterious nuclear mutations or microbes while spreading those that are advantageous, such that naturally neutral evolutionary processes are clearly in play. In their opinion, holobiont theory redefines the concepts of an individual, but does not change the validity of Darwin's theory of natural selection.

Microbiome variation can also lead to new adaptations in a fluctuating environment, much faster than merely waiting for some random mutation; it can even happen within an individual's lifetime, as shown above in the cases of diet change and infectious disease. As Zilber-Rosenberg and Rosenberg argued (2008): "Based solely on the host genome, animals and plants would evolve slowly because of (1) their relatively long generation times, (2) the fact that only changes in the DNA of the germ line are transmitted to the next generation and (3) often a whole set of new genes is required to introduce a beneficial phenotypic change."

The intimate relation with symbionts can, according to some studies, lead to reproductive barriers—i.e., to speciation. *Wolbachia* is known for inducing reproductive isolation between species, leading to parthenogenetic speciation (Bordenstein et al. 2001, Huigens and Stouthamer 2003). Brucker and Borderstein (2013) showed that in hybrids of two insect species (wasps of genus *Nasonia*), a species' microbiome can lead to hybrid lethality. These effects confirm, in their opinion, the co-evolution of the holobiont. In addition, females of *Drosophila melanogaster* strongly favor mating with individuals who have *Lactobacillus plantarum* in their gut (Sharon et al. 2010). Antibiotic treatment disrupts such preferences; however, inoculation with this microbe can again restore them, indicating that microbes, not diet, changed the mating preferences. This study shows that gut microbiota differences can lead to premating isolation.

Chemical communication can be influenced by communities of odor producing bacteria of scent-glands which varies across hyena clans (being more similar within the same social group; Theis et al. 2012). Another study by Theis et al. (2013) showed correlations between various traits such as sex, dominance and reproductive status, and the structure of microbial communities in scent glands. The influence of social grouping on taxonomic composition in gut microbiome was also shown in wild baboons (Tung et al. 2015). In the case of humans, we can imagine that microorganisms influence our partner preferences (Havlicek et al. 2008) as anaerobic microbes produce volatile fatty acids, alcohol and ketones, which are the active components of our scent. Similar studies lead to the theory that microbial communities can also be responsible for behavioral and ecological isolation, or post-mating reproductive isolation (cytoplasmic incompatibility, hybridization, etc.), all of which lead to speciation events (for a more detailed review, see Brucker and Borderstein 2012a).

In addition, the selective pressures are reciprocal: Brito et al. (2016) showed that human activities and behaviors provide selective pressures on mobile elements in the human microbiota. The variation in mobile elements between Fijian and North Americans reflects the variation of dietary habits. The evolutionary re-shaping of the bacterial gene pool has long been known: the resistance of bacteria, both pathogens and commensals, is rising due to the use of antibiotics. The previously mentioned example of how Japanese dietary habits influence the mobile elements in the microbiota of the Japanese population is just one of many positive examples. Summers et al. (1993) also reported that the use of mercury dental fillings has caused an increase in the mercury-resistant genes in the gene pool of human intestinal and oral microbiota. The tight relation of organism and symbiont is also considered to be a primary factor in the positive selection on immunity genes in *Drosophila* or apes which is stronger than in the rest of the genome (Brucker and Bordenstein 2012).

Criticism

However, the approach favoring the holobiont as a unit of selection has also brought some criticism. For example, Moran and Sloan (2015) argue that the intimacy of host-microbial relationships does not imply that there must be co-evolution between these two parts; the fact that microorganisms colonize a host and thus have an impact on its fitness does not mean they evolve in response to the host. Moran and Sloan provide the example of *Pseudomonas aeruginosa,* which can colonize lungs and is associated with disease, but the human lung is not the primary habitat of this soil bacterium. However, the holobiotic dimension involves commensal, mutualistic and also pathogenic relations. It is, as Theis et al. (2016) respond, the fact of the holobiont that does not necessarily mean cooperation only, as it involves cooperative as well as competitive selective systems.

Primacy of the level. Moran and Sloan claim (2015) that the adopting of a holobiotic level of selection means that this level is either sole or primary, but as Theis et al. (2016) emphasize, all literature devoted to multi-level selection holds for granted that multiple levels of selection can operate simultaneously.

Partner's fidelity. When treating the holobiont as a single unit of selection, we (almost notoriously) read that there should be a consistent transmission reflected in the stable genetic profile of both host and the microbiome. Even in Bordenstein and Theis (2015), we read: "The debatable and testable issue of the hologenome is whether nuclear genes and microbes are coinherited to a degree that evolution can operate on their interaction." In Moran and Sloan (2015), we read "coevolution appears more likely when lineages have a shared evolutionary history, as evidenced by a pattern of codiversification (also called cospeciation or parallel phylogenesis) where symbiont and host lineages show matching phylogenetic trees." They admit that codiversification does not necessarily mean coevolutionary history, but only the same geographic isolation and solely unidirectional selection (the bacterium can adaptively follow the changes in host but not vice versa). Similarly, Douglas and Werren (2016) see partner fidelity as a main condition for accepting holobiont as a unit of selection. In the case of horizontal transmission, such partner fidelity, i.e., stable association of host and symbiotic genotypes across generations, is very weak. To this we have to agree. However, especially on the species level of the microbiome, the holobiont is a fluctuating chimera (Chiu and Gilbert 2015), which by definition cannot be as stable as the host genome. Studies have also shown that human microbiome can be quite variable within an individual lifetime (David et al. 2014a), and although host genetics may play a role, it is more dependent on the environment than on genetics (Rothschild et al. 2018). Also, as David et al. (2014a, b) showed, the human microbiome is much more personalized and specific than the human genome. Human genomes overlap in over 99.5 per cent between individuals;

only 0.5 per cent of the genome is unique to an individual. Yet every individual has a specific and, to some extent, unique microbial signature, which can also become a target of evolution.

Theis et al. (2016) responded by arguing that it is similar to gene epistasis in the same genome, the interactions underlying a holobiont can be stable or transient and such instability of genome of host and microbiome transition can be the same as in genes that undergo recombination. The variation of holobionts is not provided only by genetic variation (mutation and recombination) but also by the acquisition of new strains from the environment; likewise, changes in abundance and composition of microbiome can generate new plasticity and novelty. We think that the possible plasticity of human microbiome can be an actual advantage when the host changes its environment, or the environment changes on the host. Sometimes, such changes are eventually reflected in the genome as well, but microbial communities provide rapid and plastic response.

Some bacterial species can also cover for each other functionally, so differences in the composition does not imply differences in function. We believe that coevolution can be defined by the reciprocal selective pressures of both parties, hosts and symbionts, but does not necessarily have to involve common evolutionary trees or even genetic stability on either side. *Holobionts are not easily convertible to hologenomes.* Also, many hosts, such as *Vibrio fischeri* or corals, rely on the stability of the environment—and arguments that favor a genetics approach, consistently and methodologically underestimate the factor of environment in holobiotic evolution.

Phylosymbiosis. All this is also connected to phylosymbiosis, a term introduced by Bordenstein and Theis (2015) to reflect the ability of symbionts to match host phylogeny (Bordenstein and Theis 2015, Theis et al. 2016). It represents the "concordance between a host phylogeny (evolutionary relationships) and microbial community dendrogram (ecological relationships) based on the degree of shared taxonomy and/or abundance of members of the community" (Theis et al. 2016). Moran and Sloan (2015) criticize this term because, in their opinion, the relationships of host phylogeny to symbiont community composition can be explained by various other phenomena such as codiversification (as in the case of organelles), and also more related species can be colonized by a similar set of bacteria. The symbiont does not need to have a shared or even similar history with their host. The response on this criticism (Theis et al. 2016) is that phylosymbiosis does not *a priori* imply coevolution, cospeciation, cocladogenesis, or codiversification because the latter vocabulary implies concordant splitting of new species from a common ancestral one. Phylosymbiosis avoids these assumptions because "it does not presume that microbial communities are stable or even vertically transmitted from generation to generation." (Brucker and Borderstein 2012a, b) Rather, "phylosymbiosis refers to a pattern in which changes in separate parts of the holobiont (host and microbiota) are related in a concordant manner. It is also a stepping stone from population genetics to community genetics." (Theis et al. 2016) They argue that phylosymbiosis is apparent in cases when the holobiont assembly is non stochastic (stochasticity implies that every microbe has the same chance to colonize the host), but deterministic. They provide controlled studies showing such patterns of phylosymbiosis in different species of *Nasonia* wasp and *Hydra* (Fraune and Bosch 2007, Brucker and Borderstein 2013).

Symbionts as abiotic environment. Moran and Sloan (2015) further argue that the host adapts to its symbionts in the same way that it adapts to the abiotic components of the environment, and that no selection on the symbiotic population need be involved. However, symbionts are not abiotic components of the environment. They are able to communicate with the host through shared pathways, and rely on historically grounded ways of communication (as in case with IgA in Donaldson et al. 2018). Communication and meaning attribution is a fundamentally new quality of life, which does not exist in abiotic nature. These qualities depend on experience and memory, which is the primary difference from the abiotic world, and these qualities are obviously illustrated by holobiotic interactions, where differing ways of communication are observable, leading to cooperation, cheating or parasitism.

11
Life as Interpretation

[T]he fundamental question to ask about any evolutionary change is 'What does this change maintain unchanged?' [...] *The survival of any population, animal or human, depends upon social interactions characterized by some minimum degree of orderliness, but orderliness in social systems depends, in turn, upon communication which must meet some minimum standard of reliability if the recipients of the message are to be willing to accept the information they receive as sufficiently reliable to depend upon.*

—R.A. Rappaport 2010, 7 and 15

All macrosocial forms of organization require the integration of activities by individuals and teams of individuals. [...] This imposes a local rationality on those working in them. It may not be very efficient. It may have grown up as the accumulation of practices that were once convenient but are now hard to justify. It may be swept aside tomorrow by the arrival of a new boss. [...] The point is that the current integrational structure—however 'good' or 'bad' it may be—is what imposes (some) limits on the 'meaning' of (some of) the actions of individuals operating within that framework, i.e., on the signs that are important to understand if you are working in that organization.

—R. Harris (2009, 162)

This book is built on four assumptions:

(1) There exists a 'Norm' established (negotiated) at the very beginnings of biosphere, and recognized since by all its inhabitants since.

(2) Particular 'lineages'[30] of life develop their own variants of living, with their own memory and experience of the past, yet—thanks to having recognized the Norm—all such variants at least partially overlap. The overlaps represent platforms for mutual understanding between lineages.

[30] Remember: lineages in quotes mean that, in fact, evolution is reticulate to a great extent.

(3) The memory and experience of a lineage represents a much larger pool of understanding, i.e., capacity for living, than can be 'lived' or 'digested' by a single individual during its lifetime; moreover, the pool is changeable—thanks to the contribution of each new generation of individuals.

(4) There is no 'basic level of description' of the living. As many authors, mentioned throughout this book, argued, life, its development, heredity, etc., cannot be described from the sole level of description (usually genetic one), but is primarily epigenetic, thus there are many sources of variation such as meaning attribution, symbiotic interactions, environmental interactions and contingency.

It follows that we acknowledge the internal forces that steer the evolution of particular lineages, forces that came into the existence only with life and are specific to it. Such forces operate on the background of the common 'Norm' that came to be installed (negotiated) right at the dawn of the biosphere (Chapter 1). In other words, Darwin's concept of Descent with Modification has a much richer fundament than random mutation, and also involves an interpretative, internal component based on memory and experience—of the lineage *but also* of the individual. Effects of the changing environment may be buffered in many ways; hence, natural selection may be plastically counterbalanced, modified, or steered in many different ways. We return to the point below.

To highlight our view: we first state our opposition to views that see in the evolutionary processes a mere continuation of ever more complicated network of chemical reactions, i.e., that see life as fully explicable from chemistry. This so called 'NASA definition of life' reads: "Life is a self-sustaining chemical system capable of Darwinian evolution."

Some authors even dispose with the concept of 'life'[31] as scientifically unsound and useless. Cleland and Chyba (2002) state: "Water is an example of what philosophers call 'natural kinds'. Natural kinds differ from non-natural kinds in that nature, rather than human convention, determines their membership. [...] Perhaps life is not a natural kind. If it is not, how we define it will forever remain a matter of no more than a linguistic choice. But if life is a natural kind, we need a theoretical framework for biology that will support a deeper understanding of life than can be provided by the features that we currently use to recognize it on Earth. [...] Indeed, it is hard to imagine what could better help us to understand the nature of life than the synthesis of candidate living systems in the laboratory or the discovery of independent extraterrestrial biologies." In our opinion, acknowledging the historical character of life and its capacity for self-interpretation may bring better results than searching the astral depths, even if you cannot create a 'natural kind' from historical processes.

We strongly maintain that the emergence of life was something brand new in our corner of the universe. To support our view, and to explain the difference, Chapter 2 gives a summary of prebiotic evolution. During this period, the syntheses of organic compounds were canalized into regular metabolic pathways, and new structures

[31] See, e.g., Editorial, Nature 447, 1031, 2007; doi.org/10.1038/4471031a.

emerged—first organomineral, later purely organic. Yet, this abiospheric work was purely 'physical', in the sense in which dissipative structures are taken as such.

No cellular closures existed yet, equipped with codes and symbolic communication, aimed internally or towards the others—thus no 'life'. The concept by Kauffman (2000) of sets of autonomous agents mutually communicating and co-constructing the biosphere, proceeding into the adjacent possible, is close to our understanding of what appeared at the singularity of life origin.

Chapter 4 is the real beginning of our version of the history of life: the establishment of a Norm of symbolic communication with the world. This inevitably leads to three possible views of living things: (1) as programmed machines executing their programs in accord with and depending on external and internal cues; or (2) as 'artificial intelligences' cumulating, sorting, putting into new contexts, etc., that which it can access and use. The problem with these two widely accepted views is that both need some kind of constructor. (3) This brings us to a third, biosemiotic, definition of life as a "complex dynamic system[32] born, endowed with semiosis, and with history" (Markoš 2014, 2016b). However short, the definition is redundant, only semiotic and historical beings can be born; semiosis requires history and cannot emerge *de novo*; and historical awareness ('historiography') requires an active, interpretative approach to one's history. Moreover, the definition conceals an implicit precondition that living beings are corporeal and cannot exist otherwise. This definition is not hindered by the fact that living beings delegate some functions to automatic, deterministic modes of functioning.

The Nine-Word Game

Life can be characterized as a complex dynamic system, so first we should define the uniqueness to life that allows it to be delimitated from other such systems. The table below of three triads of attributes of complex dynamic systems may help to orient the problem (after Markoš and Das 2016).

Table 11.1: Complex dynamic systems—properties.

1. Functionality	causality	code/program	semiosis
2. Coming into existence	creation	emergence 'from nothing'	birth
3. Development	steady state	deterministic evolution	history

1. Functioning. (i) *Deterministic causal systems* exhibit behavior explainable from the analysis of their parts, from causal relations between the parts, and from the energy gradients that drive their dynamics. Parts and relations are invariable, and the system behaves 'according to laws of physics'. Example: cellestial mechanics. Mechanical machines (see 2ii below) also belong to this category—except for the fact that they cannot be deduced from physics alone: they need a constructor.

[32] Under such a system, we understand an entity to be composed of a great number of small constituents—molecules or living beings—with an energy flow throughout the system and, possibly, also throughout its components.

(ii) *Programmable systems* must be equipped with codes and programs that determine their behavior. The code (program) is not deducible from physics, it must be introduced 'from outside'. Once present, however, it can be taken as another invariant, necessary for system description. Cues from the environment are received as *signals* (and defined for the system in advance), or as triggers for action (also defined in advance). Examples: computer, cashier's desks, typewriter, feedback control systems.

(iii) The behavior of *semiotic systems* depends on the use of memory, experience, and overall context to confront some present state of affairs. Such systems defy purely physical description; they work with *signs* that cannot be defined in advance (as in the relations in physics or in program execution): the interpretation of signs and attribution of *meaning* is always a matter of momentary contexts. For the realm of semiotics, the only semiotic system is a culture (or semiosphere), with human beings as a necessary component; biosemiotics also recognizes the whole biosphere and all living beings as such. Artificial intelligence may eventually belong here, but so far it is a mere extension, a construct of other semiotic beings—humans.

2. Coming to existence. (i) Complex dynamic systems may be produced (constructed) by an *external agency*. If we leave out of consideration some Deity capable of 'creation from nothing', such an agency is life. Here belong machines created by human civilization (programmable or not). Many functions of living bodies, e.g., organs or organelles, can be understood (modeled and studied) as such constructs.

(ii) There is another group of dynamic systems that repeatedly rise from 'nothing'—from a microscopic singularity amplified by steep external energy gradients. This is the case of *dissipative systems*, such as hurricanes, gyres, flames, stars and growing crystals. From the view of our analysis, it is important to stress the external character of the forming principle: components of the system ('molecules') are mutually indistinguishable, and indifferent (blind, passive) as to system's macroscopic parameters. The evolution of dissipative systems (as to their shape, extent, or development) cannot be exactly calculated in advance, or reconstructed from their extant state (consider the precarious quality of predictions of hurricane appearance, power and trajectory). Their multiple incidences, however, allow for statistical predictions of their behavior. As dissipative systems appear *de novo*, they bear no memory of previous occurrences of similar systems of the category.

(iii) Dynamic systems of the third category never arise *de novo* (as constructs or dissipative systems): they are *born* (or bud) from similar systems, as a continuation of a lineage. Such are living beings (species, populations) and their manifestations—cultures, life styles, overall appearances ('phenotype'), languages, religions; in short, all the ways how the things are done by living things according to tradition, heritage, fashion or necessity. Their evolution is dependent both on external forces (physical or not—e.g., natural selection or decision-making) as well as internal (heredity, culture, habits). The same holds for the components of such systems: they are unique, they are born and die, and represent an internal agency (and *not* indifferent particles), influencing the development of the system actively, and *not* merely by the fact of their presence.

3. Development in time. (i) A steady state is a circular behavior in a dynamic system, keeping it in the 'same' setting thanks to energy through-flow. The only 'development' of such a system is the wearing down of its parts. Examples: mechanical machines such as electrical motors, clockworks, metabolic cycles in cells.

(ii) Deterministic development is usually understood as the behavior of items belonging to physics or thermodynamics; very often, however, the concept is used in other realms where the development can be anticipated with some probability, such as ontogenesis, or the development of disease. Examples: ballistic curve, time-course of dissipative phenomena, chicken's development in the egg, or the flu. Components of the system may be, either completely or to a high extent, indifferent towards the system's behavior, due to the effect of strong external forces, such as canalized energy gradients in dissipative systems, cleavage and organogenesis in animal development, or an organism's overall reaction to a pathogen.

(iii) From the view of our investigation, *history* is of primary importance (see also Chapter 4 and Fig. 4.4, p. 59). As the living components (not molecules) of living ensembles are born, and the ensembles themselves develop from similar ensembles, both 'levels'—i.e., individual and biosphere—can draw on the memory and experience of countless generations of predecessors, as they plan, anticipate, and negotiate their future development. This 'drawing on' is, of course, prone to over-interpretation, misinterpretation, shifts of meaning, forgetting, etc. Evolution is isomorphic with history, as we may have known since Darwin: "All the foregoing rules and aids and difficulties in classification are explained [...] on the view that the natural system founded on descent with modification, that the characters which naturalists consider as showing true affinity between any two or more species, are those which have been inherited from a common parent, and, in so far, all true classification is genealogical; that community of descent is the hidden bond which naturalists have been unconsciously seeking, and not some unknown plan of creation, or the enunciation of general propositions, and the mere putting together and separating objects more or less alike." (Darwin 1859) This brings us back to the view of human cultures, a kind of living community, not dissimilar to species, ecological assemblages, etc.

Species as Cultures

The trajectories in Fig. 11.1 show the general pattern of evolution of live lineages, be they human cultures or biological communities.[33] In the short period of its emergence, such a system is very plastic, malleable by either external or internal forces, and subject to contingencies causing its canalization in different directions. With time, the system becomes more and more resilient to such contingent forces, and enters its elastic phase: it stretches when exposed to external pressure, but returns to its initial

[33] We stress that such evolution takes place across many generations, i.e., not in *ad hoc* assemblies such as cornfields, battlefields, POW camps, or traffic jams. Such formations may also show a long-lasting order, but one that is, to a great extent, imposed from the outside.

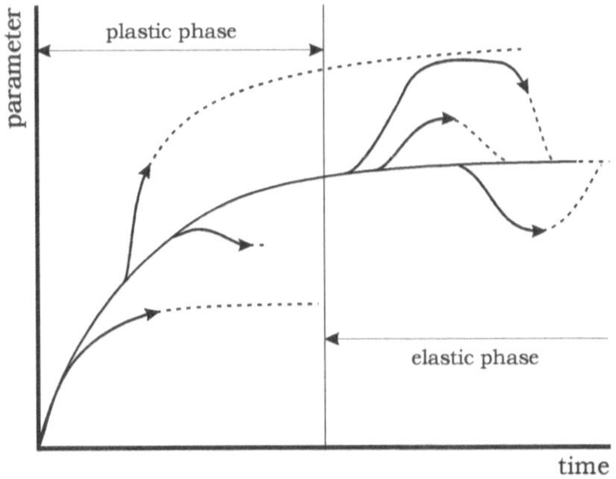

Fig. 11.1: Evolution of a lineage. In the plastic phase, its features may readily diverge induced by different external or internal cues. The elastic phase is resistant to forcing, and after its cessation the parameters return to previous values. The time axis is highly skewed: in fact, the plastic phase takes only about one per cent of the lineage's lifetime. (After Lotman (2009). Flegr (2008, 2010), and Rappaport (2010).)

configuration shortly after such forces cease. If such stresses exceed some threshold, the system may decay, collapse, or return (as a whole, or its fragments) to a plastic phase, and begin anew, with a new quality.

It is conspicuous that the curve is isomorphic for the evolution of cultures (Lotman 2009) and religions (Rappaport 2010), as well as life appearances (Flegr 2008, 2010); we have stressed the fact in previous publications (Markoš 2004, 2014, 2016, Švorcová 2016, Švorcová et al. 2018). It would appear that human cultural phenomena might represent but special cases of a general rule of the evolution of life, cases enabled by emergence of language. What follows expands the concept.

In an initial phase, the community and its members test 'what is allowed' under new conditions, and because the rules are still flexible, there is a greater abundance of possible trajectories (a view quite isomorphic with the 'phenotype first' approach, Chapter 5). In subsequent phases, the rules become 'frozen', and more stable heuristics (or simplexity rules—see Chapter 7) enter the relationships that create and maintain the community. Such heuristics lead to habits that save time in everyday tasks. The cultural *umwelt* (as discussed in Chapter 4, and as seen in Fig. 4.3, p. 58) takes shape as a single point inside a much broader potential store or responses in the community's memory and experience. This sort of evolution is not 'frozen' literally—the whole process allows the re-building of sophisticated ways of living, based on the minutiae built above the ground of such buffering. After all, this had been a common strategy ever since the dawn of life—the *whole* biosphere recognizes the 'Norm' established at that time. The Universal Sacred Postulates defined by Rappaport (2010; see Chapter 1 and the epigraph to this chapter) for religious communities may serve as a parable for much broader phenomena.

The idea of the analogy between three incommensurable authors—Flegr, Rappaport, and Lotman—points at the obvious fact that an individual (be it a cell or a multicellular organism) is born into a pre-existing community of its kin, who stipulate the ways how to interpret the world (Markoš and Das 2016). "Genetic script, experience, presuppositions, automatisms, behavioral patterns, etc. allow the new member of the community to behave along heuristic shortcuts. This saves time for explorative behavior, experimentation, and for teasing out possible deviations from the rules; all this is fully within the competence (and responsibility) of an individual—at its own risk, or its benefit. It follows that modeling a living being as nothing but a duality of genotype and phenotype is short of a third factor—belonging to a community observing such and such cultural habits." (Markoš 2016) We believe that cultural and biological evolution can be put under a common denominator. What unites them is the simultaneous duration of the society and alternation of its members, individuals that are not produced by it but *born into* it. The individuals are both indoctrinated—limited—by the existing rules, and also attempt to broaden their reach—expand their living—through their individual *idia phronesis* (art of acting or living) within given contexts. This fact—individually expressed—may influence day-to-day interpretation of the common rules by the other individuals in the society. But the interpretative capacity of species-cultures *always* greatly surpasses (as to their informational content) the genetic script stored within DNA (Markoš et al. 2013). Such may be the deepest meaning of processes recently described by sciences such as evo-devo and epigenetics.

We can broaden this view to include the interpretation of written, i.e., frozen, texts of sacred books, constitutions, or genes: "The US Constitution has always been praised as well-balanced basic law, a shining example for any democratic country—and needing only minor amendment over more than two centuries. Yet there were American communities in history that could easily accommodate slavery, racial segregation, the extermination of Indians, or McCarthism, without having any doubts about their being good and constitution-obeying citizens. In other words, written documents have many interpretative frames—and these can be thought of as being akin to phenotypes." (Markoš et al. 2009)

The view may be fully acceptable even for current versions of Neodarwinism: "Natural selection operates at the level of the individual, with the retention of mutations that benefit that individual. However, current understanding from theoretical ecology suggests that the most important level of organization for a species is the population. That is, communities are more important than the individual, at least in the present day, in maintaining ecological stability. If the processes that drive evolution operate at the level of an individual, but the success is determined at the population level, then evolutionary development can only be effective if individuals are able to interact with other individuals of the same species to ensure the success of the population." (Joint et al. 2007)

E. Danchin provides another example of such views: "Dispersers, for instance, bring not only their genes into their new population, but also their phenotype, which brings key information on the conditions that prevail outside of the population. They also bring their cultural habits (e.g., dialects), so that high immigration rates can lead to cultural meltdown in a single generation, which is equivalent to the loss of a genetic structuring. Such cultural meltdown should affect the inclusive heritability of a local population and, thus, its evolutionary dynamics." (Danchin 2013) Danchin argues that heritability and variation are only loosely correlated with genetic markers; the rest is a function of epigenetic factors such as environment, education, cultural habits and rituals, imprinting, cell (and body) structures, extracellular matrix, etc. (Fig. 11.2), i.e., roughly the themes of this book. The principal idea is, again, that living beings are born with zillions of such factors having already been negotiated before the fact of their birth (see gratuity, Chapter 7; Markoš et al. 2013). Yet interpretation, i.e., an attribution of *meaning* that functions as a rule to bind the community, that allows the community to function as one, is the job of the whole ensemble. The role of natural selection is that of top-level censorship over both individual and community doings.

In many Darwinian models, living beings can be compared to chess pieces placed in their (chessboard) environment and moved by external forces. Their 'niche' is given by strict rules imposed from outside and their evolution is likewise forced from outside. We feel that the parable should be broadened to games with living

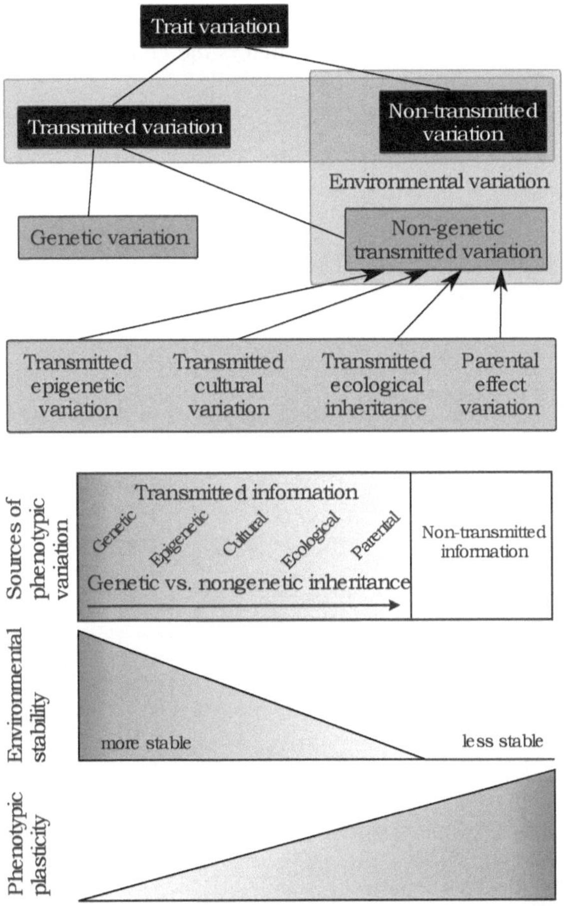

Fig. 11.2: Transmission of traits. **Above:** different forms of hereditary information (and its variation) participating on the appearance (phenotype) of an individual. Transmitted genetic variation (transfer of genes in a form of 'DNA avatar' from parents to offspring) is complemented by various forms of epigenetic, cultural, ecological, etc. endowment. **Below:** 'strength' of transmission of different kinds of traits. Genetic sources of phenotypic inheritance and/or variation are more reliable than other forms; they are also more resilient towards challenges from the environment. In contrast, phenotypic plasticity is copious in each generation, but its transmission is less reliable. (After Danchin (2013).)

players (such as ice hockey or basketball) to allow semiosis to enter the field. As we understand it, the game is without beginning and end; throughout the game, players are born, ushered into teams, and die out of them. The model of autonomous agents that create their biosphere (Kauffman 2000), and also the concept of being together (Heidegger 1995), comply with such views (see also Markoš et al. 2009). Our goal should be to properly digest the concept.

We are biologists, so we might easily be accused of the vitalist heresy. Yet we argue that models developed by the humanities may point towards a general theory of evolution valid for all life. What human cultures have in common with the biosphere is the semiotic character of individuals as well as communities, a character born of similar entities, which maintain the continuity of lineage. This allows, both individuals and communities, an interpretative approach to their history—rooted in memory and experience, yet also furthering their unique, individual, and creative approach to their genetic endowment, which is the creation of novelty and the fact of evolution.

References

Aagaard, K., Ma, J., Antoby, K.M., Ganu, R., Petrosino, J. and Versalovic, J. 2014. The placenta harbors a unique microbiome. Sci. Trans. Med. 6: 237ra65. Doi: 10.1126/scitranslmed.3008599.

Abedin, M. and King, N. 2008. The premetazoan ancestry of cadherins. Science 319: 946–8. Doi: 10.1126/science.1151084.

Abrescia, N.G., Bamford, D.H., Grimes, J.M. and Stuart, D.I. 2012. Structure unifies the viral universe. Annu. Rev. Biochem. 81: 795–822. Doi: 10.1146/annurev–biochem–060910–095130.

Adamala, K. and Szostak, J.W. 2013. Nonenzymatic template–directed RNA synthesis inside model protocells. Science 342: 1098–1100. Doi: 10.1126/science.1241888.

Adamala, K., Engelhart, A.E. and Szostak, J.W. 2015. Generation of functional RNAs from inactive oligonucleotide complexes by non-enzymatic primer extension. J. Am. Chem. Soc. 137: 483–489. Doi: 10.1021/ja511564d.

Adams, D.G. and Duggan, P.S. 2012. Signalling in cyanobacteria–plant symbioses. pp. 93–121. *In*: Perotto, S. and Baluška, F. [eds.]. Signaling and Communication in Plant Symbiosis. Springer, Berlin, Heidelberg. Doi: 10.1007/978-3-642-20966-6_5.

Adams, J.B., Johansen, L.J., Powell, L.D., Quig, D. and Rubin, R.A. 2011. Gastrointestinal flora and gastrointestinal status in children with autism—comparisons to typical children and correlation with autism severity. BMC Gastroenterol. 11: 22. Doi: 10.1186/1471-230X-11-22.

Agrawal, A.A., Laforsch, C. and Tollrian, R. 1999. Transgenerational induction of defences in animals and plants. Nature 401: 60–63. Doi: 0.1038/43425.

Ahnert, S.E., Marsh, J.A., Hernández, H., Robinson, C.V. and Teichmann, S.A. 2015. Principles of assembly reveal a periodic table of protein complexes. Science 350: aaa2245. Doi: 10.1126/science.aaa2245.

Aksoy, S. 2003. Symbiosis in tsetse. pp. 53–65. *In*: Bourtzis, K. and Miller, T.A. [eds.]. Insect Symbiosis. CRC Press, Boca Raton.

Allshire, R.C. and Madhani, H.D. 2018. Ten principles of heterochromatin formation and function. Nat. Rev. Mol. Cell Biol. 19: 229–244. Doi: 10.1038/nrm.2017.119.

Almeida, M., Pintacuda, G., Masui, O., Koseki, Y., Gdula, M., Cerase, A. et al. 2017. PCGF3/5-PRC1 initiates polycomb recruitment in X chromosome inactivation. Science 356: 1081–1084. Doi: 10.1126/science.aal2512.

Alvarez–Venegas, R. 2014. Bacterial SET domain proteins and their role in eukaryotic chromatin modification. Frontiers Genet. 5: 65. Doi: 0.3389/fgene.2014.00065.

Amato, A., Kooistra, W.H., Ghiron, J.H., Mann, D.G., Pröschold, T. and Montresor, M. 2007. Reproductive isolation among sympatric cryptic species in marine diatoms. Protist 158: 93–207. Doi: 10.1016/j.protis.2006.10.001.

Anway, M.D., Leathers, C. and Skinner, M.K. 2006. Endocrine disruptor vinclozolin induced epigenetic transgenerational adult-onset disease. Endocrinology 147: 5515–5523. Doi: 10.1210/en.2006-0640.

Ardissone, A.N., de la Cruz, D.M., Davis-Richardson, A.G., Rechcigl, K.T., Li, N., Drew, J.C. et al. 2014. Meconium microbiome analysis identifies bacteria correlated with premature birth. PLoS One 9: e90784. Doi: 10.1371/journal.pone.0090784.

Aristotle. 1972 [ca -340]. De Partibus Animalium I and De Generatione Animalium I. Translated with Notes by D.M. Balme. Oxford University Press, Oxford.

Arrieta, M.-C., Stiemsma, L.T., Dimitriu, P.A., Thorson, L., Russell, S., Yurist-Doutsch, S. et al. 2015. Early infancy microbial and metabolic alterations affect risk of childhood asthma. Sci. Transl. Med. 7: 307ra152. Doi: 10.1126/scitranslmed.aab2271.

Asano, Y., Hiramoto, T., Nishino, R., Aiba, Y., Kimura, T., Yoshihara, K. et al. 2012. Critical role of gut microbiota in the production of biologically active, free catecholamines in the gut lumen of mice. Am. J. Physiol. Gatrointest. Liver. Physiol. 303: G1288–95. Doi: 10.1152/ajpgi.00341.2012.

Auerbach, E. 1957[1946]. Mimesis. The Representation of Reality in Western Literature. Doubleday and Co., New York.

Azad, M.B., Konya, T., Persaud, R.R., Guttman, D.S., Chari, R.S., Field, C.J. et al. 2015. Impact of maternal intrapartum antibiotics, method of birth and breastfeeding on gut microbiota during the first year of life: a prospective cohort study. BJOG Int. J. Obstet. Gynaecol. 123: 983–993. Doi: 10.1111/1471-0528.13601.

Bäckhed, F., Roswall, J., Peng, Y., Feng, Q., Jia, H., Kovatcheva-Datchary, P. et al. 2015. Dynamics and stabilization of the human gut microbiome during the first year of life. Cell Host Microbe. 17: 690–703. Doi: 10.1016/j.chom.2015.04.004.

Baer, K.E. von. 1828. Über Entwickelungsgeschichte der Thiere: Beobachtung und Reflexion. Kornigsberg, Borntrager.

Baker, M. 2017. Lines of communication. Nature 549: 322–324. Doi: 10.1038/549322a.

Baldwin, M.J. 1896. A new factor in evolution. Amer. Naturalist. 30: 441–451.

Ball, S.G., Colleoni, C., Kadouche, D., Ducatez, M., Arias, M.C. and Tirtiaux, C. 2015. Toward an understanding of the function of Chlamydiales in plastid endosymbiosis. Biochim. Biophys. Acta 1847: 495–504. Doi: 10.1016/j.bbabio.2015.02.007.

Ballard, W.W. 1981. Morphogenetic movements and fate maps of vertebrates. Am. Zool. 21: 391–399.

Banani, S.F., Lee, H.O., Hyman, A.A. and Rosen, M.K. 2017. Biomolecular condensates: organizers of cellular biochemistry. Nat. Rev. Mol. Cell. Biol. 18: 285–298. Doi: 10.1038/nrm.2017.7.

Bandea, C. 2009. The origin and evolution of viruses as molecular organisms. Nat. Precedings hdl: 10101/npre.2009.3886.1.

Banerjee, S., Isaacman-Beck, J., Schneider, V.A. and Granato, M. 2013. A novel role for Lh3 dependent ECM modifications during neural crest cell migration in zebrafish. PLoS ONE 8.1: e54609. Doi: 10.1371/journal.pone.0054609.

Barbieri, M. 2003. The Organic Codes. An Introduction to Semantic Biology. Cambridge University Press, Cambridge, UK.

Barbieri, M. 2015. Code Biology. A New Science of Life. Springer.

Barge, L.M., Doloboff, I.J., Russell, M.J., VanderVelde, D., White, L.M., Stucky, G.D. et al. 2014. Pyrophosphate synthesis in iron mineral films and membranes simulating prebiotic submarine hydrothermal precipitates. Geochim. Cosmochim. Acta 128: 1–12. Doi: 10.1016/j.gca.2013.12.006.

Barrett, E., Ross, R.P., O'Toole, P.W., Fitzgelard, G.F. and Stanton, C. 2012. γ-Aminobutyric acid production by culturable bacteria from the human intestine. J. Appl. Microbiol. 113: 411–417. Doi: 10.1111/j.1365-2672.2012.05344.x.

Barton, E.S., White, D.W., Cathelyn, J.S., Brett-McClellan, K.A., Engle, M., Diamond, M.S. et al. 2007. Herpesvirus latency confers symbiotic protection from bacterial infection. Nature 447: 326–329. Doi: 10.1038/nature05762.

Beadle, G. and Tatum, E.L. 1941. Genetic control of biochemical reactions in neurospora. Proc. Natl. Acad. Sci. USA 27: 499–506.

Beatty, J. 1994. The proximate/ultimate distinction in the multiple careers of Ernst Mayr. Biology Philos. 9: 333–356.

Be'er, A., Zhang, H.P., Florin, E.L., Payne, S.M., Ben-Jacob, E. and Swinney, H.L. 2009. Deadly competition between sibling bacterial colonies. Proc. Natl. Acad. Sci. USA 106: 428–433. Doi: 10.1073/pnas.0811816106.

Be'er, A., Ariel, G., Kalisman, O., Helmanc, Y., Sirota-Madic, A., Zhang, H.P. et al. 2010. Lethal protein produced in response to competition between sibling bacterial colonies. Proc. Natl. Acad. Sci. USA 107: 6258–6263. Doi: 10.1073/Proc. Nat. Acad. Sci. USA .1001062107.

Behie, S.W., Zelisko, P.M. and Bidochka, M.J. 2012. Endophytic insect–parasitic fungi translocate nitrogen directly from insects to plants. Science 336: 1576–1577. Doi: 0.1126/science.1222289.

Beisson, J. and Sonneborn, T.M. 1965. Cytoplasmic inheritance of the organization of the cell cortex in *Paramecium aurelia*. Proc. Natl. Acad. Sci. USA 53: 275–282.

Bejerano, G., Pheasant, M., Makunin, I., Stephen, S., Kent, W.J., Mattick, J.S. et al. 2004. Ultraconserved elements in the human genome. Science 304: 1321–5. Doi: 10.1126/science.1098119.

Bell, E.A., Boehnke, P., Harrison, T.M. and Mao, W.L. 2015. Potentially biogenic carbon preserved in a 4.1 billion-year-old zircon. Proc. Nat. Acad. Sci. USA 112: 14518–14521. Doi: 10.1073/Proc. Nat. Acad. Sci. USA .1517557112.

Bell, P.J. 2009. The viral eukaryogenesis hypothesis a key role for viruses in the emergence of eukaryotes from a prokaryotic world environment. Ann. NY Acad. Sci. 1178: 91–105. Doi: 10.1111/j.1749–6632.2009.04994.x.

Bengtson, S., Sallstedt, T., Belivanova, V. and Whitehouse, M. 2017. Three-dimensional preservation of cellular and subcellular structures suggests 1.6 billion-year-old crown-group red algae. PLOS Biology 15: e2000735. Doi: 10.1371/journal.pbio.2000735.

Ben-Jacob, E., Becker, I., Shapira, Y. and Levine, H. 2004. Bacterial linguistic communication and social intelligence. Trends Microbiol. 12: 366–372. Doi: 10.1016/j.tim.2004.06.006.

Ben-Jacob, E. and Levine, H. 2005. Self-engineering capabilities of bacteria. J. R. Soc. Interface 3: 197–214. Doi: 10.1098/rsif.2005.0089.

Berendsen, R.L., Pieterse, C.M.J. and Bakker, P. 2012. The rhizosphere microbiome and plant health. Trends. Plant. Sci. 17: 478–486. Doi: 0.1016/j.tplants.2012.04.001.

Bergson, H. 2001[1913]. Time and free will. An essay on the immediate data of consciousness. Dover publications inc., New York.

Bergson, H. 2004[1912]. Matter and Memory. Dover Philosophical Classics.

Bergson, H. 2012[1946]. Creative mind: An introduction to metaphysics. Dover publications inc., New York.

Bernal, J.D. 1951. The physical basis of life. Routledge and Kegan Paul, London.

Berthoz, A. 2012. Simplexity. Simplifying Principles for a Complex World. Yale University Press.

Bhattacharya, D., Yoon, H.S. and Hackett, J.D. 2004. Photosynthetic eukaryotes unite: endosymbiosis connects the dots. BioEssays 26: 50–60. Doi: 10.1002/bies.10376.

Biggar, K.K. and Li, S.S.-C. 2015. Non-histone protein methylation as a regulator of cellular signalling and function. Nat. Rev. Mol. Cell. Biol. 16–17. Doi: 10.1038/nrm3915.

Biller, S.J., Schubotz, F., Roggensack, S.E., Thompson, A.W., Summons R.E. and Chisholm, S.W. 2014. Bacterial vesicles in marine ecosystems. Science 343: 183–186. Doi: 10.1126/science.1243457.

Bininda-Emonds, O.R., Jeffery, J.E. and Richardson, M.K. 2003. Inverting the hourglass: quantitative evidence against the phylotypic stage in vertebrate development. Proc. R. Soc. Lond. B 270: 341–6. Doi: 10.1098/rspb.2002.2242.

Bintu, L., Yong, J., Antebi, Y.E., McCue, K., Kazuki, Y., Uno, N. et al. 2016. Dynamics of epigenetic regulation at the single-cell level. Science 351: 720–724. Doi: 10.1126/science.aab2956.

Birey, F., andersen, J., Makinson, C.D., Islam, S., Wei, W., Huber, N. et al. 2017. Assembly of functionally integrated human forebrain spheroids. Nature 545: 54–59. Doi: 10.1038/nature22330.

Blain, J.C. and Szostak, J.W. 2014. Progress towards synthetic cells. Annu. Rev. Biochem. 83: 615–640. Doi: 10.1146/annurev–biochem–080411–124036.

Blöchl, E., Keller, M., Wachtershäuser, G. and Stetter, K.O. 1992. Reactions depending on iron sulfide and linking geochemistry with biochemistry. Proc. Nat. Acad. Sci. USA 89: 8117–8120.

Boddy, A.M., Fortunato, A., Wilson Sayres, M. and Aktipis, A. 2015. Fetal microchimerism and maternal health: a review and evolutionary analysis of cooperation and conflict beyond the womb. Bioessays 37: 1106–1118. Doi: 10.1002/bies.201500059.

Bohacek, J. and Mansuy, I.M. 2017. A guide to designing germline-dependent epigenetic inheritance experiments in mammals. Nat. Methods 14: 243–249. Doi: 10.1038/nmeth.4181.

Bokulich, N.A., Chung, J., Battaglia, T., Henderson, N., Jay, M., Li, H. et al. 2016. Antibiotics, birth mode, and diet shape microbiome maturation during early life. Sci. Transl. Med. 343: 343ra82. Doi: 10.1126/scitranslmed.aad7121.

Boles, B.R., Thoendel, M. and Singh, P.K. 2004. Self-generated diversity produces 'insurance effects' in biofilm communities. Proc. Natl. Acad. Sci. USA 101: 16630–16635. Doi: 10.1073/Proc. Nat. Acad. Sci. USA .0407460101.

Bonduriansky, R. and Day, T. 2018. Extended heredity. A New Understanding of Inheritance and Evolution. Princeton University Press, Princeton.

Booth, A. 2014. Symbiosis, selection, and individuality. Biol. Phil. 29: 657–673. Doi: 10.1007/s10539-014-9449.

Bordenstein, S.R., O'Hara, F.P. and Werren, J.H. 2001. *Wolbachia*-induced incompatibility precedes other hybrid incompatibilities in *Nasonia*. Nature 409: 707–710. Doi: 10.1038/35055543.

Bordenstein, S.R. and Theis, K.R. 2015. Host biology in light of the microbiome: ten principles of holobionts and hologenomes. PLoS Biol. 13: e1002226. Doi: 10.1371/journal.pbio.1002226.

Borgia, A., Borgia, M.B., Bugge, K., Kissling, V.M., Heidarsson, P.O., Fernandes, C.B. et al. 2018. Extreme disorder in an ultrahigh-affinity protein complex. Nature 555: 61–66. doi: 10.1038/nature25762.

Bothe, H., Tripp, J. and Zehr, J.P. 2010. Unicellular cyanobacteria with a new mode of life: the lack of photosynthetic oxygen evolution allows nitrogen fixation to proceed. Arch. Microbiol. 92: 783–790. Doi: 10.1007/s00203-010-0621-5.

Bourtzis, K. and Miller, T.A. [eds.]. 2003. Insect Symbiosis. I–III. CRC Press, Boca Raton.

Bouskra, D., Brézillon, C., Bérard, M., Werts, C., Varona, R., Boneca, I.G. et al. 2008. Lymphoid tissue genesis induced by commensals through NOD1 regulates intestinal homeostasis. Nature 456: 507–510. Doi: 10.1038/nature07450.

Boveri, T. 1910. Die Potenzen der *Ascaris*-Blastomeren bei abgeänderter Furchung, zugleich ein Beitrag zur Frage qualitativ-ungleicher Chromosomen-Teilung. Festschrift für Richard Hertwig, Vol. 3. Gustav Fischer Jena.

Bowler, P.J. 1983. The eclipse of Darwinism: Anti-Darwinian evolution theories in the decades around 1900. Johns Hopkins University Press, London.

Boyle, A.P., Araya, C.L., Brdlik, C., Cayting, P., Cheng, C., Cheng, Y. et al. 2014. Comparative analysis of regulatory information and circuits across distant species. Nature 512: 453–6. Doi: 10.1038/nature13668.

Braendle, C., Miura, T., Bickel, R., Shingleton, A.W., Kambhampati, S. and Stern, D.L. 2003. Developmental origin and evolution of bacteriocytes in the aphid–buchnera symbiosis. PLoS Biol. 1: E21. Doi: 10.1371/journal.pbio.0000021.

Braithwaite, V.A. and Odling-Smee, L. 1999. The paradox of the stickleback: different yet the same. Trends Ecol. Evol. 14: 460–461.

Branciamore, S., Gallori, E., Szathmáry, E. and Czárán, T. 2009. The origin of life: Chemical evolution of a metabolic system in a mineral honeycomb? J. Mol. Evol. 69: 458–469. Doi: 10.1007/s00239–009–9278–6.

Bravo, J.A., Forsythe, P., Chew, M.V., Escaravage, E., Savignac, H.M., Dinan, T.G. et al. 2011. Ingestion of *Lactobacillus* strain regulates emotional behavior and central GABA receptor expression in a mouse via the vagus nerve. Proc. Natl. Acad. Sci. USA 108: 16050–16055. Doi: 10.1073/Proc. Nat. Acad. Sci. USA .1102999108.

Breslow, R. and Cheng, Z.-L. 2009. On the origin of terrestrial homochirality for nucleosides and amino acids. Proc. Nat. Acad. Sci. USA 106: 9144–9146. Doi: 10.1073/Proc. Nat. Acad. Sci. USA .0904350106 .

Bright, M. and Burghelesi, S. 2010. A complex journey: transmission of microbial symbionts. Nature Revs. Microbiol. 8: 218–230. Doi:10.1038/nrmicro2262.

Brink, R. 1973. Paramutation. Annu. Rev. Genet. 7: 29–152.

Brinkman, F.S.L., Blanchard, J.L., Cherkasov, A., Av-Gay, Y., Brunham, R.C., Fernandez, R.C. et al. 2002. Evidence that plant-like genes in *Chlamydia* species reflect an ancestral relationship between Chlamydiaceae, cyanobacteria, and the chloroplast. Genome Res. 12: 1159–1167. Doi: 10.1101/gr.341802.

Brito, I.L., Yilmaz, S., Huang, K., Xu, L., Jupiter, S.D., Jenkins, A.P. et al. 2016. Mobile genes in the human microbiome are structured from global to individual scales. Nature 535: 435–439. Doi: 10.1038/nature18927.

Brockhurst, M.A., Buckling, A., Racey, D. and Gardner, A. 2008. Resource supply and the evolution of public-goods cooperation in bacteria. BMC Biol. 6: 20. Doi: 10.1186/1741-7007-6-20.

Brockman, J. 1995. The third culture. Beyond the scientific revolution. Simon and Schuster, New York.

Bronner, M.E. and Simões-Costa, M. 2016. The neural crest migrating into the 21st century. Curr. Top. Dev. Biol. 116: 115–134. Doi: 0.1016/bs.ctdb.2015.12.003.

Browne, H.P., Forster, S.C., Anonye, B.O., Kumar, N., Neville, B.A., Stares, M.D. et al. 2016. Culturing of 'unculturable' human microbiota reveals novel taxa and extensive sporulation. Nature 533: 543–546. Doi: 10.1038/nature17645.

Brucker, R.M. and Bordenstein, S.R. 2012a. Speciation by symbiosis. Trends Ecol. Evol. 27: 443–451. Doi: 10.1016/j.tree.2012.03.011.

Brucker, R.M. and Bordenstein, S.R. 2012b. The roles of host evolutionary relationships (genus: *Nasonia*) and the development in structuring microbial communities. Evolution 66: 349–62. Doi: 10.1111/j.1558-5646.2011.01454.x.

Brucker, R.M. and Bordenstein, S.R. 2013. The hologenomic basis of speciation: gut bacteria cause hybrid lethality in the genus *Nasonia*. Science 341: 667–669. Doi: org/10.1126/science.1240659.

Brun, Y.V. and Shimkets, L.J. [eds.]. 2000. Bacterial Development. ASM Press, Washington, DC.

Brüssow, H. 2009. The not so universal tree of life or the place of viruses in the living worlds. Phil. Trans. R. Soc. B 364: 2663–2274. Doi: 10.1098/rstb.2009.0036.

Brykczynska, U., Hisano, M., Erkek, S., Ramos, L., Oakeley, E.J., Roloff, T.C. et al. 2010. Repressive and active histone methylation mark distinct promoters in human and mouse spermatozoa. Nat. Struct. Mol. Biol. 17: 679–687. Doi: 10.1038/nsmb.1821.

Buckeridge, M.S. 2018. The evolution of the glycomic codes of extracellular matrices. BioSystems 164: 112–120. Doi: 10.1016/j.biosystems.2017.10.003.

Buffington, S.A., Di Prisco, G.V., Auchtung, T.A., Ajami, N.J., Petrosino, J.F. and Costa-Mattioli, M. 2016. Microbial reconstitution reverses maternal diet-induced social and synaptic deficit in offspring. Cell 165: 1762–1775. Doi: 10.1016/j.cell.2016.06.001.

Burghelesi, S. 2016. Bacterial cell biology outside the streetlight. Environ. Microbiol. 18: 2305–2318. Doi: 10.1111/1462-2920.13406.

Burton, A. and Torres–Padilla, M.-E. 2014. Chromatin dynamics in the regulation of cell fate allocation during early embryogenesis. Nature Revs. Mol. Cell. Biol. 15: 722–734. Doi: 10.1038/nrm3885.

Butler, S. 2009[1878]. Life and Habit. Cambridge University Press, Cambridge, UK.

Butler, S. 2014[1910]. Unconscious Memory. The Project Gutenberg eBook.

Cairns–Smith, A.G. 1982. Genetic Takeover and the Mineral Origins of Life. Cambridge University Press, Cambridge.

Cairns–Smith, A.G. 1985. Seven Clues to the Origin of Life. Cambridge University Press, Cambridge.

Cameron, R.A., Oetersin, K.J. and Davidson, E.H. 1998. Developmental gene regulation and the evolution of large animal body plans. Amer. Zool. 38: 609–620.

Caporaso, J.G., Lauber, C.L., Costello, E.K., Berg-Lyons, D., Gonzalez, A., Stombaugh, J. et al. 2011. Moving pictures of the human microbiome. Genome Biol. 12: R50. Doi: 10.1186/gb-2011-12-5-r50.

Carroll, S.B. 2005. Endless forms most beautiful. The new science of evo devo and the making of animal kingdom. WW Norton and Co., London.

Carroll, S.B., Grenier, J.K. and Weatherbee, S.D. 2006. From DNA to diversity. Molecular genetics and the evolution of animal design. Blackwell Science, Maden.

Čepl, J., Blahůšková, A., Cvrčková, F. and Markoš, A. 2014. Ammonia produced by bacterial colonies promotes growth of ampicillin sensitive *Serratia* sp. by means of antibiotic inactivation. FEMS Microbiol. Lett. 354: 126–32. Doi: 10.1111/1574-6968.12442.

Čepl, J., Scholtz, V. and Scholtzová, J. 2015. The fitness change and the diversity maintenance in the growing mixed colony of two *Serratia rubidaea* clones. Arch. Microbiol. 2015: 301–6. Doi: 10.1007/s00203-015-1177-1.

Čepl, J., Blahůšková, A., Neubauer, Z. and Markoš, A. 2016. Variations and heredity in bacterial colonies. Commun. Integrative Biol. 9: e1261228. Doi: 10.1080/19420889.2016.1261228.

Čepl, J.J., Pátková, I., Blahůšková, A., Cvrčková, F. and Markoš, A. 2010. Patterning of mutually interacting bacterial bodies: close contacts and airborne signals. BMC Microbiol. 10: 39. Doi: 10.1186/1471-2180-10-139.

Chaikeeratisak, V., Nguyen, K., Khanna, K., Brilot, A.F., Erb, M.L. and Coker, J.K. et al. 2017. Assembly of a nucleus-like structure during viral replication in bacteria. Science 355: 194–197. DOI: 10.1126/science.aal2130.

Chamberlain, N.L., Hill, R., Kapan, D.D., Gilbert, L.E. and Kronforst, M.R. 2009. Polymorphic butterfly reveals the missing link in ecological speciation. Science 326: 847–850. 10.1126/science.1179141.

Chen, L., Xiao, S., Pang, K., Zhou, C. and Yuan, X. 2014. Cell differentiation and germ–soma separation in Ediacaran animal embryo-like fossils. Nature 516: 238–241. Doi: 10.1038/nature13766.

Chen, Q., Yan, M., Cao, Z., Li, X., Zhang, Y., Shi, J. et al. 2016. Sperm tsRNAs contribute to intergenerational inheritance of an acquired metabolic disorder. Science 351: 397–400. Doi: 10.1126/science.aad7977.

Chen, T. and Dent, S.Y.R. 2014. Chromatin modifiers and remodellers: regulators of cellular differentiation. Nature Revs. Genet. 15: 93–106. Doi: 10.1038/nrg3607.

Chen, W., Koide, R.T., Adams, T.S., DeForest, J.L., Cheng, L. and Eissenstat, D.M. 2016. Root morphology and mycorrhizal symbioses together shape nutrient foraging strategies of temperate trees. Proc. Natl. Acad. Sci. USA 113: 8741–6. Doi: 10.1073/Proc. Nat. Acad. Sci. USA .1601006113.

Cheng, X., Hui, J.H., Lee, Y.Y., Wan Law, P.T. and Kwan, H.S. 2015. A 'developmental hourglass' in fungi. Mol. Biol. Evol. 32: 1556–66. Doi: 10.1093/molbev/msv047.

Chiu, L. and Gilbert, S.F. 2015. The Birth of the holobiont: multi-species birthing through mutual scaffolding and niche construction. Biosemiotics 8: 191–210. Doi: 10.1007/s12304-015-9232-5.

Cho, H.J., Jönsson, H., Campbell, K., Melke, P., Williams, J.W., Jedynak, B. et al. 2007. Self-organization in high-density bacterial colonies: efficient crowd control. PLoS Biol. 5: e302. Doi: 10.1371/journal.pbio.0050302.

Cho, I., Yamanishi, S., Cox, L., Methé, B.A., Zavadil, J., Gao, Z. et al. 2012. Antibiotics in early life alter the murine colonic microbiome and adiposity. Nature 488: 621–626. Doi: 10.1038/nature11400.

Chou, S., Daugherty, M.D., Peterson, S.B., Biboy, J., Yang, Y., Jutras, B.L. et al. 2015. Transferred interbacterial antagonism genes augment eukaryotic innate immune function. Nature 518: 98–101. Doi: 10.1038/nature13965.

Chuang, J.H. and Li, H. 2004. Functional bias and spatial organization of genes in mutational hot and cold regions in the human genome. PLoS Biol. 2: E29. Doi: 10.1371/journal.pbio.0020029.

Chuong, E.B. 2013. Retroviruses facilitate the rapid evolution of the mammalian placenta. Bioessays 35: 853–861. Doi: 10.1002/bies.201300059.

Chuong, E.B., Elde, N.C. and Feschotte, C. 2016. Regulatory evolution of innate immunity through co-option of endogenous retroviruses. Science 351: 1083–1087. Doi: 10.1126/science.aad5497.

Clarke, K.J., Finlay, B.J., Esteban, G., Guhl, B.E. and Embley, T.M. 1993. *Cyclidium porcatum* n. sp.: a free-living anaerobic scuticociliate containing a stable complex of hydrogenosomes, eubacteria and archaeobacteria. Eur. J. Protistol. 29: 262–70. Doi: 10.1016/S0932-4739(11)80281-6.

Cleland, C.E. and Chyba, C.F. 2002. Defining 'life'. Orig. Life Evol. Biosphere 32: 387–393.

Cohen, L.J., Esterhazy, D., Kim, S.H., Lemetre, C., Aguilar, R.R., Gordon, E.A. et al. 2017. Commensal bacteria make GPCR ligands that mimic human signalling molecules. Nature 549: 48–53. Doi: 0.1038/nature23874.

Coleman, R.T. and Struhl, G. 2017. Causal role for inheritance of H3K27me3 in maintaining the OFF state of a Drosophila HOX gene. Science 356: eaai8236. Doi: 10.1126/science.aai8236.

Collado, M.C., Rautava, S., Aakko, J., Isolauri, E. and Salminen, S. 2016. Human gut colonization may be initiated *in utero* by distinct microbial communities in the placenta and amniotic fluid. Sci. Rep. 6: 23129. Doi: 10.1038/srep23129.

Collantes-Fernandez, E., Arrighi, R.B., Alvarez-García, G., Weidner, J.M., Regidor-Cerrillo, J., Boothroyd, J.C. et al. 2012. Infected dendritic cells facilitate systemic dissemination and transplacental passage of the obligate intracellular parasite *Neospora caninum* in mice. PLoS ONE 7: e32123. 10.1371/journal.pone.0032123.

Conway Morris, S. 2003. Life's Solution: Inevitable Humans in a Lonely Universe. Cambridge University Press, Cambridge.

Correns, C.G. 1900. Mendels Regel über das Verhalten der Nachkommenschaft der Rassenbastarde. Ber. Deutsch. Bot. Ges. 18: 158–168.

Costanzo, M., Van der Sluis, B., Koch, E.N., Baryshnikova, A., Pons, C., Tan, G. et al. 2016. A global genetic interaction network maps a wiring diagram of cellular function. Science 353: aafl420. Doi: 10.1126/science.aafl420.

Cowley, S.J. 2016. Changing the *idea* of language: Nigel love's perspective. Language Sci. 61: 43–55. Doi: 10.1016/j.langsci.2016.09.008.

Cox, L.M., Yamanishi, S., Sohn, J., Alekseyenko, A.V., Leung, J.M., Cho, I. et al. 2014. Altering the intestinal microbiota during a critical development window has lasting metabolic consequences. Cell 158: 705–721. Doi: 10.1016/j.cell.2014.05.052.

Coyte, K., Schluter, J. and Foster, K.R. 2015. The ecology of the microbiome: Networks, competition, and stability. Science 350: 663–666. 10.1126/science.aad2602.

Crick, F.H.C. 1958. On protein synthesis. pp. 138–163. *In*: Sanders, F.K. [ed.]. Symp. Soc. Exp. Biol. XII: The Biological Replication of Macromolecules. Cambridge University Press.

Crick, F.H.C. 1963. On the genetic code. Science 139: 461–464.

Crick, F. 1970. Central dogma of molecular biology. Nature 227: 561–3.

Crisp, A., Boschetti, C., Perry, M., Tunnacliffe, A. and Micklem, G. 2015. Expression of multiple horizontally acquired genes is a hallmark of both vertebrate and invertebrate genomes. Genome Biol. 16: 50. Doi: 10.1186/s13059–015–0607–3.

Cronin, J. and Reisse, J. 2005. Chirality and the origin of homochirality. pp. 473–515. *In*: Gargaud, M., Barbier, B., Martin, H. and Reisse, J. [eds.]. Lectures in Astrobiology, Vol. 1. Springer, Berlin.

Cropley, J.E., Suter, C.M., Beckman, K.B. and Martin, D.I.K. 2006. Germ-line epigenetic modification of the murine Avy allele by nutritional supplementation. PNAS 103: 17308–17312. Doi: 10.1073/pnas.0607090103.

Curtis, B.A., Tanifuji, G., Burki, F., Gruber, A., Irimia, M., Maruyama, S. et al. 2012. Algal genomes reveal evolutionary mosaicism and the fate of nucleomorphs. Nature 492: 59–65. Doi: 10.1038/nature11681.

Cvrčková, F. and Markoš, A. 2005. Beyond bioinformatics: Can similarity be measured in the digital world? J. Biosemiotics 1: 93–112.

Danchin, E. 2013. Avatars of information: towards an inclusive evolutionary synthesis. Trends. Ecol. Evol. 28: 351–358. Doi: 10.1016/j.tree.2013.02.010.

Danchin, E. and Pocheville, A. 2014. Inheritance is where physiology meets evolution. J. Physiol. 592: 2307–2317. Doi: 10.1113/jphysiol.2014.272096.

Dann, G.P., Liszczak, G.P., Bagert, J.D., Müller, M.M., Nguyen, U.T.T., Wojcik, F. et al. 2017. ISWI chromatin remodellers sense nucleosome modifications to determine substrate preference. Nature 548: 607–611. Doi: 10.1038/nature23671.

Danovaro, R., Dell'Anno, A., Corinaldesi, C., Magagnini, M., Noble, R., Tamburini, C. et al. 2008. Major viral impact on the functioning of benthic deep-sea ecosystems. Nature 454: 1084–1087. Doi: 10.1038/nature07268.

Darwin, C. 1859. The origin of species by means of natural selection. John Murray, London.

Darwin, C. 1868. The variation of animals and plants under domestication. London.

Darwin, C. 1876. The origin of species by means of natural selection, 6th ed. John Murray, London.

Davankov, V.A. 2009. Inherent hoomochirality of primary particles and meteorite impacts as possibles of prebiotic molecular chirality. Rus. J. Phys. Chem. 83: 1247–1256. Doi: 10.1134/S0036024409080019.

David, L.A. and Alm, E.J. 2011. Rapid evolutionary innovation during an Archaean genetic expansion. Nature 469: 93–96. Doi: 10.1038/nature09649.

David, L.A., Materna, A.C., Friedman, J., Campos-Baptista, M.I., Blackburn, M.C., Perrotta, A. et al. 2014a. Host lifestyle affects human microbiota on daily timescales. Genome Biol. 15: R89. Doi: 10.1186/gb-2014-15-7-r89.

David, L.A., Maurice, C.F., Carmody, R.N., Gootenberg, D.B., Button, J.E., Wolfe, B.E. et al. 2014b. Diet rapidly and reproducibly alters the human gut microbiome. Nature 505: 559–563. Doi: 10.1038/nature12820.

Davidson, E.H., Peterson, K.J. and Cameron, R.A. 1995. Origin of Bilaterian body plans: evolution of developmental regulatory mechanisms. Science 270: 1319–1325. Doi: 10.1126/science.270.5240.1319.

Davidson, E.H. 2006. The Regulatory Genome. Gene Regulatory Networks in Development and Evolution. Academic Press, Amsterdam.

Dawkins, R. 1976. The Selfish Gene. Oxford University Press, Oxford.

Dawkins, R. 1982. The Extended Phenotype. Oxford University Press, Oxford.

Dawkins, R. 1995. River out of Eden. Basic, New York.

de Duve, C. 1991. Blueprint for a cell: the nature and origin of life. Carolina Biological Supply Co.

De la Maza, L.M. and Carter, B.J. 1981. Inhibition of adenovirus oncogenicity in hamsters by adeno-associated virus DNA. J. Nat. Cancer Inst. 67: 1323–1326.

de Vries, H. 1900. Sur la loi de disjonction des hybrides. Compt. Rendus. Acad. Sci. Paris 130: 845–847.

de Vrieze, A., de Groot, P.F., Kootte, R.S., Knaapen, M., van Nood, E. and Nieuwdorp, M. 2013. Fecal transplant: a safe and sustainable clinical therapy for restoring intestinal microbial balance in human disease? Best. Pract. Res. Clin. Gastroenterol. 27: 127–137. Doi: 10.1016/j.bpg.2013.03.003.

Dekker, J., Belmont, A.S., Guttman, M., Leshyk, V.O., Lis, J.T., Lomvardas, S. et al. 2017. The 4D nucleome project. Nature 549: 219–226. Doi: 10.1038/nature23884.

Deleuze, G. 1994[1968]. Difference and repetition. Columbia University Press, New York.

den Besten, G., van Eunen, K., Groen, A.K., Venema, K., Reijngoud, D.J. and Bakker, B.M. 2013. The role of short-chain fatty acids in teh interplay between diet, gut microbiota, and host energy metabolism. J. Lipid. Res. 54: 2325–40. Doi: 10.1194/jlr.R036012.

Denison, R.F. and Kiers, E.T. 2011. Life histories of symbiotic rhizobia and mycorrhizal fungi. Curr. Biol. 21: R775–R785. Doi: 10.1016/j.cub.2011.06.018.

Denton, M. and Marshall, C. 2001. Laws of form revisited. Nature 410: 417. Doi: 10.1038/35068645.

Denton, M.J., Marshall, C.J. and Ledge, J. 2002. Protein folds as platonic forms: New support for the pre-Darwinian conception of evolution by natural law. J. Theor. Biol. 219: 325–342.

Depew, D.J. and Weber, B.H. 1996. Darwinism Evolving. Systems Dynamics and the Genealogy of Natural Selection. MIT Press, Cambridge, Ma.

Desai, M.M. and Walczak, A.M. 2015. Flexible gene pools. Science 348: 977–978. Doi: 10.1126/science.aab3957.

Deschamps, P. 2015. Primary endosymbiosis: have cyanobacteria and Chlamydiae ever been roommates? Acta Soc. Bot. Pol. 83: 291–302. Doi: 10.5586/asbp.2014.048.

Deterding, K., Tegtmeyer, B., Cornberg, M., Hadem, J., Potthoff, A., Böker, K.H. et al. 2006. Hepatitis A virus infection suppresses hepatitis C virus replication and may lead to clearance of HCV. J. Hepatol. 45: 770–778. Doi: 10.1016/j.jhep.2006.07.023.

Dey, G., Thattai, M. and Baum, B. 2016. On the archaeal origins of eukaryotes and the challenges of inferring phenotype from genotype. Trends Cell. Biol. 26: 476–485. Doi: 10.1016/j.tcb.2016.03.009.

Diaz, B.G. and Resssler, K.J. 2014. Parental olfactory experience influences behavior and neural structure in subsequent generations. Nat. Neurosci. 17: 89–96. Doi: 10.1038/nn.3594.

Ding, Y., Tang, Y., Kwok, C.K., Zhang, Y., Bevilacqua, P.C. and Assmann, S.M. 2014. *In vivo* genome-wide profiling of RNA secondary structure reveals novel regulatory features. Nature 505: 696–700. Doi: 10.1038/nature12756.

Dittmer, T.A. and Misteli, T. 2011. The lamin protein family. Genome. Biol. 12: 222. Doi: 10.1186/gb-2011-12-5-222.

Docampo, R. and Moreno, S.N.J. 2011. Acidocalcisomes. Cell Calcium 50: 113–119. Doi: 10.1016/j.ceca.2011.05.012.

Dodd, M.S., Papineau, D., Grenne, T., Slack, J.F., Rittner, M., Pirajno, F. et al. 2017. Evidence for early life in Earth's oldest hydrothermal vent precipitates. Nature 543: 60–64. Doi: 10.1038/nature21377.

Domazet-Lošo, T. and Tautz, D. 2010. A phylogenetically based transcriptome age index mirrors ontogenetic divergence patterns. Nature 468: 815–818. Doi: 10.1038/nature09632.

Dominguez-Bello, M.G., Costello, E.K., Contreras, M., Magris, M., Hidalgo, G., Fierer, N. et al. 2010. Delivery mode shapes the acquisition and structure of the initial microbiota across multiple body habitats in newborns. Proc. Nat. Acad. Sci. USA 107: 11971–11975. Doi: 10.1073/pnas.1002601107.

Donaldson, G.P., Lee, S.M. and Mazmanian, S.K. 2016. Gut biogeography of the bacterial microbiota. Nature Revs. Microbiol. 14: 20–32. Doi: 10.1038/nrmicro3552.

Donaldson, G.P., Ladinsky, M.S., Yu, K.B., Sanders, J.G., Yoo, B.B., Chou, W.-C. et al. 2018. Gut microbiota utilize immunoglobulin A for mucosal colonization. Science 360: 795–800. Doi: 10.1126/science.aaq0926.

Douglas, A.E. 2010. The Symbiotic Habit. Princenton University Press.

Douglas, A.E. 2014. The molecular basis of bacterial-insect symbiosis. J. Mol. Biol. 426: 3830–3837. Doi: 10.1016/j.jmb.2014.04.005.

Douglas, A.E. 2015. Multiorganismal insects: Diversity and function of resident microorganisms. Annu. Rev. Entomol. 60: 17–34. Doi: 10.1146/annurev-ento-010814-020822.

Douglas, A.E. and Werren, J.H. 2016. Holes in the hologenome: why host-microbe symbioses are not holobionts. mBio 7: e02099-15. Doi:10.1128/mBio.02099-15.

Drew, K., Lee, C., Huizar, R.L., Tu, F., Borgeson, B., McWhite, C.D. et al. 2017. Integration of over 9,000 mass spectrometry experiments builds a global map of human protein complexes. Mol. Syst. Biol. 8: 13: 932. Doi: 10.15252/msb.20167490.

Driesch, H. 1914. The history and theory of vitalism. Macmillan, London.

Drobner, E., Huber, H., Wächtershäuser, G., Rose, D. and Stetter, K.O. 1990. Pyrite formation linked with hydrogen evolution under anaerobic conditions. Nature 346: 742–744. Doi: 10.1038/346742a0.

Drost, H.G., Bellstädt, J., Ó'Maoiléidigh, D.S., Silva, A.T., Gabel, A., Weinholdt, C. et al. 2016. Post-embryonic hourglass patterns mark ontogenetic transitions in plant development. Mol. Biol. Evol. 33: 1158–63. Doi: 10.1093/molbev/msw039.

Drost, H.G., Janitza, P., Grosse, I. and Quint, M. 2017. Cross-kingdom comparison of the developmental hourglass. Curr. Opin. Genet. Dev. 45: 69–75. Doi: 10.1016/j.gde.2017.03.003.

Du, Z., Zheng, H., Huang, B., Ma, R., Wu, J., Zhang, X. et al. 2017. Allelic182 reprogramming of 3D chromatin architecture during early mammalian development. Nature 547: 232–235. Doi: 10.1038/nature23263.

Duboule, D. 1994. Temporal colinearity and the phylogenetic progression: a basis for the stability of a vertebrate bauplan and the evolution of morphologies through heterochrony. Dev. Suppl. 135–142.

Eco, U. 1994a. Six Walks in the Fictional Woods. Harvard University Press, Cambridge, MA.

Eco, U. 1994b. The Limits of Interpretation. Indiana University Press, Bloomington.

Eigen, M. and Winkler-Oswatitsch, R. 1992. Steps Towards Life. A Perspective on Evolution. Oxford University Press, Oxford.

Eigen, M. 2013. From Strange Simplicity to Complex Familiarity. Oxford University Press, Oxford.

Eisenstein, M. 2016. Microbiome: Bacterial broadband. Nature 533: S104–S106. Doi: 10.1038/533S104a.

Embley, T.M. and Finlay, B.J. 1993. Systematic and morphological diversity of endosymbiotic methanogens in anaerobic ciliates. Anton. van Leeuw. 64: 261–271. Doi: 10.1007/BF00873086.

Engel, P. and Moran, N.A. 2013. The gut microbiota of insects—diversity in structure and function. FEMS Microbiol. Rev. 37: 699–735. Doi: 10.1111/1574-6976.12025.

Engels, F. 1972 [from manuscripts dated 1873–1882]. Dialectics of Nature. Progress Publishers, Moscow.

Entrevan, M., Schuettengruber, B. and Cavalli, G. 2016. Regulation of genome architecture and function by polycomb proteins. Trends Cell. Biol. 26: 11–525. Doi: 10.1016/j.tcb.2016.04.009.

Erez, Z., Steinberger-Levy, I., Shamir, M., Doron, S., Stokar-Avihail, A., Peleg, Y. et al. 2017. Communication between viruses guides lysis-lysogeny decisions. Nature 541: 488–493. Doi: 10.1038/nature21049.

Feagin, J.E., Abraham, J.M. and Stuart, K. 1988. Extensive editing of the cytochrome c oxidase III transcripts in *Trypanosoma brucei*. Cell 53: 413–422. Doi: 10.1016/0092-867488)90161-4.

Felten, J., Martin, F. and Legué, V. 2012. Signalling in ectomycorrhizal symbiosis. pp. 123–142. *In*: Perotto, S. and Baluška, F. [eds.]. Signaling and Communication in Plant Symbiosis. Springer, Berlin, Heidelberg. Doi: 10.1007/978-3-642-20966-6_6.

Fenchel, T. and Finlay, B.J. 1991. The biology of free-living anaerobic ciliates. Eur. J. Protistol. 26: 201–215.

Fisher, R.A. 1918. The correlation between relatives on the supposition of Mendelian inheritance. Trans. R. Soc. Edinburgh 52: 399–433.

Fisher, R.A. 1930. The Genetical Theory of Natural Selection. At The Clarendon Press.

Flegr, J. 2008. Frozen evolution, or, that's not the way it is, Mr Darwin. Prague: Charles University.

Flegr, J. 2010. Elastic, not plastic species: Frozen plasticity theory and the origin of adaptive evolution in sexually reproducing organisms. Biol. Direct 5: 2. Doi: 10.1186/1745-6150-5-2.

Flegr, J. 2013. Microevolutionary, macroevolutionary, ecological and taxonomical implications of punctuational theories of adaptive evolution. Biol. Direct 8: 1. Doi: 10.1186/1745-6150-8-1.

Flyamer, I.M., Gassler, J., Imakaev, M., Brandão, H.B., Ulianov, S.V., Abdennur, N. et al. 2017. Single-nucleus Hi-C reveals unique chromatin reorganization at oocyte-to-zygote transition. Nature 544: 110–114. Doi: 10.1038/nature21711.

Forterre, P. 2010a. Defining life: the virus viewpoint. Orig. Life Evol. Biosph. 40: 51–160. Doi: 10.1007/s11084–010–9194–1.

Forterre, P. 2010b. Giant viruses: conflicts in revisiting the virus concept. Intervirology 53: 362–37. Doi: 10.1159/000312921.

Forterre, P. 2013. The virocell concept and environmental microbiology. ISME J. 7: 233–236. Doi: 10.1038/ismej.2012.110.

Forterre, P. and Gaïa, M. 2016. Giant viruses and the origin of modern eukaryotes. Curr. Opin. Microbiol. 31: 44–49. Doi: 10.1016/j.mib.2016.02.001.

Franche, C. and Bogusz, D. 2012. Signalling and communication in the actinorhizal symbiosis. pp. 73–92. *In*: Perotto, S. and Baluška, F. [eds.]. Signaling and Communication in Plant Symbiosis. Springer, Berlin, Heidelberg. Doi: 978-3-642-20966-6_4.

François, A., Grebert, D., Rhimi, M., Mariadassou, M., Naudon, L., Rabot, S. et al. 2016. Olfactory epithelium changes in germfree mice. Sci. Rep. 6: 24687. Doi: 1038/srep24687.

Frankel, J. 1989. Pattern Formation. Ciliate Studies and Models. Oxford University Press, Oxford.

Fraune, S. and Bosch, T.C. 2007. Long-term maintenance of species-specific bacterial microbiota in the basal metazoan *Hydra*. Proc. Natl. Acad. Sci. USA 104: 13146–13151. Doi:10.1073/pnas.0703375104.

Freud, S. 1991. Gesammelte Werke. Imago Publishing Co. London.

Fuhrman, J.A. 1999. Marine viruses and their biogeochemical and ecological effects. Nature 399: 541–548. Doi: 10.1038/21119.

Funkhouser, L.J. and Bordenstein, S.R. 2013. Mom knows best: the universality of maternal microbial transmission. PLoS Biol. 11: e1001631. Doi: 10.1371/journal.pbio.1001631.

Gabius, H.-J. 2018. The sugar code: Why glycans are so important. BioSystems 164: 102–111. Doi: 10.1016/j.biosystems.2017.07.003.

Galis, F. and Metz, J.A.J. 2001. Testing the vulnerability of the phylotyic stage: On modularity and evolutionary conservation. J. Exp. Zool. 291: 195–204.

Gama, J.A., Reis, A.M., Domingues, I., Mendes-Soares, H., Matos, A.M. and Dionisio, F. 2013. Temperate bacterial viruses as double-edged swords in bacterial warfare. PLoS One 8: e59043. Doi: 10.1371/journal.pone.0059043.

Gapp, K., Jawaid, A., Sarkies, P., Bohacek, J., Pelczar, P., Prados, J. et al. 2014. Implication of sperm RNAs in transgenerational inheritance of the effects of early trauma in mice. Nat. Neurosci. 17: 667–669. Doi: 10.1038/nn.3695.

Garcia-Seisdedos, H., Empereur-Mot, C., Elad, N. and Levy, E.D. 2017. Proteins evolve on the edge of supramolecular self-assembly. Nature 548: 244–247. Doi: 0.1038/nature23320.

Gardner, A. and West, S.A. 2004. Spite. Curr. Biol. 16: R662. Doi: 10.1016/j.cub.2006.08.015.

Gast, R.J., Sanders, R.W. and Caron, D.A. 2009. Ecological strategies of protists and their symbiotic relationships with prokaryotic microbes. Trends Microbiol. 17: 563–569. Doi: 10.1016/j.tim.2009.09.00.

Gatenby, R.A. 2017. Is the genetic paradigm of cancer complete? Radiology 284: 1–3.

Gavelis, G.S., Hayakawa, S., White, R.A. 3rd, Gojobori, T., Suttle, C.A., Keeling, P.J. et al. 2015. Eye-like ocelloids are built from different endosymbiotically acquired components. Nature 523: 204–207. Doi: 10.1038/nature14593.

Genre, A. 2012. Signalling and the re-structuring of plant cell architecture in AM symbiosis. pp. 51–71. *In*: Perotto, S. and Baluška, F. [eds.]. Signaling and Communication in Plant Symbiosis. Springer, Berlin, Heidelberg. Doi: 10.1007/978-3-642-20966-6_3.

Gibson, G. 2008. The environmental contribution to gene expression profiles. Nat. Rev. Genet. 9: 575–581. Doi: 10.1038/nrg2383.

Gilbert, J.A., Martin Blaser, M.A., Caporaso, Jansson, J.K., Lynch, S.V. and Knight, R. 2018. Current understanding of the human microbiome. Nat. Med. 24: 392–400. Doi: 10.1038/nm.4517.

Gilbert, S.F., Opitz, J.M. and Raff, R.A. 1996a. Resynthesizing evolutionary and developmental biology. Dev. Biol. 175: 357–372.

Gilbert, S.F., Opitz, J.M. and Raff, R.A. 1996b. Reply to Lipshitz. Dev. Biol. 175: 618–619.

Gilbert, S.F., Sapp, J. and Tauber, A.I. 2012. A symbiotic view of life: we have never been individuals. Q. Rev. Biol. 87: 325–341. Doi: 10.1086/668166.

Gilbert, S.F. 2014. A holobiont birth narrative: the epigenetic transmission of the human microbiome. Frontiers Genet. 5: 282. Doi: 10.3389/fgene.2014.00282.

Gilbert, S.F. and Epel, D. 2015. Ecological developmental biology: The environmental regulation of development, health, and evolution. 2nd edition. Sinauer Associates, Inc.

Gilbert, W.V., Bell, T.A. and Schaening, C. 2016. Messenger RNA modifications: Form, distribution, and function. Science 352: 1408–12. Doi: 10.1126/science.aad8711.

Gilmore, M.S., Rauch, M., Ramsey, M.M., Himes, P.R., Varahan, S., Manson, J.M. et al. 2015. Pheromone killing of multidrug-resistant *Enterococcus faecalis* V583 by native commensal strains. Proc. Natl. Acad. Sci. USA 112: 7273–7278. Doi: 10.1073/pnas.1500553112.

Ginsburg, S. and Jablonka, E. 2010. The evolution of associative learning: A factor in the Cambrian explosion. J. Theor. Biol. 266: 11–20. Doi: 10.1016/j.jtbi.2010.06.017.

Ginzburg, C. 1992a. The Cheese and the Worms: The Cosmos of a Sixteenth-century Miller. Johns Hopkins University Press.

Ginzburg, C. 1992b. The Night Battles: Witchcraft and Agrarian Cults in the Sixteenth and Seventeenth Centuries. Johns Hopkins University Press.

Gissis, S.B. and Jablonka, E. 2011. Transformations of Lamarckism. From Subtle Fluids to Molecular Biology, Vienna Series in Theoretical Biology. Cambridge: MIT Press.

Gladyshev, E.A., Meselson, M. and Arkhipova, I.R. 2008. Massive horizontal gene transfer in bdelloid Rotifers. Science 320: 1210–1213. Doi: 10.1126/science.1156407.

Glanssdorff, N., Xu, Y. and Labedan, B. 2008. The last universal common ancestor: emergence, constitution and genetic legacy of an elusive forerunner. Biol. Direct 3: 29. Doi: 10.1186/1745–6150–3–29.

Gold, T. 1999. The deep hot biosphere. Springer, New York.

Goldford, J.E., Hartman, H., Smith, T.F. and Segrè, D. 2017. Remnants of an ancient metabolism without phosphate. Cell 168: 1–9. Doi: 10.1016/j.cell.2017.02.001.

Goldschmidt, R. 1940. Material Basis of Evolution. Yale University Press.

Gomez de Agüero, M., Ganal-Vonarburg, S.C., Fuhrer, T., Rupp, S., Uchimura, Y., Steinert, A. et al. 2016. The maternal microbiota drives early postnatal innate immune development. Science 351: 1296–1302. Doi: 10.1126/science.aad2571.

Goodrich, J.K., Waters, J.L., Poole, A.C., Sutter, J.L., Koren, O., Blekhman, R. et al. 2014. Human genetics shape the gut microbiome. Cell 159: 789–799. Doi: 10.1016/j.cell.2014.09.053.

Goodwin, B.C. 1985. Developing organisms as self-organizing fields. *In*: Antonelli, P.L. [ed.]. Mathematical Essays on Growth and the Emergence of Form. University Alberta Press.

Goodwin, B., Sibatani, A. and Webster, G. [eds.]. 1989. Dynamic Structures in Biology. Edinburgh University Press, Edinburgh.

Gopalakrishnan, V., Spencer, C.N., Nezi, L., Reuben, A., andrews, M.C., Karpinets, T.V. et al. 2018. Gut microbiome modulates response to anti-PD-1 immunotherapy in melanoma patients. Science 359: 97–103. Doi: 10.1126/science.aan4236.

Gorby, Y.A., Yanina, S., McLean, J.S., Rosso, K.M., Moyles, D., Dohnalkova, A. et al. 2006. Electrically conductive bacterial nanowires produced by *Shewanella oneidensis* strain MR–1 and other microorganisms. Proc. Nat. Acad. Sci. USA 103: 1138–11363. Doi/10.1073/Proc. Nat. Acad. Sci. USA .0604517103.

Gosalbes, M.J., Llop, S., Vallès, Y., Moya, A., Ballester, F. and Francino, M.P. 2012. Meconium microbiota types dominated by lactic acid or enteric bacteria are differentially associated with maternal eczema and respiratory problems in infants. Clinical et Experimental Allergy 43: 198–211. Doi: 10.1111/cea.12063.

Goto, Y., Obata, T., Kunisawa, J., Sato, S., Ivanov, I.I., Lamichhane, A. et al. 2014. Innate lymphoid cells regulate intestinal epithelial cell glycosylation. Science 345: 1254009. Doi: 10.1126/science.1254009.

Goto, Y., Lamichhane, A., Kamioka, M., Sato, S., Honda, K., Kunisawa, J. et al. 2015. IL-10-producing CD4+ T cells negatively regulates fucosylation of epithelial cells in the gut. Sci. Rep. 5: 15918. Doi: 10.1038/srep15918.

Gould, S.J. 1977. Ontogeny and Phylogeny. Harvard University Press, Cambridge, MA.

Gould, S.J. 1987. Freud's phylogenetic fantasy. Natural History 96: 10.

Gould, S.J. 2002. The Structure of Evolutionary Theory. Harvard University Press, Cambridge, MA.

Grbić, M., Nagy, L.M., Carroll, S.B. and Strand, M. 1996. Polyembryonic development: insect pattern formation in a cellularized environment. Development 122: 795–804.

Green, J.J. and Elisseef, J.H. 2016. Mimicking biological functionality with polymers for biomedical applications. Nature 540: 386–394. Doi: 10.1038/nature21005.

Greene, E. 1989. A diet-induced developmental polymorphism in a caterpillar. Science 243: 643–646. Doi: 10.1126/science.243.4891.643.

Grow, E.J., Flynn, R.A., Chavez, S.L., Bayless, N.L., Wossidlo, M., Wesche, D.J. et al. 2015. Intrinsic retroviral reactivation in human preimplantation embryos and pluripotent cells. Nature 522: 221–225. Doi: 10.1038/nature14308.

Guerrero-Bosagna, C. 2017. Evolution with no reason: A neutral view on epigenetic 479 changes, genomic variability, and evolutionary novelty. BioScience 67: 469–476.

Guo, H., Zhu, P., Yan, L., Li, R., Hu, B., Lian, Y. et al. 2014. The DNA methylation landscape of human early embryos. Nature 511: 606–10. Doi: 10.1038/nature13544.

Gurwitsch, A.G. 1904. Morphologie und biologie der Zelle. Gustav Fischer, Jena.

Haaber, J., Leisner, J.J., Cohn, M.T., Catalan-Moreno, A., Nielsen, J.B., Westh, H et al. 2016. Bacterial viruses enable their host to acquire antibiotic resistence genes from neighbouring cells. Nature Comun. 7: 13333. Doi: 10.1038/ncomms13333.

Haeckel, E. 1874. Anthropogenie oder Entwicklungsgeschichte des Menschen. Engelmann, Leipzig.

Haeckel, E. 1876a [1868]. History of creation: or the development of the earth and its inhabitants by the action of natural causes. D. Appelton and Co., New York.

Haeckel, E. 1876b. Perigenesis der Plastidule oder die Wellenzeugung der Lebenstheilchen. Reimer, Berlin.

Hall, B.K. 1999. The neural crest in development and evolution. Springer, New York.

Hall, B.K. 2001. Organic selection: proximate environmental effects on the evolution of morphology and behaviour. Biol. Philos. 16: 215–237.

Hall, B.K., Pearson, R.D. and Müller, G.B. 2003. Environment, Development and Evolution. Towards a Synthesis. Vienna Series in Theoretical Biology. MIT Press, London.

Hammoud, S.S., Nix, D.A., Zhang, H., Purwar, J., Carrell, D.T. and Cairns, B.R. 2009. Distinctive chromatin in human sperm packages genes for embryo development. Nature 460: 473–478.

Hanna, C.W. and Kelsey, G. 2017. Genomic imprinting beyond DNA methylation: a role for maternal histones. Genome Biol. 18: 177. Doi: 10.1186/s13059-017-1317-9.

Hansen, T.H., Gøbel, R.J., Hansen, T. and Pedersen, O. 2015. The gut microbiome in cardio-metabolic health. Genome Medicine 7: 33. Doi: 10.1186/s13073-015-0157-z.

Harcourt, E.M., Kietrys, A.M. and Kool, E.T. 2017. Chemical and structural effects of base modifications in messenger RNA. Nature 541: 339–346. Doi: 0.1038/nature21351.

Hardin, G. 1968. The tragedy of the commons. Science 162: 1243–1248. Doi: 10.1126/science.162.3859.1243.

Harms, A., Maisonneuve, E. and Gerdes, K. 2016. Mechanisms of bacterial persistence during stress and antibiotic exposure. Science 354: aaf4268. Doi: 10.1126/science.aaf4268.

Harris, R. 2009. Rationality and the literate mind. Routlege.

Harrison, S.E., Sozen, B., Christodoulou, N., Kyprianou, C. and Zernicka-Goetz, M. 2017. Assembly of embryonic and extraembryonic stem cells to mimic embryogenesis *in vitro*. Science 356, pii: eaal1810. Doi: 10.1126/science.aal1810.

Hartl, F.U., Bracher, A. and Hayer-Hartl, M. 2011. Molecular chaperones in protein folding and proteostasis. Nature 475: 324–32. Doi: 10.1038/nature10317.

Harumuto, T. and Lemaitre, B. 2018. Male-killing toxin in a bacterial symbiont of Drosophila. Nature 557: 252–255. Doi: 10.1038/s41586-018-0086-2.

Hassenkam, T., Andersson, M.P., Dalby, K., Mackenzie, D.M.A. and Rosing, M.T. 2017. Elements of Eoarchean life trapped in mineral inclusions. Nature 548: 78–81. Doi: 0.1038/nature23261.

Hausser, J. and Zavolan, M. 2014. Identification and consequences of miRNA-target interactions—beyond repression of gene expression. Nat. Rev. Genet. 15: 599–612. Doi: 10.1038/nrg3765.

Havlicek, J., Saxton, T.K., Roberts, S.C., Jozifkova, E., Lhota, S., Valentova, J. et al. 2008. He sees, she smells? Male and female reports of sensory reliance in mate choice and non-mate choice contexts. Personality and Individual Differences 45: 565–570. Doi:10.1016/j.paid.2008.06.019.

Hehemann, J.H., Correc, G., Barbeyron, T., Helbert, W., Czjzek, M. and Michel, G. 2010. Transfer of carbohydrate-active enzymes from marine bacteria to Japanase gut microbiota. Nature 464: 908–912. Doi: 10.1038/nature08937.

Heidegger, M. 1995[1982]. The Fundamental Concepts of Metaphysics. Indiana University Press, Bloomington.

Heikkilä, M.P. and Saris, P.E.J. 2003. Inhibition of *Staphylococcus aureus* by the commensal bacteria of human milk. J. Appl. Microbiol. 95: 471–478. Doi.org/10.1046/j.1365-2672.2003.02002.x.

Hein, J.E. and Blackmond, D.G. 2012. On the origin of single chirality of amino acids and sugars in biogenesis. Accounts. Chem. Res. 45: 2045–2054. Doi: 10.1021/ar200316n.

Heliconius Genome Consortium. 2012. Butterfly genome reveals promiscuous exchange of mimicry adaptations among species. Nature 487: 94–98. 10.1038/nature11041.

Hendrickson, D.G., Kelley, D., Tenen, D., Bernstein, B. and Rinn, J.L. 2016. Widespread RNA binding by chromatin-associated proteins. Genome Biol. 16: 17: 8. Doi: 10.1186/s13059-016-0878-3.

Hering, E. 1897. On memory and the specific energies of the nervous system. Open Court Publishing Co., Chicago.

Hering, E. 1921[1870]. Über das Gedächtnis als allgemeine Funktion der organisierten Materie. Vortrag gehalten in der feierlichen Sitzung der Kaiserlichen Akademie der Wissenschaften in Wien am XXX. Mai MDCCCLXX. Akademische Verlagsgesellschaft, Leipzig.

Hernando-Herraez, I., Heyn, H., Fernandez-Callejo, M., Vidal, E., Fernandez-Bellon, H., Prado-Martinez, J. et al. 2015. The interplay between DNA methylation and sequence divergence in recent human evolution. Nucl. Acid. Res. 43: 8204–8214. Doi: 10.1093/nar/gkv693.

Herz, H.-M., Morgan, M., Gao, X., Jackson, J., Rickels, R., Swanson, S.K. et al. 2014. Histone H3 lysine-to-methionine mutants as a paradigm to study chromatin signaling. Science 345: 1065–1070. Doi: 10.1126/science.1255104.

Hofmeister, B.T., Lee, K., Rohr, N.A., Hall, D.K. and Schmitz, R.J. 2017. Stable inheritance of DNA methylation allows creation of epigenotype maps and the study of epiallele inheritance patterns in the absence of genetic variation. Genome Biol. 18: 155. Doi: 10.1186/s13059-017-1288-x.

Holmes, E.C. 2011. What does virus evolution tell us about virus origins? J. Virol. 85: 5247–5251. Doi: 10.1128/JVI.02203-10.

Holoch, D. and Moazed, D. 2015. RNA-mediated epigenetic regulation of gene expression. Nat. Rev. Genet. 16: 71–84. Doi: 10.1038/nrg3863.

Holzmann, M., Berney, C. and Hohenegger, J. 2006. Molecular identification of diatom endosymbionts in nummulitid Foraminifera. Symbiosis 42: 93–101.

Hom, E.F.Y. and Murray, A.W. 2014. Niche engineering demonstrates a latent capacity for fungal-algal mutualism. Science 345: 94–98. Doi: 10.1126/science.1253320.

Hooper, L.V., Stappenbeck, T.S., Hong, C.V. and Gordon, J.I. 2003. Angiogenins: a new class of microbicidal proteins involved in innate immunity. 4: 269–73. Doi: 10.1038/ni888.

Horgan, J. 1998. The end of science. Facing the limits of knowledge in the twilight of scientific age. Little, Brown and Co., London.

Horie, M., Honda, T., Suzuki, Y., Kobayashi, Y., Daito, T., Oshida, T. et al. 2010. Endogenous non-retroviral RNA virus elements in mammalian genomes. Nature 463: 84. Doi: 10.1038/nature08695.

Hormann, M., Heubeck, C., Bontognali, T.R.R., Bouvier, A.-S., Baumgartner, L.P. and Airo, A. 2016. Evidence for cavity-dwelling microbial life in 3.22 Ga tidal deposits. Geology 44: 51–54. Doi: 10.1130/G37272.1.

Horwitz, R. and Johnson, G.T. 2017. Whole cell maps chart a course for 21st-century cell biology. Science 356: 806–807. Doi: 10.1126/science.aan5955.

Hosokawa, T., Kikuchi, Y., Nikoh, N., Shimada, M. and Fukatsu, T. 2006. Strict host-symbiont cospeciation and reductive genome evolution in insect gut bacteria. PLoS Biol 4: e337. Doi: 10.1371/journal.pbio.0040337.

Hosokawa, T., Kikuchi, Y., Shimada, M. and Fukatsu, T. 2008. Symbiont acquisition alters behavior of stinkbug nymphs. Biol. Lett. 4: 45–48. Doi: 10.1098/rsbl.2007.0510.

Hou, J., Jun, S.R., Zhang, C. and Kim, S.H. 2005. Global mapping of the protein structure space and application in structure-based inference of protein function. Proc. Nat. Acad. Sci. USA 102: 3651–3656. Doi: 10.1073/Proc. Nat. Acad. Sci. USA .0409772102.

Hsiao, E.Y., McBride, S.W., Hsien, S., Sharon, G., Hyde, E.R., McCue, T. et al. 2013. Microbiota modulate behavioral and physiological abnormalities associated with neurodevelopmental disorders. Cell 155: 1451–1463. Doi: 10.1016/j.cell.2013.11.024.

Hsu, P.-J., Shi, H. and He, C. 2017. Epitranscriptomic influences on development and disease. Genome Biol. 18: 197. DOI 10.1186/s13059-017-1336-6.

Huang, P.S., Boyken, S.E. and Baker, D. 2016. The coming of age of *de novo* protein design. Nature 537: 320–327. Doi: 10.1038/nature19946.

Huber, C. and Wächtershäuser, G. 1997. Activated acetic acid by carbon fixation on (Fe,Ni)S under primordial conditions. Science 276: 245–247. Doi: 10.1126/science.276.5310.245.

Hug, L.A., Baker, B.J., Anantharaman, K., Brown, C.T., Probst, A.J., Castelle, C.J. et al. 2016. A new view of the tree of life. Nat. Microbiol. 1: 16048. Doi: 10.1038/nmicrobiol.2016.48.

Huigens, M.E. and Stouthamer, R. 2003. Parthenogenesis associated with *Wolbachia*. Insect Symbiosis 1: 247–266.

Human Microbiome Project Consortium. 2012. Structure, function and diversity of the healthy human microbiome. Nature 486: 207–214. Doi: 10,1038/nature11234.

Humphrey, J.D., Dufresne, E.R. and Schwartz, M.A. 2014. Mechanotransduction and extracellular matrix homeostasis. Nat. Rev. Mol. Cell. Biol. 15: 802–12. Doi: 10.1038/nrm3896.

Hutchinson, C.A. 3rd, Chuang, R.Y., Noskov, V.N., Assad-Garcia, N., Deerinck, T.J., Ellisman, M.H. et al. 2016. Design and synthesis of a minimal bacterial genome. Science 351: aaa6253. Doi: 10.1126/science.aad6253.

Huttlin, E.L., Bruckner, R.J., Paulo, J.A., Cannon, J.R., Ting, L., Baltier, K. et al. 2017. Architecture of the human interactome defines protein communities and disease networks. Nature 545: 505–509. Doi: 0.1038/nature22366.

Huxley, J. 1974 [1942]. Evolution: The Modern Synthesis. Allen and Unwin.

Inaba, M., Buszczak, M. and Yamashita, Y.M. 2015. Nanotubes mediate niche-stem-cell signaling in the *Drosophila* testis. Nature 523: 329–32. Doi: 10.1038/nature14602.

Inoue, A., Jiang, L., Lu, F., Suzuki, T. and Zhang, Y. 2017. Maternal H3K27me3 controls DNA methylation-independent imprinting. Nature 547: 19–424. Doi: 10.1038/nature23262.

Irie, N. and Kuratani, S. 2014. The developmental hourglass model: a predictor of the basic body plan? Development 141: 4649–55. Doi: 10.1242/dev.107318.

Irmler, I., Schmidt, K. and Starck, J.M. 2004. Developmental variability during early embryonic development of zebra fish, *Danio rerio*. J. Exp. Zool. B. Mol. Dev. Evol. 302: 446–57. Doi: 10.1002/jez.b.2101.

Ishikawa, H. 2003. Insect symbiosis: An introduction. pp. 1–22. *In*: Bourtzis, K. and Miller, T.A. [eds.]. Insect Symbiosis. CRC Press, Boca Raton.

Jablonka, E. and Lamb, M.J. 1995. Epigenetic Inheritance and Evolution. The Lamarckian Dimension, Oxford University Press, Oxford.

Jablonka, E. and Raz, G.D. 2009. Transgenerational epigenetic inheritance: Prevalence, mechanisms, and implications for the study of heredity and evolution. Q. Rev. Biol. 84: 131–176.

Jacob, F. and Monod. J. 1961a. Genetic regulatory mechanisms in the synthesis of proteins. J. Mol. Biol. 3: 318–356.

Jacob, F. and Monod. J. 1961b. On the regulation of gene activity. Cold Spring Harb. Symp. Quant. Biol. 26: 193–211.

Jägersten, G. 1972. Evolution of the Metazoan Life Cycle. A Comprehensive Theory. Academic Press, London.

Jarosz, D.F. and Lindquist, S. 2010. Hsp90 and environmental stress transform the adaptive value of natural genetic variation. Science 330: 1820–1824. DOI: 10.1126/science.1195487.

Javaux, E.J., Marshall, C.P. and Bekker, A. 2010. Organic-walled microfossils in 3.2-billion-year-old shallow–marine siliciclastic deposits. Nature 463: 934–938. Doi: 10.1038/nature08793.

Jeffery, I.B., O'Toole, P.W., Öhman, L., Claesson, M.J., Deane, J., Quigley, E.M. et al. 2012. An irritable bowel syndrome subtype defined by species-specific alterations in fecal microbiota. Gut 61: 997–1006. Doi: 10.1136/gutjnl-2011-301501.

Jiang, D. and Berger, F. 2017. DNA replication-coupled histone modification maintains Polycomb gene silencing in plants. Science 357: 1146–1149. Doi: 10.1126/science.aan4965.

Jiao, L. and Liu, X. 2015. Structural basis of histone H3K27 trimethylation by an active polycomb repressive complex 2. Science 350: aac4383. Doi: 10.1126/science.aac4383.

Jih, G., Iglesias, N., Currie, M.A., Bhanu, N.V., Paulo, J.A., Gygi, S.P. et al. 2017. Unique roles for histone H3K9me states in RNAi and heritable silencing of transcription. Nature 247: 463–467. Doi: 10.1038/nature23267.

Jimenez, E., Fernández, L., Marín, M.L., Martín, R., Odriozola, J.M., Nueno-Palop, C. et al. 2005. Isolation of commensal bacteria from umbilical cord blood of healthy neonates born by cesarean section. Curr. Microbiol. 51: 270–274. Doi: 10.1007/s00284-005-0020-3.

Jimenez, E., Marín, M.L., Martín, R., Odriozola, J.M., Olivares, M. and Xaus, J. 2008. Is meconium from healthy newborns actually sterile? Res. Microbiol. 159: 187–193. Doi: 10.1016/j.resmic.2007.12.007.

Johannsen, W. 1909. Elemente der exakten Erblichkeitslehre. Gustav Fischer, Jena.

Johnson, M.D., Oldach, D., Delwiche, C.F. and Stoecker, D.K. 2007. Retention of transcriptionally active cryptophyte nuclei by the ciliate *Myrionecta rubra*. Nature 445: 426–428. Doi: 10.1038/nature05496.

Johnson, T.C., Scholz, C.A., Talbot, M.R., Kelts, K., Ricketts, R.D., Ngobi, G. et al. 1996. Late Pleistocene desiccation of Lake Victoria and rapid evolution of cichlid fishes. Science 273: 1091–1093.

Joint, I., Downie, J.A. and Williams, P. 2007. Bacterial conversations: talking, listening and eavesdropping. An introduction. Phil. Trans. R. Soc. B. 362: 1115–1117. Doi: 10.1098/rstb.2007.2038.

Kaiser, J. 2016. Malignant messengers. Science 352: 164–166.

Kalinka, A.T., Varga, K.M., Gerrard, D.T., Preibisch, S., Corcoran, D.L., Jarrells, J. et al. 2010. Gene expression divergence recapitulates the developmental hourglass model. Nature 468: 811–814. Doi: 0.1038/nature09634.

Kang, D.W., Park, J.G., Ilhan, Z.E., Wallstrom, G., Labaer, J., Adams, J.B. et al. 2013. Reduced incidence of *Prevotella* and other fermenters in intestinal microflora of autistic children. PloS One 8: e68322. Doi: 10.1371/journal.pone.0068322.

Karlsson, F., Tremaroli, V., Nielsen, J. and Bäckhed, F. 2013. Assessing the human microbiota in metanolic diseases. Diabetes 62: 3341–9. Doi: 10.2337/db13-0844.

Kashtan, N., Roggensack, S.E., Rodrigue, S., Thompson, J.W., Biller, S.J., Coe, A. et al. 2014. Single-cell genomics reveals hundreds of coexisting subpopulations in wild *Prochlorococcus*. Science 344: 416–420. Doi: 10.1126/science.1248575.

Kassam, Z., Lee, C.H., Yuan, Y. and Hunt, R.H. 2013. Fecal microbiota transplantation for *Clostridium difficile* infection: systematic review and meta-analysis. Am. J. Gastroenterol. 108: 500–508. Doi: 10.1038/ajg.2013.59.

Kaucka, M., Ivashkin, E., Gyllborg, D., Zikmund, T., Tesarova, M., Kaiser, J. et al. 2016. Analysis of neural crest–derived clones reveals novel aspects of facial development. Sci. Adv. e1600060. Doi: 10.1126/sciadv.160006.

Kauffman, S. 1993. The Origins of Order. Self-organization and Selection in Evolution. New York: Oxford University Press.

Kauffman, S. 2000. Investigations. Oxford University Press, New York.

Kawashima, T. and Berger, F. 2014. Epigenetic reprogramming in plant sexual reproduction. Nat. Rev. Genet. 15: 613–24. Doi: 10.1038/nrg3685.

Keefe, A.D. and Szostak, J.W. 2001. Functional proteins from a random-sequence library. Nature 410: 715–718.

Keese, P.K. and Gibbs, A. 1992. Origins of genes: Big bang or continuous creation. Proc. Nat. Acad. Sci. USA 89: 9489–9493.

Keller, L. and Surette, M.G. 2006. Communication in bacteria: an ecological and evolutionary perspective. Nature Revs. Microbiol. 4: 249–258. Doi: 10.1038/nrmicro1383.

Kelley, D.S., Karson, J.A., Früh-Green, G.L., Yoerger, D.R., Shank, T.M., Butterfield, D.A. et al. 2005. A serpentinite-hosted ecosystem: The lost city hydrothermal field. Science 307: 1428–1434. Doi: 10.1126/science.1102556.

Kelsey, G., Stegle, O. and Reik, W. 2017. Single-cell epigenomics: Recording the past and predicting the future. Science 358: 69–75. Doi: 10.1126/science.aan6826.

Kenrick, P. and Strullu-Derrien, C. 2014. The origin and early evolution of roots. Plant Physiol. 166: 570–580. Doi: 10.1104/pp.114.244517.

Kernbauer, E., Ding, Y. and Cadwell, K. 2014. An enteric virus can replace the beneficial function of commensal bacteria. Nature 516: 94–98. Doi: 10.1038/nature13960.

Kerr, B. and Godfrey-Smith, P. 2002. Individualist and multi-level perspectives on selection in structured populations. Biol. Philo. 17: 477–517.

Kerr, B., Riley, M.A., Feldman, M.W. and Bohannan, B.J.M. 2002. Local dispersal promotes biodiversity in a real game of rock-paper-scissors. Nature 418: 171–174. Doi: 10.1038/nature00823.

Kiers, E.T., Duhamel, M., Beesetty, Y., Mensah, J.A., Franken, O., Verbruggen, E. et al. 2011. Reciprocal rewards stabilize cooperation in the mycorrhizal symbiosis. Science 333: 880–882. Doi: 10.1126/science.1208473.

Kiers, E.T. and West, S.A. 2015. Evolving new organisms via symbiosis. Science 348: 392–394. Doi: 10.1126/science.aaa96052.

Kim, G., LeBlanc, M.L., Wafula, E.K., de Pamphilis, C.W. and Westwood, J.H. 2014. Genomic-scale exchange of mRNA between a parasitic plant and its hosts. Science 345: 808–811. Doi: 10.1126/science.1253122.

Kim, H.K., Fuchs, G., Wang, S., Wei, W., Zhang, Y., Park, H. et al. 2017. A transfer-RNA-derived small RNA regulates ribosome biogenesis. Nature 552: 57–62. Doi: 10.1038/nature25005.

Kim, K.S., Lee, S. and Ryu, C.M. 2013. Interspecific bacterial sensing through airborne signals modulates locomotion and drug resistance. Nat. Commun. 4: 1809. Doi: 10.1038/ncomms2789.

Kimura, M. 1983. The Neutral Theory of Molecular Evolution. Cambridge University Press, New York.

Kistner, D.H. 1968. Revision of the African species of the termitophilous tribe Corotocini (Coeloptera: Stapylinidae). I. A new genus and species from Ovamboland and its zoogeographic significance. J. New York Entomol. Soc. 76: 213–221.

Kistner, D.H. 1990. The integration of foreign insect into termite societies or why do termites tolerate foreign insects in their societies? Sociobiology 17: 191–215.

Klein, T., Siegwolf, R.T. and Körner, C. 2016. Belowground carbon trade among tall trees in a temperate forest. Science 352: 342–344. Doi: 10.1126/science. aad6188.

Kleisner, K. and Markoš, A. 2005. Semetic rings: towards the new concept of mimetic resemblances. Theor. Biosci. 123: 209–22. Doi: 10.1016/j.thbio.2004.09.001.

Kleisner, K. and Markoš, A. 2009. Mutual understanding and misunderstanding in biological systems mediated by self-representational meaning of organisms. Semiotike—Sign Syst. Stud. 37: 299–310.

Koenig, J.E., Spor, A., Scalfone, N., Fricker, A.D., Stombaugh, J., Knight, R. et al. 2011. Succession of microbial consortia in the developing infant gut microbiome. Proc. Natl. Acad. Sci. USA 108: 4578–4585. Doi: 10.1073/pnas.1000081107.

Kollar, E.J. and Fisher, C. 1980. Tooth induction in chick epithelium: Expression of quiescent genes for enamel synthesis. Science 207: 993–995.

Komárek, S. 2003. Mimicry, aposemantism and related phenomena. LINCOM EUROPA, München.

Koonin, E.V., Senkevich, T.G. and Dolja, V.V. 2006. The ancient virus world and evolution of cells. Biol. Direct 1: 9. Doi: 10.1186/1745–6150–1–29.

Koren, O., Goodrich, J.K., Cullender, T.C., Spor, A., Laitinen, K., Bäckhed, H.K. et al. 2012. Host remodeling of the gut microbiome and metabolic changes during pregnancy. Cell 150: 470–480. Doi: 10.1016/j.cell.2012.07.008.

Krautkramer, K.A., Kreznar, J.H., Romano, K.A., Vivas, E.I., Barrett-Wilt, G.A., Rabaglia, M.E. et al. 2016. Diet-microbiota interactions mediate global epigenetic programming in multiple host tissues. Mol. Cell. 64: 982–992. Doi: 10.1016/j.molcel.2016.10.025.

Kremer, N., Philipp, E.E., Carpentier, M.C., Brennan, C.A., Kraemer, L., Altura, M.A. et al. 2013. Initial symbiont contact orchestrates host-organ-wide transcriptional changes that prime tissue colonization. Cell Host Microbe 14: 183–194. Doi: 10.1016/j.chom.2013.07.006.

Kreysing, M., Keil, L., Lanzmich, S. and Braun, D. 2015. Heat flux across an open pore enables the continuous replication and selection of oligonucleotides towards increasing length. Nature Chem. 7: 203–208. Doi: 10.1038/nchem.2155.

Krupovič, M. and Bamford, D.H. 2010. Order to the viral universe. J. Virol. 84: 12476–12479. Doi: 10.1128/JVI.01489–10.

Ku, C., Nelson-Sathi, S., Roettger, M., Sousa, F.L., Lockhart, P.J., Bryant, D. et al. 2015. Endosymbiotic origin and differential loss of eukaryotic genes. Nature 524: 427–432. Doi: 10.1038/nature14963.

Kulesa, B.M. and Gammill, L.S. 2010. Neural crest migration: Patterns, phases and signals. Dev. Biol. 344: 566–568. Doi: 10.1016/j.ydbio.2010.05.005.

Kunte, K., Zhang, W., Tenger-Trolander, A., Palmer, D.H., Martin, A., Reed, R.D. et al. 2014. Doublesex is a mimicry supergene. Nature 507: 229–232. 10.1038/nature13112.

Kurokawa, K., Itoh, T., Kuwahara, T., Oshima, K., Toh, H., Toyoda, A. et al. 2007. Comparative metagenomics revealed commonly enriched gene sets in human gut microbiomes. DNA Res. 14: 169–181. Doi: 10.1093/dnares/dsm018.

Kuzmin, E., VanderSluis, B., Wang, W., Tan, G., Deshpande, R., Chen, Y. et al. 2018. Systematic analysis of complex genetic interactions. Science 360: eaao1729. Doi: 10.1126/science.aao1729.

Lacey, R. and Danziger, D. 1999. The year 1000. What life was like at the turn of the first millenium. Little, Brown and Co., GB.

Lake, J.A. 2009. Evidence for an early prokaryotic endosymbiosis. Nature 460: 967–971. Doi: 10.1038/nature08183.

Laland, K.N., Odling Smee, J. and Myles, S. 2010. How culture shaped the human genome: bringing genetics and the human sciences together. Nat. Rev. Genet. 11: 137–48. Doi: 10.1038/nrg2734.

Laland, K.N., Sterelny, K., Odling-Smee, J., Hoppitt, W. and Uller, T. 2011. Cause and effect in biology revisited: Is Mayr's proximate-ultimate dichotomy still useful? Science 334: 1512–1516. Doi: 10.1126/science.1210879.

Laland, K.N., Uller, T., Feldman, M.W., Sterelny, K., Müller, G.B., Moczek, A. et al. 2015. The extended evolutionary synthesis: its structure, assumptions and predictions. Proc. Biol. Sci. 282: 20151019. Doi: 10.1098/rspb.2015.1019.

Lamarck, J.B.P.A.D.M. 2012 [1809]. Zoological Philosophy. London: Forgotten Books.

Lamb, M.J. 2011. Attitudes to soft inheritance in great Britain, 1930s–1970s. pp. 109–120. *In*: Gissis, S.B. and Jablonka, E. [eds.]. Transformations of Lamarckism. From Subtle Fluids to Molecular Biology, Vienna Series in Theoretical Biology. Cambridge: MIT Press.

Laming, S.R., Duperron, S., Gaudron, S.M., Hilário, A. and Cunha, M.R. 2015. Adapted to change: The rapid development of symbiosis in newly settled, fast-maturing chemosymbiotic mussels in the deep sea. Mar. Environ. Res. 112: 100–112. Doi: 10.1016/j.marenvres.2015.07.014.

Lander, A.D. 2010. The edges of understanding. BMC Biology 8: 40. Doi: 10.1186/1741-7007-8-40.

Lane, N. and Martin, W.F. 2012. The origin of membrane bioenergetics. Cell 151: 1406–1416. Doi: 10.1016/j.cell.2012.11.050.

Langeberg, L.K. and Scott, J.D. 2015. Signalling scaffolds and local organization of cellular behaviour. Nature Revs. Mol. Cell. Biol. 16: 232–244. Doi: 10.1038/nrm3933.

Larson, A.G., Elnatan, D., Keenen, M.M., Trnka, M.J., Johnston, J.B., Burlingame, A.L. et al. 2017. Liquid droplet formation by HP1α suggests a role for phase separation in heterochromatin. Nature 547: 236–240. Doi: 10.1038/nature22822.

Lauder, A.P., Roche, A.M., Sherrill-Mix, S., Bailey, A., Laughlin, A.L., Bittinger, K. et al. 2016. Comparison of placenta samples with contamination controls does not provide evidence for a distinct placenta microbiota. Microbiome 4: 29. Doi: 10.1186/s40168-016-0172-3.

Lawrence, R.M. and Lawrence, R.A. 2004. Breast milk and infection. Clin. Perinatol. 31: 501–528. Doi: 10.1016/j.clp.2004.03.019.

Lax, S., Smith, D.P., Hampton-Marcell, J., Owens, S.M., Handley, K.M., Scott, N.M. et al. 2014. Longitudinal analysis of microbial interaction between humans and the indoor environment. Science 345: 1048–1052. Doi: 10.1126/science.1254529.

Le Chatelier, E., Nielsen, T., Qin, J., Prifi, E., Hildebrand, F., Falony, G. et al. 2013. Richness of human microbiome correlates with metabolic markers. Nature 500: 541–546. Doi: 10.1038/nature12506.

Le Douarin, N. 1973. A biological cell labeling technique and its use in experimental embryology. Dev. Biol. 30: 217–222.

Le Lievre, C.S. and Le Douarin, N.M. 1975. Mesenchymal derivatives of the neural crest: Analysis of chimaeric quail and chick embryos. J. Embryol. Exp. Morphol. 34: 25–154.

Lee, Y.K. and Mazmanian, S.K. 2010. Has microbiota played a critical role in the evolution of the adaptive immune system? Science 330: 1768–73. Doi: 10.1126/science.1195568.

Lehninger, A.L. 1975. Biochemistry. The molecular basis of cell structure and function. Worth publishers, New York.

Lei, L. and Spradling, A.C. 2016. Mouse oocytes differentiate through organelle enrichment from sister cyst germ cells. Science 352: 95–99. Doi: 10.1126/science.aad2156.

Lenoir, A., Háva, J., Hefetz, A., Dahbi, A., Cerdá, X. and Boulay, R. 2013. Chemical integration of *Thorictus myrmecophilous* beetles into *Cataglyphis* ant nests. Biochem. Syst. Ecol. 51: 335–342. Doi: 10.1016/j.bse.2013.10.002.

Levin, M., Anavy, L., Cole, A.G., Winter, E., Mostov, N., Khair, S. et al. 2016. The mid-developmental transition and the evolution of animal body plans. Nature 531: 637–641. Doi: 10.1038/nature16994.

Linewaver, H. 2006. We have not detected extraterrestrial life or have we? pp. 445–457. *In*: Seckbach, J. [ed.]. Life in Extreme Habitats and Astrobiology. Springer, Dordrecht.

Lipshitz, H.D. 1996. Resynthesis or revisionism? Dev. Biol. 175: 616–618.

Liscovitch-Brauer, N., Alon, S., Porath, H.T., Elstein, B., Unger, R., Ziv, T. et al. 2017. Trade-off between transcriptome plasticity and genome evolution in cephalopods. Cell 169: 191–202. Doi: 10.1016/j.cell.2017.03.025.

Liu, J., Martinez-Corral, R., Prindle, A., Lee, D.D., Larkin, J., Gabalda-Sagarra, M. et al. 2017. Coupling between distant biofilms and emergence of nutrient time-sharing. Science 356: 638–642. Doi: 10.1126/science.aah4204.

Liu, N., Lee, C.H., Swigut, T., Grow, E., Gu, B., Bassik, M.C. et al. 2018. Selective silencing of euchromatic L1s revealed by genome-wide screens for L1 regulators. Nature 553: 228–232. Doi: 10.1038/nature25179.

Lombardo, M.P. 2008. Access to mutualistic endosymbiotic microbes: an underappreciated benefit to group living. Behav. Ecol. Sociobiol. 62: 479–497.

Lönnerdal, B. 2013. Bioactive proteins in breast milk. J. Paediatr. Child Health. 1: 1–7. Doi: 10.1111/jpc.12104.

Lorenz, K. 1966. On agression. Routledge, London.

Lotman, Y.M. 2009[1992]. Culture and explosion. Berlin: Mouton de Gruyter 2009.

Lu, C., Jain, S.U., Hoelper, D., Bechet, D., Molden, R.C., Ran, L. et al. 2016. Histone H3K36 mutations promote sarcomagenesis through altered histone methylation landscape. Science 352: 844–9. Doi: 10.1126/science.aac7272.

Luczynski, P., McVey Neufeld, K.-A., Oriach, C.S., Clarke, G., Dinan, T.G. and Cryan, J.F. 2016. Growing up in a Bubble: Using germ-free animals to assess the influence of the gut microbiota on brain and behavior. Int. J. Neuropsychopharm. 19: pyw020. Doi:10.1093/ijnp/pyw020.

Luef, B., Frischkorn, K.R., Wrighton, K.C., Holman, H Y., Birarda, G., Thomas, B.C. et al. 2015. Diverse uncultivated ultra-small bacterial cells in groundwater. Nat. Commun. 6: 6372. Doi: 10.1038/ncomms7372.

Luginbuehl, L.H., Menard, G.N., Kurup, S., Van Erp, H., Radhakrishnan, G.V., Breakspear, A. et al. 2017. Fatty acids in arbuscular mycorrhizal fungi are synthesized by the host plant. Science 356: 1175–1178. Doi: 10.1126/science.aan0081.

Luo, C., Keown, C.L., Kurihara, L., Zhou, J., He, Y., Li, J. et al. 2017. Single-cell methylomes identify neuronal subtypes and regulatory elements in mammalian cortex. Science 357: 600–604. Doi: 10.1126/science.aan3351.

Lyell, C. 1830. Principles of Geology. John Murray, London.

Maden, B.E.H. 1995. No soup for starters? Autotrophy and the origin of metabolism. Trends Biochem. Sci. 20: 337–341.

Mahamid, J., Pfeffer, S., Schaffer, M., Villa, E., Danev, R., Cuellar, L.K. et al. 2016. Visualizing the molecular sociology at the HeLa cell nuclear periphery. Science 351: 969–72. Doi: 10.1126/science.aad8857.

Makova, K.D. and Hardison, R.C. 2015. The effects of chromatin organization on variation in mutation rates in the genome. Nat. Rev. Genet. 16: 213–223. Doi: 10.1038/nrg3890.

Maldonado-Contreras, A., Goldfarb, K.C., Godoy-Vitorino, F., Karaoz, U., Contreras, M., Blaser, M.J. et al. 2011. Structure of the human gastric bacterial community in relation to *Helicobacter pylori-* status. ISME J. 5: 574–579. Doi: 10.1038/ismej.2010.149.

Malthus, T. 1798. An Essay on the Principle of Population, as it affects the future Improvement of Society, with Remarks on the Speculations of Mr. Godwin, M. Condorcet and Other Writers. London.

Mangold, H. and Spemann, H. 1924. Über Induktion von Embryonalanlagen durch Implantation artfremder Organisatoren. Archiv für Mikroskopische Anatomie und Entwicklungsmechanik 100: 599–638.

Maraldi, N.M. 2018. The lamin code. BioSystems 164: 68–75. Doi: 10.1016/j.biosystems.2017.07.006.

Margulis, L. and Fester, R. 1991. Symbiosis as a Source of Evolutionary Innovation: Speciation and Morphogenesis. Cambridge, MA: The MIT Press.

Margulis, L. 1996. Archaeal-eubacterial mergers in the origin of Eukarya: phylogenetic classification of life. Proc. Natl. Acad. Sci. USA 93: 1071–1076.

Marijuán, P.C., Navarro, J. and Del Moral, R. 2018. How prokaryotes 'encode' their environment: Systemic tools for organizing the information flow. BioSystems 164: 26–38. Doi: 10.1016/j.biosystems.2017.10.002.

Markmann, K., Radutoiu, S. and Stougaard, J. 2012. Infection of *Lotus japonicus* roots by *Mesorhizobium loti*. pp. 31–50. *In*: Perotto, S. and Baluška, F. [eds.]. Signaling and Communication in Plant Symbiosis. Springer, Berlin, Heidelberg. Doi: 10.1007/978-3-642-20966-6_2.

Markoš, A. 2002. Readers of the Book of Life. Contextualizing Developmental Evolutionary Biology. Oxford University Press, New York.

Markoš, A. 2004. In the quest for novelty: Kauffman's biosphere and Lotman's semiosphere. Sign Syst. Stud. 32: 309–327.

Markoš, A. and Švorcová, J. 2009. Recorded vs. organic memory. Interaction of two worlds as demonstrated by the chromatin dynamics. Biosemiotics 2: 131–149. Doi: 10.1007/s12304-009-9045-5.

Markoš, A., Grygar, F., Hajnal, L., Kleisner, K., Kratochvíl, Z. and Neubauer, Z. 2009. Life as its own designer. Darwins Origin and Western thought. Springer, Dodrecht.

Markoš, A., Švorcová, J. and Lhotský, J. 2013. Living as languaging. Distributed knowledge in living beings. pp. 71–92. *In*: Cowley, S.J. and Vallée-Tourangeau, F. [eds.]. Cognition Beyond the Brain. Computation, Interactivity and Human Artifice. Springer.

Markoš, A. 2014. Biosphere as a semiosphere. Variations on Lotman. Sign Syst. Stud. 42: 487–498.

Markoš, A. 2016a. Evoluční tápání [Evolutionary groping, in Czech]. Pavel Mervart, Červený Kostelec.

Markoš, A. 2016b. The birth and life of species–cultures. Biosemiotics 9: 73–84. Doi: 10.1007/s12304-015-9252-1.

Markoš, A. and Das, P. 2016. Levels or domains of life? Biosemiotics 9: 319–330. Doi: 10.1007/s12304-016-9271-6.

Marks, J. 2003. What it Means to be 98% Chimpanzee: Apes, People, and their Genes. University of California Press.

Martienssen, R. 1996. Epigenetic phenomena: Paramutation and gene silencing in plants. Curr. Biol. 6: 10–813.

Martijn, J., Vosseberg, J., Guy, L., Offre, P. and Ettema, T.J.G. 2018. Deep mitochondrial origin outside the sampled alphaproteobacteria. Nature 557: 101–105. Doi: 10.1038/s41586-018-0059-5.

Martik, M.L. and Bronner, M.E. 2017. Regulatory logic underlying diversification of the neural crest. Trends Genet. 33: 715–727. Doi: 10.1016/j.tig.2017.07.015.

Martín, R., Langa, S., Reviriego, C., Jiménez, E., Marín, M.L., Olivares, M. et al. 2004. The commensal microflora of human milk: new perspectives for food bacteriotherapy and probiotics. Trends Food Sci. Technol. 15: 121–127. Doi: 10.1016/j.tifs.2003.09.010.

Martin, W. and Müller, M. 1998. The hydrogen hypothesis for the first eukaryote. Nature 392: 37–41. Doi: 10.1038/32096.

Martin, W. and Embley, T.M. 2004. Evolutionary biology: Early evolution comes full circle. Nature 431: 134–137. Doi: 10.1038/431134a.

Martin. W. and Russell, M.J. 2007. On the origin of biochemistry at an alkaline hydrothermal vent. Phil. Trans. R.Soc. B 362: 1887–1925. Doi: 10.1098/rstb.2006.1881.

Maruyama, M. and Parker, J. 2017. Deep-time convergence in rove beetle symbionts of army ants. Curr. Biol. 27: 920–926. Doi: 10.1016/j.cub.2017.02.030.

Massana, R. 2011. Eukaryotic picoplankton in surface oceans. Annu. Rev. Microbiol. 65: 91–110. Doi: 10.1146/ann urev–micro–090110–102903.

Matsuda, R. 1987. Animal Evolution in Changing Environments, with Special Reference to Abnormal Metamorphosis. John Wiley and Sons.

Matthews, B., De Meester, L., Jones, C.G., Ibelings, B.W., Bouma, T.J., Nuutinen, V. et al. 2014. Under niche construction: an operational bridge between ecology, evolution, and ecosystem science. Ecol. Monogr. 84: 245–263.

Mattiroli, F., Bhattacharyya, S., Dyer, P.N., White, A.E., Sandman, K., Burkhart, B.W. et al. 2017. Structure of histone-based chromatin in Archaea. Science 357: 609–612. Doi: 10.1126/science. aaj1849.

Maynard Smith, J. and Szathmary, E. 1995. The Major Transitions in Evolution. Oxford University Press, Oxford, United Kingdom.

Mayr, E. 1961. Cause and effect in biology. Science 134: 1501–1506.

Mayr, E. 1993. Proximate and ultimate causations. Biol. Phil. 1: 93–94.

Maze, I., Noh, K.-M., Soshnev, A.A. and Allis, D. 2014. Every amino acid matters: essential contributions of histone variants to mammalian development and disease. Nat. Rev. Genet. 15: 259–71. Doi: 10.1038/nrg3673.

Mazmanian, S.K., Round, J.L. and Kasper, D.L. 2008. A microbial symbiosis factor prevents intestinal inflammatory disease. Nature 453: 620–5.

McCollom, T. and Seewald, J. 2007. Abiotic synthesis of organic compounds in deep-sea hydrothermal environments. Chem. Rev. 107: 382–401. Doi: 10.1021/cr0503660.

McCormack, J.E., Faircloth, B.C., Crawford, N.G., Gowaty, P.A., Brumfield, R.T. and Glenn, T.C. 2012. Ultraconserved elements are novel phylogenomic markers that resolve placental mammal phylogeny when combined with species-tree analysis. Genome Res. 22: 746–54. Doi: 10.1101/gr.125864.111.

McCracken, W., Aihara, E., Martin, B., Crawford, C.M., Broda, T., Treguier, J. et al. 2017. Wnt/β-catenin promotes gastric fundus specification in mice and humans. Nature 541: 182–187. Doi: 10.1038/nature21021.

McCutcheon, J.P. and von Dohlen, C.D. 2011. An interdependent metabolic patchwork in the nested symbiosis of mealybugs. Curr. Biol. 21: 1366–72. Doi: 10.1016/j.cub.2011.06.051.

McFall-Ngai, M., Heath-Heckman, E.A.C., Gillette, A.A., Peyer, S.M. and Harvie, E.A. 2012. The secret languages of coevolved symbioses: Insights from the *Euprymna scolopes-Vibrio fischeri* symbiosis: Seminars. Immunol. 24: 3–8.

McFall-Ngai, M., Hadfield, M.G., Bosch, T.C.G, Carey, H.V., Domazet-Lošo, T. and Douglas, A.E. 2013. Animals in a bacterial world, a new imperative for the life sciences. Proc. Acad. Sci. USA 110: 3229–3236. Doi: 10.1073/pnas.1218525110.

McGinnis, W., Levine, M.S., Hafen, E., Kuroiwa, A. and Gehring, W.J. 1984. A conserved DNA sequence in homoeotic genes of the Drosophila Antennapedia and bithorax complexes. Nature 308: 428–33. Doi: 10.1038/308428a0.

McGlynn, S.E., Chadwick, G.L., Kempes, C.P. and Orphan, V.J. 2015. Single cell activity reveals direct electron transfer in methanotrophic consortia. Nature 526: 531–535. Doi: 10.1038/nature15512.

McGuire, B.A., Carroll, P.B., Loomis, R.A., Finneran, I.A., Jewell, P.R., Remijan, A.J. et al. 2016. Discovery of the interstellar chiral molecule propylene oxide (CH_3CHCH_2O). Science 352: 1449–1452. Doi: 10.1126/science.aae0328.

McMennamin, M.A.S. and McMennamin, D.L.S. 1990. The Emergence of Animals. The Cambrian Breakthrough. Columbia University Press, New York.

Meadows, D.L., Meadows, D.J., Randers, J. and Behrens, W.W. 3rd. 1972. The Limits to Growth. University Books, New York.

Meaney, M.J. 2001. Maternal care gene expression and the transmission of individual differences in stress reactivity across generations. Annual. Rev. Neurosci. 24: 161–192. Doi: 10.1146/annurev.neuro.24.1.1161.

Meaney, M.J. and Szyf, M. 2005. Environmental programming of stress responses through DNA methylation: life at the interface between a dynamic environment and a fixed genome. Dial. Clin. Neurosci. 7: 103–123.

Medvedev, Z.A. 1969. The rise and fall of T.D. Lysenko. Columbia University Press, New York.

Melentijevic, I., Toth, M.L., Arnold, M.L., Guasp, R.J., Harinath, G., Nguyen, K.C. et al. 2017. *C. elegans* neurons jettison protein aggregates and mitochondria under neurotoxic stress. Nature 542: 367–371. Doi: 10.1038/nature21362.

Mentel, M. and Martin, W.F. 2008. Energy metabolism among eukaryotic anaerobes in light of Proterozoic ocean chemistry. Phil. Trans. R. Soc. B 363: 2717–2729. Doi: 10.1098/rstb.2008.0031.

Mentel, M., Röttger, M., Leys, S., Tielens, A.G. and Martin, W.F. 2014. Of early animals, anaerobic mitochondria, and a modern sponge. Bioessays 36: 924–932. Doi: 10.1002/bies.201400060.

Merleau-Ponty, M. 2003. Nature: Course Notes from the Collège de France. Northwestern University Press Evanston, Illinois.

Mi, S., Lee, X., Li, X., Veldman, G.M., Finnerty, H. and Racie, L. 2000. Syncytin is a captive retroviral envelope protein involved in human placental morphogenesis. Nature 403: 785–789.

Miller, J.B. and Oldroyd, G.E.D. 2012. The role of diffusible signals in the establishment of rhizobial and mycorrhizal symbioses. pp. 1–30. *In*: Perotto, S. and Baluška, F. [eds.]. Signaling and Communication in Plant Symbiosis. Springer-Verlag Berlin Heidelberg. Doi: 10.1007/978-3-642-20966-6_1.

Miller, S.L. 1953. A production of amino acids under possible primitive earth conditions. Science 117: 528–529.

Miller, S.L. and Urey, H.C. 1959. Organic compound synthesis on the primitive Earth. Science 130: 245–251.

Minoux, M., Holwerda, S., Vitobello, A., Kitazawa, T., Kohler, H., Stadler, M.B. et al. 2017. Gene bivalency at Polycomb domains regulates cranial neural crest positional identity. Science 355: eaal2913. Doi: 10.1126/science.aal2913.

Mo, G.C.H., Ross, B., Hertel, F., Manna, P., Yang, X., Greenwald, E. et al. 2017. Genetically encoded biosensors for visualizing live-cell biochemical activity at super-resolution. Nature Methods 14: 427–434. Doi: 0.1038/nmeth.4221.

Monod, J. 1972. Chance and Necessity. Vintage Books, New York.

Moore, D. 1998. Fungal Morphogenesis. Cambridge University Press, New York.

Moran, N.A. and Sloan, D.B. 2015. The hologenome concept: helpful or hollow? PLoS Biology 13: e1002311. Doi: 10.1371/journal.pbio.1002311.

Moreira, D. and Lopez-Garcia, P. 2009. Ten reasons to exclude viruses from the tree of life. Nature Revs. Microbiol. 7: 306–311. Doi: 10.1038/nrmicro2108.

Morgan, D.K. and Whitelaw, E. 2008. The case for transgenerational epigenetic inheritance in humans. Mam. Genome 19: 394–397. Doi: 10.1007/s00335-008-9124-y.

Morgan, K.O. [ed.]. 1999. The Oxford History of Britain. Oxford University Press.

Morgan, T.H. 1919. The Physical Basis of Heredity. Lippincott Company, Philadelphia and London.

Morgan, T.H. 1932a. The Scientific Basis of Evolution. W.W. Norton and Co., New York.

Morgan, T.H. 1932b. The rise of genetics. Science 76: 261–7.

Morgan, T.H. 1934. Embryology and Genetics. Columbia University Press, New York.

Morowitz, H. and Smith, E. 2007. Energy flow and the organization of life. Complexity 13: 51–59. Doi: 10.1002/cplx.20191.

Mouw, J.K., Ou, G. and Weaver, V.M. 2014. Extracellular matrix assembly: a multiscale deconstruction. Nat. Rev. Mol. Cell. Biol. 15: 771–85. Doi: 10.1038/nrm3902.

Müller, G.B. and Newman, S.A. 2005. The innovation triad: An evodevo agenda. J. Exp. Zool. B. Mol. Dev. Evol. 15: 487–503. Doi: 10.1002/jez.b.21081.

Müller, G.B. 2007. Evo-devo: extending the evolutionary synthesis. Nat. Rev. Genet. 8: 943–949. Doi: 10.1038/nrg2219.

Muller, H.J. 1922. Variation due to change in the individual gene. Am. Nat. 56: 32–50.

Muller, H.J. 1926. The gene as the basis of life. Proc. Internat. Cong. Plant Sci. 1: 897–921.

Müller, M. 1992. Energy metabolism of ancestral eukaryotes: A hypothesis based on the biochemistry of amitochondriate parasitic protists. BioSystems 28: 33–40.

Nachmanovitch, S. 1990. Free play. Improvisation in Life and Art. Penguin Putnam, New York.

Nagano, T., Lubling, Y., Várnai, C., Dudley, C., Leung, W., Baran, Y. et al. 2017. Cell-cycle dynamics of chromosomal organization at single-cell resolution. Nature 547: 61–67. Doi: 10.1038/nature23001.

Nägeli, K. von. 1884. Mechanisch-physiologische Theorie der Abstammungslehre. R. Oldenbourg Leipzig.

Nahum, J.R., Harding, B.N. and Kerr, B. 2011. Evolution of restraint in a structured rock-paper-scissors community. Proc. Natl. Acad. Sci. 108: 10831–10838. Doi: 10.1073/Proc. Nat. Acad. Sci. USA .1100296108.

Naik, S., Larsen, S.B., Gomez, N.C., Alaverdyan, K., Sendoel, A., Yuan, S. et al. 2017. Inflammatory memory sensitizes epithelial stem cells to tissue damage. Nature 550: 475–480. Doi: 0.1038/nature24271.

Nakayama, T. and Inagaki, Y. 2011. Genomic divergence within non-photosynthetic cyanobacterial endosymbionts in rhopalodiacean diatoms. Sci. Rep. 7: 13075. Doi: 10.1038/s41598-017-13578-8.

Nasir, A., Forterre, P., Kim, K.M. and Caetano-Anollés, G. 2014. The distribution and impact of viral lineages in domains of life. Front. Microbiol. 5: 194. Doi: 10.3389/fmicb.2014.00194.

Nasir, A. and Caetano-Anollés, G. 2015. A phylogenomic data–driven exploration of viral origins and evolution. Sci. Adv. 1: e15005. Doi: 10.1126/sciadv.1500527.

Natchiar, S.K., Myasnikov, A.G., Kratzat, H., Hazemann, I. and Klaholz, B.P. 2017. Visualization of chemical modifications in the human 80S ribosome structure. Nature 551: 472–477. Doi: 10.1038/nature24482.

Nelson, C. 2004. Selector genes and the genetic control of developmental modules. pp. 17–33. *In*: Schlosser, G. and Wagner, G.P. [eds.]. Modularity in Development and Evolution. The University of Chicago Press, Chicago.

Neubauer, Z. 1989. Pojem morfogenetického pole. pp. 191–234. *In*: Kůrka, P. [ed.]. Geometrie živého. Čs. vědeckotechnol. spol.

Neuman, Y. 2008. Reviving the Living: Meaning Making in Living Systems. Elsevier.

Newman, S.A. and Müller, G.B. 2000. Epigenetic mechanisms of character origination. J. Exp. Zool. 288: 304–317. Doi: 10.1002/1097-010X.

Newman, S.A., Forgacs, G. and Müller, G.B. 2006. Before programs: The physical origination of multicellular forms. Int. J. Dev. Biol. 50: 289–299. Doi: 10.1387/ijdb.052049sn.

Newman, S.A. and Bhat, R. 2009. Dynamical patterning modules: a pattern language for development and evolution of multicellular forms. Int. J. Dev. Biol. 53: 693–705. Doi: 10.1387/ijdb.072481sn.

Newman, S.A. and Bhat, R. 2011. Lamarck's dangerous idea. pp. 157–170. *In*: Gissis, S.B. and Jablonka, E. [eds.]. Transformations of Lamarckism. From Subtle Fluids to Molecular Biology, Vienna Series in Theoretical Biology. Cambridge: MIT Press.

Nezami, B.G. and Srinivasan, S. 2013. Enteric nervous system in the small intestine: Pathophysiology and clinical implications. Curr. Gastroenterol. Rep. 12: 358–365. Doi:10.1007/s11894-010-0129-9.

Nicholson, D.J. and Dupré, J. 2018. A manifesto for a processual philosophy of biology. *In*: Nicholson, D.J. and Dupré, J. [eds.]. Everything Flows: Towards a Processual Philosophy of Biology. Oxford University Press, Oxford.

Nikoh, N., Hosokawa, T., Oshima, K., Hattori, M. and Fukatsu, T. 2011. Reductive evolution of bacterial genome in insect gut environment. Genome Biol. Evol. 3: 702–714. Doi: 10.1093/gbe/evr064.

Nitschke, W. and Russell, M.J. 2013. Beating the acetyl coenzyme S–pathway to the origin of life. Phil. Trans. R. Soc. B 368: 20120258. Doi: 10.1098/rstb.2012.0258.

Noffke, N., Christian, D., Wacey and Hazen, R.M. 2013. Microbially induced sedimentary structures recording an ancient ecosystem in the ca. 3.48 billion-year-old dresser formation, pilbara, western Australia. Astrobiology 13: 1103–1124. DOI: 10.1089/ast.2013.1030.

Nixon-Abell, J., Obara, C.J., Weigel, A.V., Li, D., Legant, W.R., Xu, C.S. et al. 2016. Increased spatiotemporal resolution reveals highly dynamic dense tubular matrices in the peripheral ER. Science 354: aaf3928. Doi: 10.1126/science.aaf3928.

Nowack, E.C.M. 2014. *Paulinella chromatophora*—rethinking the transition from endosymbiont to organelle. Acta Soc. Bot. Pol. 83: 387–397. Doi: 10.5586/asbp.2014.049.

Nowack, E.C. and Melkonian, M. 2010. Endosymbiotic associations within protists. Philos. Trans. R. Soc. Lond. B. Biol. Sci. 365: 699–712. Doi: 10.1098/rstb.2009.0188.

Nowak, M.A. 2006. Five rules for the evolution of cooperation. Science 301: 1560–1563. Doi: 10.1126/science.1133755.

Ntarlagiannis, D., Atekwana, E.A., Hill, E.A. and Gorby, J. 2007. Microbial nanowires: Is the subsurface hardwired? Geophys. Res. Lett. 34: L17305. Doi: 10.1029/2007GL030426.

Nussbaumer, A.D., Fisher, C.R. and Bright, M. 2006. Horizontal endosymbiont transmission in hydrothermal vent tubeworms. Nature 441: 345–8. Doi: 10.1038/nature04793.

Nutman, A.P., Bennett, V.C., Friend, C.R., Van Kranendonk, M.J. and Chivas, A.R. 2016. Rapid emergence of life shown by discovery of 3,700-million-year-old microbial structures. Nature 537: 535–538. Doi: 10.1038/nature19355.

O'Mahony, S.M., Felice, V.D., Nally, K., Savignac, H.M. and Claesson, M.J. et al. 2014. Disturbance of the gut microbiota in early-life selectively affects visceral pain in adulthood without impacting cognitive or anxiety-related behaviors in male rats. Neuroscience 277: 885–890. Doi: 10.1016/j.neuroscience.2014.07.054.

O'Toole, G., Kaplan, H.B. and Kolter, R. 2000. Biofilm formation as microbial development. Annu. Rev. Microbiol. 54: 49–79. Doi: 10.1146/annurev.micro.54.1.49.

Odling-Smee, F.J., Laland, K.N. and Feldman, M.F. 2003. Niche Construction: the Neglected Process in Evolution. Princeton University Press.

Odum, E.P. 1971. Fundamentals of Ecology. Saunders, Philadelphia.

Ohkuma, M. 2008. Symbioses of flagellates and prokaryotes in the gut of lower termites. Trends. Microbiol. 16: 345–352. Doi:10.1016/j.tim.2008.04.004.

Okasha, S. 2013. Evolution and the Levels of Selection. Clarendon Press, Oxford.

Ong-Abdullah, M., Ordway, J.M., Jiang, N., Ooi, S-E., Kok, S.-Y., Sarpa, N. et al. 2015. Loss of *Karma* transposon methylation underlies the mantled somaclonal variant of oil palm. Nature 525: 533–537. Doi: 0.1038/nature15365.

Ooi, S.L. and Henikoff, S. 2007. Germline histone dynamics and epigenetics. Curr. Opin. Cell Biol. 19: 257–26. Doi: 10.1016/j.ceb.2007.04.015.

Oparin, A.I. 2013 [1924]. Genesis and Evolutionary Development of Life. Academic Press.

Orgel, L. 2008. The implausibility of metabolic cycles on the prebiotic Earth. PLoS Biol. 6: e18. Doi: 0.1371/journal.pbio.0060018.

Osswald, M., Jung, E., Sahm, F., Solecki, G., Venkataramani, V., Blaes, J. et al. 2015. Brain tumour cells interconnect to a functional and resistant network. Nature 528: 93–8. Doi: 10.1038/nature16071.

Ostdiek, G. 2016. Me, Myself, and Semiotic Function: Finding the "I" in Biology. Biosemiotics 9: 435–450. Doi:10.1007/s12304-016-9268-1.

Osterwalder, M., Barozzi, I., Tissières, V., Fukuda-Yuzawa, Y., Mannion, B.J., Afzal, S.Y. et al. 2018. Enhancer redundancy provides phenotypic robustness in mammalian development. Nature 554: 239–243. Doi: 10.1038/nature25461.

Otis, L. 1994. Organic Memory: History and the Body in the Late Nineteenth and Early Twentieth Centuries. University of Nebraska Press, Nebraska.

Ou, H.D., Phan, S., Deerinck, T.J., Thor, A., Ellisman, M.H. and O'Shea, C.C. 2017. ChromEMT: Visualizing 3D chromatin structure and compaction in interphase and mitotic cells. Science 357: eaag0025. Doi: 10.1126/science.aag0025.

Oyama, S. 2000. The Ontogeny of Information: Developmental Systems and Evolution. Duke University Press, Durham.

Oyama, S., Griffithsm, P.E. and Gray, R.D. 2001. Cycles of Contingency: Developmental Systems and Evolution. MIT Press, Cambridge.

Pais, R., Lohs, C., Wu, Y., Wang, J. and Aksoy, S. 2008. The obligate mutualist *Wigglesworthia glossinidia* influences reproduction, digestion, and immunity processes of its host, the tsetse fly. Appl. Environ. Microbiol. 74: 5965–74. Doi: 10.1128/AEM.00741-08.

Papke, T.P. and Gogarten, J.P. 2012. How bacterial lineages emerge. Science 336: 45–46. Doi: 10.1126/science.1219241.

Park, M.G., Kim, M. and Kang, M. 2013. A dinoflagellate *Amylax triacantha* with plastids of the cryptophyte origin: phylogeny, feeding mechanism, and growth and grazing responses. J. Eukaryot. Microbiol. 60: 363–376. Doi: 10.1111/jeu.12041.

Parker, A. and Kornfield, I. 1997. Evolution of the mitochondrial DNA control region in the mbuna (Cichlidae) species flock of Lake Malawi, East Africa. J. mol. Evol. 45: 70–83.

Parker, J. 2016. Myrmecophily in beetles (Coleoptera): evolutionary patterns and biological mechanisms. Myrmecol. News 22: 65–108.

Paşca, S.P. 2018. The rise of three-dimensional human brain cultures. Nature 553: 437–445. Doi: 10.1038/nature25032.

Patel, B., Percivalle, C., Ritson, D.J., Duffy, C.D. and Sutherland, J.D. 2015. Common origins of RNA, protein and lipid precursors in a cyanosulfidic protometabolism. Nature Chem. 7: 301–307. Doi: 10.1038/nchem.2202.

Pátková, I., Čepl, J., Rieger, T., Blahůšková, A., Neubauer, Z. and Markoš, A. 2012. Developmental plasticity of bacterial colonies and consortia in germ-free and gnotobiotic settings. BMC Microbiol. 12: 78. Doi: 0.1186/1471-2180-12-178.

Paungfoo-Lonhienne, C., Rentsch, D., Robatzek, S., Webb, R.I., Sagulenko, E., Näsholm, T. et al. 2010. Turning the table: plants consume microbes as a source of nutrients. PLoS ONE 5: e11915. Doi: 10.1371/journal.pone.0011915.

Pearson, C.J., Lemons, D. and McGinnis, W. 2005. Modulating Hox gene functions during animal body patterning. Nat. Rev. Genet. 6: 893–904. Doi: 10.1038/nrg1726.

Pedersen, H.K., Gudmundsdottir, V., Nielsen, H.B., Hyotylainen, T., Nielsen, T., Jensen, B.A. et al. 2016. Human gut microbes impact host serum metabolome and insulin sensitivity. Nature 535: 376–381. Doi: 10.1038/nature18646.

Pelechano, V., Wei, W. and Steinmetz, L.M. 2013. Extensive transcriptional heterogeneity revealed by isoform profiling. Nature 497: 127–131. Doi: 10.1038/nature12121.

Penders, J., Thijs, C., Vink, C., Stelma, F.F., Snijders, B., Kummeling, I. et al. 2006. Factors influencing the composition of the intestinal microbiota in early infancy. Pediatrics 118: 511–521. Doi: 10.1542/peds.2005-2824.

Peng, H., Shi, J., Zhang, Y., Zhang, H., Liao, S., Li, W. et al. 2012. A novel class of tRNA-derived small RNAs extremely enriched in mature mouse sperm. Cell. Res. 22: 1609–1612. Doi: 10.1038/cr.2012.141.

Peng, J., Narasimhan, S., Marchesi, J.R., Benson, A., Wong, F.S. and Wen, L. 2014. Long term effect of gut microbiota transfer on diabetes development. J. Autoimmun. 53: 85–94. Doi: 10.1016/j.jaut.2014.03.005.

Pengelly, A.R., Copur, Ö., Jäckle, H., Herzig, A. and Müller, J. 2013. A histone mutant reproduces the phenotype caused by loss of histone-modifying factor polycomb. Science 339: 698–699. Doi: 10.1126/science.1231382.

Penney, H.D., Hassall, C., Skevington, J.H., Abbott, K.R. and Sherratt, T.N. 2012. A comparative analysis of the evolution of imperfect mimicry. Nature 483: 461–464. Doi: 10.1038/nature10961.

Pereira, P.A.B., Aho, V.T.E., Paulin, L., Pekkonen, E., Auvinen, P. and Scheperjans, F. 2017. Oral and nasal microbiota in Parkinson's disease. Parkinsonism Rela. Disord. 38: 61–67. Doi: 10.1016/j.parkreldis.2017.02.026.

Perez-Muñoz, M.E., Arrieta, M.-C., Ramer-Tait, A.E. and Walter, J. 2017. A critical assessment of the "sterile womb" and "in utero colonization" hypotheses: implications for research on the pioneer infant microbiome. Microbiome 5: 48. http://Doi.org/10.1186/s40168-017-0268-4.

Perris, D. and Perissinotto, D. 2000. Role of extracellular matrix during neural crest migration. Mechs. Dev. 95: 3–21.

Perry, J.R., Peng, L., Barry, N.A., Cline, G.W., Zhang, D., Cardone, R.L. et al. 2016. Acetate mediates a microbiome-brain-β-cell axis to promote metabolic syndrome. Nature 534: 213–217. Doi: 10.1038/nature18309.

Petersen, J.M., Zielinski, F.U., Pape, T., Seifert, R., Moraru, C., Amann, R. et al. 2011. Hydrogen is an energy source for hydrothermal vent symbioses. Nature 476: 176–180. Doi: 10.1038/nature10325.

Pfeffer, C., Larsen, S., Song, J., Dong, M., Besenbacher, F., Meyer, R.L. et al. 2012. Filamentous bacteria transport electrons over centimetre distances. Nature 491: 218–221. Doi:10.1038/nature11586.

Piersma, T. and van Gils, J.A. 2011. The Flexible Phenotype. A Body-centered Integration of Ecology, Physiology, and Behavior. Oxford University Press, Oxford.

Pigliucci, M. 2009. An extended synthesis for evolutionary biology. Ann. N. Y. Acad. Sci. 1168: 218–228. Doi: 10.1111/j.1749-6632.2009.04578.x.

Pigliucci, M. and Müller, G.B. 2010. Evolution: The Extended Synthesis. Cambridge MIT Press, Cambridge.

Pittis, A.A. and Gabaldón, T. 2016. Late acquisition of mitochondria by a host with chimaeric prokaryotic ancestry. Nature 531: 101–104. Doi.org/10.1038/nature16941.

Piunti, A. and Shilatifard, A. 2016. Epigenetic balance of gene expression by Polycomb and COMPASS families. Science 352: aad9780. Doi: 10.1126/science.aad9780.

Pombo, A. and Dillon, N. 2015. Three-dimensional genome architecture: players and mechanisms. Nat.Rev. Mol. Cell. Biol. 16: 245–57. Doi: 10.1038/nrm3965.

Popkun, G. 2016. The physics of life. Nature 529: 16–18.

Por, F.D. 2011. Lamarck's 'pouvoir de la nature' demystified: a thermodynamic foundation to Lamarck's concept of progressive evolution. pp. 373–376. *In*: Gissis, S.B. and Jablonka, E. [eds.]. Transformations of Lamarckism. From Subtle Fluids to Molecular Biology, Vienna Series in Theoretical Biology. Cambridge: MIT Press.

Portin, P. 2012. Does epigenetic inheritance revolutionize the foundations of the theory of evolution? Topics. Curr. Genet. 5: 49–59.

Portin, P. and Wilkins, A. 2017. The evolving definition of the term 'gene'. Genetics 205: 1353–1364. doi: 10.1534/genetics.116.196956.

Powers, R.E., Wang, S., Liu, T.Y. and Rapoport, T.A. 2017. Reconstitution of the tubular endoplasmic reticulum network with purified components. Nature 543: 257–260. Doi: 10.1038/nature21387.

Powner, M.W., Gerland, B. and Sutherland, J.D. 2009. Synthesis of activated pyrimidine nucleotides in prebiotically plausible conditions. Nature 459: 239–242. Doi: 10.1038/nature08013.

Prakash, K. and Fournier, D. 2018. Evidence for the implication of the histone code in building the genome structure. BioSystems 167: 49–59. Doi: 10.1016/j.biosystems.2017.11.

Proctor, L.M. and Fuhrman, J.A. 1990. Viral mortality of marine bacteria and cyanobacteria. Nature 343: 60–62.

Pross, A. 2004. Causation and the origin of life. Metabolism or replication first? Orig. Life and Evol. Biosphere 34: 307–321. Doi: 10.1023/B: RIG.0000016446.51012.bc.

Ptashne, M. 1987. A Genetic Switch, Third Edition: Phage Lambda Revisited. Cold Spring Harbor Laboratory Press.

Qiao, Y., Wu, M., Fenf, Y., Zhou, Z., Chen, L. and Chen, F. 2018. Alterations of oral microbiota distinguish children with autism spectrum disorders from healthy controls. Sci. Rep. 8: 1597. Doi: 10.1038/s41598-018-19982-y.

Qin, J., Li, Y., Cai, Z., Li, S., Zhu, J., Zhang, F et al. 2012. A metagenome-wide association study of gut microbiota in type 2 diabetes. Nature 490: 55–61. Doi: 10.1038/nature11450.

Quadrato, G., Nguyen, T., Macosko, E.Z., Sherwood, J.L., Min Yang, S., Berger, D.R et al. 2017. Cell diversity and network dynamics in photosensitive human brain organoids. Nature 545: 48–53. Doi: 10.1038/nature22047.

Quint, M., Drost, H.G., Gabel, A., Ullrich, K.K., Bönn, M. and Grosse, I. 2012. A transcriptomic hourglass in plant embryogenesis. Nature 490: 8–101. Doi: 0.1038/nature11394. Doi: 10.1126/science.aan3351.

Radl, E. 1930. The History of Biological Theories. Oxford University Press, London.

Raff, R.A. 1996. The Shape of Life: Genes, Development, and the Evolution of Animal Form. University of Chicago Press, Chicago.

Rajilić-Stojanović, M., Biagi, E., Heilig, H.G., Kajander, K., Kekkonen, R.A., Tims, S. et al. 2011. Global and deep molecular analysis of microbiota signatures in faecal samples from patients with irritable bowel syndrome. Gastroenterology 5: 1792–1801. Doi: 10.1053/j.Gastro.2011.07.043.

Rakoff-Nahoum, S., Foster, K.R. and Comstock, L.E. 2016. The evolution of cooperation within the gut microbiota. Nature 533: 255–259. Doi: 0.1038/nature17626.

Rappaport, R.A. 2010[1999]. Ritual and Religion in the Making of Humanity. Cambridge University Press, Cambridge.

Rasmussen, B., Fletcher, I.R., Brocks, J.J. and Kilburn, M.R. 2008. Reassessing the first appearance of eukaryotes and cyanobacteria. Nature 455: 1101–1104. Doi: 10.1038/nature07381.

Rasmussen, S.R., Füchtbauer, W., Novero, M., Volpe, V., Malkov, N., Genre, A. et al. 2016. Intraradical colonization by arbuscular mycorrhizal fungi triggers induction of a lipochitooligosaccharide receptor. Sci. Rep. 6: 29733. Doi: 10.1038/srep29733.

Rauch, C., Jahns, P., Tielens, A.G.M., Gould, S.B. and Martin, W.F. 2017. On being the right size as an animal with plastids. Front. Plant. Sci. 8: 1402. Doi: 10.3389/fpls.2017.01402.

Raveh-Sadka, T., Thomas, B.C., Singh, A., Firek, B., Brooks, B., Castelle, C.J. et al. 2015. Gt bacteria are rarely shared by co-hospitalized premature infants, regardless of necrotizig enterocolitis development. eLife 4: e05477. Doi: 10.7554/eLife.05477.

Reed, R.D., Papa, R., Martin, A., Hines, H.M., Counterman, B.A., Pardo-Diaz, C. et al. 2011. *Optix* drives the repeated convergent evolution of butterfly wing pattern mimicry. Science 333: 1137–41. Doi: 10.1126/science.1208227.

Reguera, G. 2009. Are microbial conversations being lost in translation? Microbe 4: 506–512. Doi: 10.1128/microbe.4.506.1.

Reilly, J.N., McLaughlin, E.A., Stanger, S.J., Anderson, A.L., Hutcheon, K., Church, K. et al. 2016. Characterisation of mouse epididymosomes reveals a complex profile of microRNAs and a potential mechanism for modification of the sperm epigenome. Sci. Rep. 6: 31794. Doi: 10.1038/srep31794.

Rescigno, M., Urbano, M., Valsazina, B., Francoloni, M., Rotta, G., Bonasio, R. et al. 2001. Dendritic cells express tight junction proteins and penetrate gut epithelial monolayers to sample bacteria. Nature Immunol. 2: 367. Doi: 10.1078/0171-2985-00094.

Reyes, A., Haynes, M., Hanson, N., Angly, F.E., Heath, A.C., Rohwer, F. et al. 2010. Viruses in the faecal microbiota of monozygotic twins and their mothers. Nature 466: 334–338. Doi: 10.1038/nature09199.

Reyes-Prieto, A., Weber, A.P. and Bhattacharya, D. 2007. The origin and establishment of the plastid in algae and plants. Annu. Rev. Genet. 41: 47–168. Doi: 10.1146/annurev.genet.41.110306.130134.

Reyes-Prieto, A., Moustafa, A. and Bhattacharya, D. 2008. Multiple genes of apparent algal origin suggest ciliates may once have been photosynthetic. Curr. Biol. 18: 956–962. Doi: 10.1016/j.cub.2008.05.042.

Richardson, M.K. 1995. Heterochrony and the phylotypic period. Dev. Biol. 172: 412–421.

Richardson, M.K., Hanken, J., Gooneratne, M.L., Pieau, C., Raynaud, A., Selwood, L. et al. 1997. There is no highly conserved embryonic stage in the vertebrates: implications for current theories of evolution and development. Anat. Embryol. 196: 91–106.

Richardson, S.L. 2001. Endosymbiont change as a key innovation in the adaptive radiation of Soritida (Foraminifera). Paleobiology 27: 262–289. Doi: 10.1666/0094-8373.

Richerson, P.J. and Boyd, R. 2006. Not by Genes Alone. University of Chicago Press, London.

Ridaura, V.K., Faith, J.J., Rey, F.E., Cheng, J., Duncan, A.E., Kau, A.L. et al. 2013. Gut microbiota from twins discordant for obesity modulate metabolism in mice. Science 341: 1241214. Doi: 10.1126/science.1241214.

Riedel, C.U., Schwiertz, A. and Egert, M. 2014. The stomach and small and large intestinal microbiomes. pp. 1–19. *In*: Marchesi, J.R. [eds.]. Human Microbiota and Microbiome. Adv. Mol. Cell. Microbiol. 25. CAB International.

Rieger, T., Neubauer, Z., Blahůšková, A., Cvrčková, F. and Markoš, A. 2008. Bacterial body plans. Colony ontogeny in *Serratia marcescens*. Commun. Integr. Biol. 1: 78–87. Doi: 10.4161/cib.1.1.6547.

Rignano, E. 1911. Upon the inheritance of acquired characters. A hypothesis of heredity, development, and assimilation. The open court publishing Co., Chicago.

Rivera, M.C. and Lake, J.A. 2004. The ring of life provides evidence for a genome fusion origin of eukaryotes. Nature 431: 152–155. Doi: 10.1038/nature02848.

Rivron, N.C., Frias-Aldeguer, J., Vrij, E.J., Boisset, J.C., Korving, J., Vivié, J. et al. 2018. Blastocyst-like structures generated solely from stem cells. Nature 557: 106–111. Doi: 10.1038/s41586-018-0051-0.

Robertson, M.P. and Joyce, G.F. 2012. The origins of the RNA world. Cold Spring Harbor Perspect. Biol. 4: a003608. Doi: 10.1101/cshperspect.a003608.

Robertson, M.P. and Joyce, G.F. 2014. Highly efficient self-replicating RNA enzymes. Chem. Biol. 21: 238–245. Doi: 10.1016/chembiol.2013.12.004.

Robinson, G.E. and Barron, A.B. 2017. Epigenetics and the evolution of instincts. Science 356: 26–27. Doi: 10.1126/science.aam6142.

Rocklin, G.J., Chidyausiku, T.M., Goreshnik, I., Ford, A., Houliston, S., Lemak, A. et al. 2017. Global analysis of protein folding using massively parallel design, synthesis, and testing. Science 357: 168–175. Doi: 10.1126/science.aan0693.

Roger, L.C., Costabile, A., Holland, D.T., Hoyles, T. and McCartney, A.L. 2010. Examination of faecal *Bifidobacterium* populations in breast-fed and formula-fed infants during the first 18 months of life. Microbiology 158: 3329–3341. Doi: 10.1099/mic.0.043224-0.

Rohner,N.,Jarosz,D.F.,Kowalko,J.E.,Yoshizawa,M.,Jeffery,W.R.,Borowsky,R.L.etal.2013.Cryptic variation in morphological evolution: HSP90 as a capacitor for loss of eyes in cavefish. Science 342: 1372–1375. Doi: 10.1126/science.1240276.

Rohwer, F. and Thurber, R.V. 2009. Viruses manipulate the marine environment. Nature 459: 207–212. Doi: 10.1038/nature.

Root-Bernstein, M. and Root-Bernstein, R. 2015. The ribosome as a missing link in the evolution of life. J. Theor. Biol. 367: 130–158. Doi: 10.1016/j.jtbi.2014.11.025.

Root-Bernstein, M. and Root-Bernstein, R. 2016. The ribosome as a missing link in prebiotic evolution II: Ribosomes encode ribosomal proteins that bind to common regions of their own mRNAs and rRNAs. J. Theor. Biol. 397: 115–127. Doi/10.1016/j.jtbi.2016.02.030.

Root-Bernstein, R. 2007. Simultaneous origin of homochirality, the genetic code and its directionality. BioEssays 29: 689–698. Doi: 10.1002/bies.2060208060.

Rosen, M.J., Davison, M., Bhaya, D. and Fisher, D.S. 2015. Fine-scale diversity and extensive recombination in a quasisexual bacterial population occupying a broad niche. Science 348: 1019–1023. Doi: 10.1126/science.aaa4456.

Rosenberg, E. and Zilber-Rosenberg, I. 2013. The Hologenome Concept: Human, Animal and Plant Microbiota. Switzerland: Springer.

Rosenberg, H. 1965[1959]. The Tradition of the New. McGraw-Hill, New York.

Rosenzweig, R.F. and Adams, J. 1994. Microbial adaptation to a changeable environment: Cell-cell interactions mediate physiological and genetic differentiation. BioEssays 16: 715–717. Doi: 10.1002/bies.950161005.

Rothschild, D., Weissbrod, O., Barkan, E., Kurilshikov, A., Korem, Zeevi, D. et al. 2018. Environment dominates over host genetics in shaping human gut microbiota. Nature 555: 210–215. Doi: 10.1038/nature25973.

Röttger, R., Dettmering, C., Krüger, R., Schmaljohann, R. and Hohenegger, J. 1998. Gametes in nummulitids (Foraminifera). J. Foraminifer. Res. 28: 345–348.

Roux, W. 1883. Über die Zeit der Bestimmung der Hauptrichtungen des Froschembryo. Wilhelm Engelmann, Leipzig.

Russell, M.J., Daniel, R.M., Hall, A.J. and Sherringham, J.A. 1994. A hydrothermally precipitated catalytic iron sulphide membrane as a first step towards life. J. Mol. Evol. 39: 231–243.

Russell, M.J., Nitschke, W. and Branscomb, E. 2015. The inevitable journey to being. Phil. Trans. R. Soc. B. 368: 20120254. Doi: 10.1098/rstb.2012.0254.

Rutherford, S. and Lindquist, S. 1998. Hsp90 as a capacitor for morphological evolution. Nature 396: 336–342. Doi: 10.1038/24550.

Ruyer, R. 1974. La gnose de Princeton [The Princeton gnosis]. Fayard, Paris.

Sachs, J.L., Skophammer, R.G. and Regus, J.U. 2011. Evolutionary transitions in bacterial symbiosis. Proc. Nat. Acad. Sci. USA 108: 10800–10807. Doi/10.1073/Proc. Nat. Acad. Sci. USA .1100304108.

Saladino, R., Botta, G., Pino, S., Costanzo, G. and Di Mauro, E. 2012. From the one-carbon amide formamide to RNA all the steps are prebiotically possible. Biochimie 94: 1451–1456. Doi: 10.1016/biochi.2012.02.018.

Sanchez, T., Chen, D.T., DeCamp, S.J., Heymann, M. and Dogic, Z. 2012. Spontaneous motion in hierarchically assembled active matter. Nature 491: 431–434. Doi: 10.1038/nature11591.

Sander, K. 1983. The evolution of patterning mechanisms: gleanings form insect embryogenesis and spermatogenesis. pp. 137–159. *In*: Goodwin, B.C., Holder, N. and Wylie, C.C. [eds.]. Development and Evolution. Cambridge University Press, Cambridge.

Sapp, J. 1990. Where the Truth Lies. Franz Moewus and the Origins of Molecular Biology. Cambridge University Press.

Sapp, J. 2003. Genesis: The Evolution of Biology. Oxford University Press, Oxford.

Schiano, P., Provost, A., Clocchiatti, R. and Faure, F. 2006. Transcrystalline melt migration and Earth's mantle. Science 314: 970–974. Doi: 10.1126/science.1132485.

Schilbach, S., Hantsche, M., Tegunov, D., Dienemann, C., Wigge, C., Urlaub, H. et al. 2017. Structures of transcription pre-initiation complex with TFIIH and Mediator. Nature 551: 204–209. Doi: 10.1038/nature24282.

Schimmel, P. 2018. The emerging complexity of the tRNA world: mammalian tRNAs beyond protein synthesis. Nature. Revs. Mol. Cell. Biol. 19: 45–58. Doi: 10.1038/nrm.2017.77.

Schlosser, G. 2004. Modules in development and evolution. pp. 520–582. *In*: Schlosser, G. and Wagner, G.P. [eds.]. Modularity in Development and Evolution. The University of Chicago Press, Chicago.

Schmalhausen, I.I. 1949. Factors of evolution: the theory of stabilizing selection. Blakiston Co., Philadelphia.

Schmitz, R.J., Schultz, M.D., Lewsey, M.G., O'Malley, R.C., Urich, M.A., Libiger, O. et al. 2011. Transgenerational epigenetic instability is a source of novel methylation variants. Science 334: 369–73. Doi: 10.1126/science.1212959.

Schönknecht, G., Chen, W.H., Ternes, C.M., Barbier, G.G., Shrestha, R.P., Stanke, M. et al. 2013a. Gene transfer from bacteria and Archaea facilitated evolution of an extremophilic eukaryote. Science 339: 1207–1210. Doi: 10.1126/science.1231707.

Schönknecht, G., Weber, A.P. and Lercher, M.J. 2013b. Horizontal gene acquisitions by eukaryotes as drivers of adaptive evolution. Bioessays 36: 9–20. Doi: 10.1002/bies.201300095.

Schopf, J.W. [ed.]. 1983. Earth's Earlier Biosphere, its Origin and Evolution. Princeton University Press, Princeton.

Schopf, J.W. and Klein, C. [eds.]. 1992. The Proterozoic Biosphere: A Multidisciplinary Study. Cambridge University Press, Cambridge.

Schopf, J.W. 1993. Microfossils of the early archean apex chert: New evidence of the antiquity of life. Science 260: 640–646. Doi: 10.1126/science.260.5108.640.

Schrödinger, E. 1944. What is Life. Cambridge University Press, Cambridge, UK.

Schuchmann, K. and Muller, V. 2014. Autotrophy at the thermodynamic limit of life: a model for energy conservation in acetogenic bacteria. Nature Rev. Microbiol. 12: 809–821. Doi: 10.1038/nrmicro3365.

Schulz, H.N., Brinkhoff, T., Ferdelman, T.G., Mariné, M.H., Teske, A. and Jorgensen, B.B. 1999. Dense populations of a giant sulfur bacterium in Namibian shelf sediments. Science 284: 493–495. Doi:10.1126/science.284.5413.493.

Seenivasan, R., Sausen, N., Medlin, L.K. and Melkonian, M. 2013. *Picomonas judraskeda* Gen. et sp. nov.: The first identified member of the Picozoa phylum nov., a widespread group of Picoeukaryotes, formerly known as 'Picobiliphytes'. PLoS ONE 8.3.: e59565. Doi: 10.1371/journal.pone.0059565.

Seet, C.S., He, C., Bethune, M.T., Li, S., Chick, B., Gschweng, E.H. et al. 2017. Generation of mature T cells from human hematopoietic stem and progenitor cells in artificial thymic organoids. Nat. Methods 14: 521–530. Doi: 10.1038/nmeth.4237.

Sekelja, M., Paulsen, J. and Collas, P. 2015. 4D nucleomes in single cells: what can computational modeling reveal about spatial chromatin conformation? Genome Biol. 17: 54. Doi 10.1186/s13059–016–0923–2.

Sela, D.A., Li, Y., Lerno, L., Wu, S., Marcobal, A.M., German, J.B. et al. 2011. An infant-associated bacterial commensal utilizes breast milk sialyloligosaccharides. J. Biol. Chem. 286: 11909–11918. Doi: 10.1074/jbc.M110.193359.

Semon, R. 1904. Die Mneme. W. Engelmann, Leipzig.

Sender, R., Fuchs, S. and Milo, R. 2016. Revised estimates for the number of human and bacteria cells in the body. PLoS Biol. 14: e1002533. Doi: 10.1371/journal.pbio.1002533.

Seong, K.-H., Li, D., Shimizu, H., Nakamura, R. and Ishii, S. 2011. Inheritance of stress-induced, ATF-2 Dependent epigenetic change. Cell 145: 1049–1061. Doi: 10.1016/j.cell.2011.05.029.

Shaffer, S.M., Dunagin, M.C., Torborg, S.R., Torre, E.A., Emert, B., Krepler, C. et al. 2017. Rare cell variability and drug-induced reprogramming as a mode of cancer drug resistance. Nature 546: 431–435. Doi: 10.1038/nature22794.

Shahid, S., Kim, G., Johnson, N.R., Wafula, E., Wang. F., Coruh, C. et al. 2018. MicroRNAs from the parasitic plant *Cuscuta campestris* target host messenger RNAs. Nature 553: 82–85. Doi: 10.1038/nature25027.

Shapiro, B.J., Friedman, J., Cordero, O.X., Preheim, S.P., Timberlake, S.C., Szabó, G. et al. 2012. Population genomics of early events in the ecological differentiation of bacteria. Science 336: 48–51. Doi: 10.1126/science.1218198.

Shapiro, J.A. 1997. Multicellularity: The rule, not the exception. Lessons from *E. coli* colonies. pp. 14–49. *In*: Dworkin, M. and Shapiro, J.A. [eds.]. Bacteria as Multicellular Organisms. Oxford University Press.

Shapiro, J.A. 1998. Thinking about bacterial populations as multicellular organism. Ann. Rev. Microbiol. 52: 81–104. Doi: 10.1146/annurev.micro.52.1.81.

Shapiro, J.A. 2016. Nothing in evolution makes sense except in the light of genomics: read-write genome evolution as an active biological process. Biology 5: 27. Doi: 10.3390/biology5020027.

Sharma, U., Conine, C.C., Shea, J.M., Boskovic, A., Derr, A.G., Bing, X.Y. et al. 2016. Biogenesis and function of tRNA fragments during sperm maturation and fertilization in mammals. Science 351: 391–396. Doi: 10.1126/science.aad6780.

Sharon, G., Segal, D., Ringo, J.M., Hefetz, A., Zilber-Rosenberg, I. and Rosenberg, E. 2010. Commensal bacteria play a role in mating preference of *Drosophila melanogaster*. Proc. Natl. Acad. Sci. USA 107: 20051–20056. Doi: 10.1073/pnas.1009906107.

Shechner, D.M., Grant, R.A., Bagby, S.C., Koldobskaya, Y., Piccirilli, J.A. and Bartel, D.P. 2009. Crystal structure of the catalytic core of an RNA-polymerase ribozyme. Science 326: 1271–1275. Doi: 10.1126/science.1174676.

Sheehan, D., Moran, C. and Shanahan, F. 2015. The microbiota in inflammatory bowel disease 50: 495–507. Doi: 10.1007/s00535-015-1064-1.

Shoemark, D.K. and Allen, S.J. 2015. The microbiome and disease: reviewing the links between the oral microbiome, aging, and Alzheimer's disease. J. Alzheimers Dis. 43: 725–738. Doi: 10.3233/JAD-141170.

Sibatani, A. 1987. An attempt to structuralize biology. Riv. Biol. Biol. Forum 80: 558–564.

Siegel, J. 2008. The Emergence of Pidgin and Creole Languages. Oxford University Press, Oxford.

Siklenka, K., Erkek, S., Godmann, M., Lambrot, R., McGraw, S., Lafleur, C. et al. 2015. Disruption of histone methylation in developing sperm impairs offspring health transgenerationally. Science 350: aab2006. Doi: 10.1126/science.aab2006.

Simpson, G.G. 1944. Tempo and Mode of Evolution. Columbia University Press, New York.

Slack, J.M.W., Holland, P.W.H. and Graham, C.F. 1993. The zootype and the phylotypic stage. Nature 361: 490–492.

Sleep, N.H., Bird, D.K. and Pope, E. 2012. Paleontology of Earth's mantle. Annu. Rev. Earth. Planet. Sci. 40: 277–300. Doi: 10.1146/annurev–earth–092611–090602.

Smith, E. and Morowitz, H. 2007. Universality in intermediary metabolism. Proc. Nat. Acad. Sci. USA 101: 13168–13173. Doi: 10.1073/Proc. Nat. Acad. Sci. USA .0404922101.

Smith, M.L., Bruhn, J.N. and Anderson, J.B. 1992. The fungus *Armillaria bulbosa* is among the largest and oldest living organisms. Nature 356: 428–431. Doi: 0.1038/356428a0.

Smith, Z.D., Chan, M.M., Humm, K.C., Karnik, R., Mekhoubad. S., Regev, A. et al. 2014. DNA methylation dynamics of the human preimplantation embryo. Nature 511: 611–615. Doi: 10.1038/nature13581.

Smith, Z.D., Shi, J., Gu, H., Donaghey, J., Clement, K., Cacchiarelli, D. et al. 2017. Epigenetic restriction of extraembryonic lineages mirrors the somatic transition to cancer. Nature 549: 543–547. Doi: 10.1038/nature23891.

Smits, L.P., Bouter, K.E., de Vos, W.M., Borody, T.J. and Nieuwdorp, M. 2013. Therapeutic potential of fecal microbiota transplantation. Gastroenterology 145: 964–53. Doi: 10.1053/j.gastro.2013.08.058.

Sogin, M.L., Morrison, H.G., Hinkle, G. and Silberman, J.D. 1996. Ancestral relationships of the major eukaryotic lineages. Microbiologia SEM 12: 17–28.

Sojo, V., Herschy, B., Whicher, A., Camprubí, E. and Lane, N. 2016. The origin of life in alkaline hydrothermal vents. Astrobiology 16: 181–97. Doi: 10.1089/ast.2015.1406.

Son, J.S., Zheng, L.J., Rowehl, L.M., Tian, X., Zhang, Y., Zhu, W. et al. 2015. Comparison of fecal microbiota in children with autism spectrum disorders and neurotypical siblings in the Simons simplex collection. PloS One 10: e0137725. Doi: 10.1371/journal.pone.0137725.

Song, S.J., Lauber, C., Costello, E.K., Lozupone, C.A., Humprey, G., Berg-Lyons, D. et al. 2013. Cohabiting family members share microbiota with one another and with their dogs. eLIfe 2: e00458. Doi: 10.7554/eLife.00458.

Sonneborn, T.M. 1964. The determinants and evolution of life. The differentiation of cells. Proc. Natl. Acad. Sci. USA 51: 915–929.

Sonnenburg, E.D., Smits, S.A., Tikhonov, M., Higginbottom, S.K., WinGreen, N.S. and Sonnenburg, J.L. 2016. Diet-induced extinctions in the gut microbiota compound over generations. Nature 529: 212–215. Doi: 10.1038/nature16504.

Soo, R.M., Hemp, J., Parks, D.H., Fischer, W.W. and Hugenholtz, P. 2017. On the origins of oxygenic photosynthesis and aerobic respiration in Cyanobacteria. Science 355: 1436–1440. Doi: 10.1126/science.aal3794.

Sousa, F.L., Thiergart, T., Landan, G., Nelson-Sathi, S., Pereira, I.A., Allen, J.F. et al. 2015. Early biogenetic revolution. Phil. Trans. R. Soc. B 368: 20130088. Doi: 10.1098/rstb.2013.0088.

Soyfer, V.N. 1994. Lysenko and the Tragedy of Soviet Science. Rutgers University Press.

Spang, A., Saw, J.H., Jørgensen, S.L., Zaremba-Niedzwiedzka, K., Martijn, J., Lind, A.E. et al. 2015. Complex archaea that bridge the gap between prokaryotes and eukaryotes. Nature 521: 173–179. Doi: 10.1038/nature14447.

Specchia, V., Piacentini, L., Tritto, P., Fanti. L., D'Alessandro, R., Palumbo, G. et al. 2010. Hsp90 prevents phenotypic variation by suppressing the mutagenic activity of transposons. Nature 463: 662–665. Doi: 10.1038/nature08739.

Spemann, H. 1938. Embryonic Development and Induction. Yale University Press, London.

Spencer, H. 1864. Principles of Biology. Williams and Norgate, London.

Spribille, T., Tuovinen, V., Resl, P., Vanderpool, D., Wolinski, H., Aime, M.C. et al. 2016. Basidiomycete yeasts in the cortex of ascomycete macrolichens. Science 353: 488–92. Doi: 10.1126/science.aaf8287.

Stadler, M.R. and Eisen, M.B. 2017. Atlas…t, patterns from every cell. Science 358: 172–173. Doi: 10.1126/science.aap8493.

Stappenbeck, T.S. and Virgin, H.W. 2016. Accounting for reciprocal host-microbiome interactions in experimental science. Nature 534: 191–199. Doi: 10.1038/nature18285.

Starr, T.N., Picton, L.K. and Thornton, J.W. 2017. Alternative evolutionary histories in the sequence space of an ancient protein. Nature 549: 409–413. Doi: 10.1038/nature2390.

Stebbins, G.L. 1950. Variation and Evolution in Plants. Columbia University Press, New York.

Stephenson, M. and Rowatt, E. 1947. The production of acetylcholine by strain of *Lactobacillus plantarum*. J. Gen. Microbiol. 1: 279–298.

Stevens, T.J., Lando, D., Basu, S., Atkinson, L.P., Cao, Y., Lee, S.F. et al. 2017. 3D structures of individual mammalian genomes studied by single-cell Hi-C. Nature 544: 59–64. Doi: 10.1038/nature21429.

Stokholm, J., Thorsen, J., Chawes, B.L., Schjørring, S., Krogfelt, K.A., Bønnelykke, K. et al. 2016. Cesarean section changes neonatal gut colonization. J. Allergy Clin. Immunol. 138: 881–889. Doi: 10.1016/j.jaci.2016.01.028.

Stout, M.J., Conlon, B., Landeau, M., Lee, I., Bower, C., Zhao, Q. et al. 2013. Identification of intracellular bacteria in the basal plate of teh human placenta in term and preterm gestation. Am. J. Obstet. Gynecol. 208: 226.e1-7. Doi: 10.1016/j.ajog.2013.01.018.

Stoye, J.P. 2009. Proviral protein provides placental function. Proc. Nat. Acad. Sci. USA 106: 11827–11828. Doi: 10.1073/Proc. Nat. Acad. Sci. USA .0906295106.

Sturtevant, A.H. 1925. The effects of unequal crossing over at the *Bar* locus in *Drosophila*. Genetics 10: 117–147.

Suez, J., Korem, T., Zeevi, D., Zilberman-Schapira, G., Thaiss, C.A., Maza, O. et al. 2014. Artificial sweeteners induce glucose intolerance by altering the gut microbiota. Nature 514: 181–186. Doi: 10.1038/nature13793.

Summers, A.O., Wireman, J., Vimy, M.J., Lorscheider, F.L., Marshall, B., Levy, S.B et al. 1993. Mercury released from dental "silver" fillings provokes an increase in mercury- and antibiotic-resistant bacteria in oral and intestinal floras of primates. Antimicrob. Agents. Chemother. 37: 825–834.

Sutter, M., Greber, B., Aussignargues, C. and Kerfeld, C.A. 2017. Assembly principles and structure of a 6.5-MDa bacterial microcompartment shell. Science 356: 1293–1297. Doi: 10.1126/science.aan3289.

Szostak, J.W. 2011. An optimal degree of physical and chemical heterogeneity for the origin of life? Phil. Trans. R. Soc. B 366: 2894–2901. Doi: 10.1098/rstb.2011.0140.

Švorcová, J. 2012. The phylotypic stage as a boundary of modular memory: non mechanistic perspective. Theor. Biosci. 131: 31–42. Doi: 10.1007/s12064-012-0149-0.

Švorcová, J. 2016. Distributed heredity and development: a heterachical perspective. Biosemiotics 9: 331–343. Doi: 10.1007/s12304-016-9276-1.

Švorcová, J., Markoš, A. and Das, P. 2018. Origins of the cellular biosphere. pp. 271–290. *In*: Sahi, V.P. and Baluška, F. [eds.]. Concepts in Cell Biology—History and Evolution. Springer.

Švorcová, J. and Kleisner, K. 2018. Evolution by meanining attribution: Notes on biosemiotics interpretations of extended evolutionary synthesis. Biosemiotics 1–14 (published online). Doi: 10.1007/s12304-018-9328-9.

Taipale, M., Jarosz, D.F. and Lindquist, S. 2010. HSP90 at the hub of protein homeostasis: emerging mechanistic insights. Nature. Revs. Mol. Cell. Biol. 11: 515–528. Doi: 10.1038/nrm2918.

Takeshita, K. and Kikuchi, Y. 2017. *Riptortus pedestris* and *Burkholderia* symbiont: an ideal model system for insect-microbe symbiotic associations. Res. Microbiol. 168: 175–187. Doi: 10.1016/j.resmic.2016.11.005.

Taleb, N.N. 2008. The Black Swan. The Impact of the Highly Improbable. Penguin Books.

Taniguchi, C.M., Emanuelli, B. and Kahn, C.R. 2006. Critical nodes in signalling pathways: insights into insulin action. Nat. Rev. Mol. Cell. Biol. 7: 85–96. Doi: 10.1038/nrm1837.

Tebb, G. 1998. Kim Nasmyth: The universal truth. Curr. Biol. 8: R257–R258. Doi: 10.1016/S0960-9822(98)70165-4.

Terakawa, T., Bisht, S., Eeftens, J.M., Dekker, C., Haering, C.H. and Greene, E.C. 2017. The condensin complex is a mechanochemical motor that translocates along DNA. Science 358: 672–676. Doi: 10.1126/science.aan651.

Tessarz, P. and Kouzarides, T. 2014. Histone core modifications regulating nucleosome structure and dynamics. Nature. Revs. Mol. Cell. Biol. 15: 703–708. Doi: 10.1038/nrm3890.

The ENCODE Project Consortium. 2012. An integrated encyclopedia of DNA elements in the human genome. Nature 489: 57–74.

Theis, K.R., Schmidt, T.M. and Holekamp, K.E. 2012. Evidence for a bacterial mechanism for group-specific social odors among hyenas. Sci. Rep. 2: 615. Doi: 10.1038/srep00615.

Theis, K.R., Venkataraman, A., Dycus, J.A., Koonter, K.D., Schmitt-Matzen, E.N., Wagner, A.P. et al. 2013. Symbiotic bacteria appear to mediate hyena social odors. Proc. Natl. Acad. Sci. USA 110: 19832–19837. Doi: 10.1073/pnas.1306477110.

Theis, K.R., Dheilly, N.M., Klassen, J.L., Brucker, R.M., Baines, J.B., Bosh, T.C.G. et al. 2016. Getting the hologenome concept right: an eco-evolutionary framework for hosts and their microbiomes 1: e00028–16. DOI: 10.1128/mSystems.00028-16.

Thompson, A.W., Foster, R.A., Krupke, A., Carter, B.J., Musat, N., Vaulot, D. et al. 2012. Unicellular cyanobacterium symbiotic with a single–celled eukaryotic alga. Science 337: 1546–1550. Doi: 10.1126/science.1222700.

Thompson, D.W. 1961 [1917]. On Growth and Form. Cambridge University Press, Cambridge.

Thul, P.J., Åkesson, L., Wiking, M., Mahdessian, D., Geladaki, A., Ait Blal, H. et al. 2017. A subcellular map of the human proteome. Science 356: eaal3321. Doi: 10.1126/science.aal3321.

Timmers, A.C., Soupène, E., Auriac, M.C., de Billy, F., Vasse, J., Boistard, P. et al. 2000. Saprophytic intracellular rhizobia in alfalfa nodules. Mol. Plant. Microbe. Interact. 13: 1204–1213. Doi: 10.1094/MPMI.2000.13.11.1204.

Timmons, C.L., Shao, Q., Wang, C., Liu, L., Liu, H., Dong, X. et al. 2013. GB virus type C E2 protein inhibits human immunodeficiency virus type 1 assembly through interference with HIV-1 gag plasma membrane targeting. J. Infect. Dis. 207: 1171–1180. Doi: 10.1093/infdis/jis459.

Trainor, P.A. 2014. Neural Crest Cells. Evolution, Development and Disease. Elsevier.

Trasande, L., Blustein, J., Liu, M., Corwin, E., Cox, L.M. and Blaser, M.J. 2013. Infant antibiotic exposures and early-life body mass. Int. J. Obes. (Lond.) 37: 16–23. Doi: 10.1038/ijo.2012.132.

Tschermak, E. 1900. Über künstliche Kreuzung bei *Pisum sativum*. Ber. Deut. Bot. Ges. 18: 232–239.

Tung, J., Barreiro, L.B., Burns, M.B., Grenier, J.-C., Lynch, J., Grieneisen, L.E. et al. 2015. Social networks predict gut microbiome composition in wild baboons. eLife 4: e05224. 10.7554/eLife.05224.

Turgay, Y., Eibauer, M., Goldman, A.E., Shimi, T., Khayat, M., Ben-Harush, K. et al. 2017. The molecular architecture of lamins in somatic cells. Nature 543: 261–264. Doi: 10.1038/nature21382.

Turner, B.M. 2009. Epigenetic responses to environmental change and their evolutionary implications. Doi: 10.1098/rstb.2009.01253-18.

Uexküll, J. von 2010 [1940]. The theory of meaning. pp. 90–114. *In*: Favareau, D. [ed.]. Essential Readings in Biosemiotics. Springer, Dodrecht.

Vaidya, N., Manapat, M.L., Chen, I.A., Xulvi-Brunet, R., Hayden, E.J. and Lehman. N. 2012. Spontaneous network formation among cooperative RNA replicators. Nature 491: 72–77. Doi: 10.1038/nature11549.

Valm, A.M., Cohen, S., Legant, W.R., Melunis, J., Hershberg, U., Wait, E. et al. 2017. Applying systems-level spectral imaging and analysis to reveal the organelle interactome. Nature 546: 162–167. Doi: 10.1038/nature22369.

van Nood, E., Vrieze, A., Nieuwdorp, M., Fuentes, S., Zoetendal, E.G., de Vos, W.M. et al. 2013. Duodenal infusion of donor feces for recurrent *Clostridium difficile*. New England J. Med. 368: 407–415. Doi: 10.1056/NEJMoa120503.

Van Valen, L. 1973. A new evolutionary law. Evol. Theory 1: 1–30.

van Zeijl, A., Op den Camp, R.H., Deinum, E.E., Charnikhova, T., Franssen, H., Op den Camp, H.J. et al. 2015. *Rhizobium* lipo-chitooligosaccharide signaling triggers accumulation of cytokinins in *Medicago truncatula* roots. Mol. Plant 8: 1213–1226. Doi: 10.1016/j.molp.2015.03.010.

Vecchi, D. and Hernández, I. 2014. The epistemological resilience of the concept of morphogenetic field. pp. 79–94. *In*: Minelli, A. and Pradeu, T. [eds.]. Towards a Theory of Development. Oxford University Press, Oxford.

Veluchamy, A., Rastogi, A., Lin, X., Lombard, B., Murik, O., Thomas, Y. et al. 2015. An integrative analysis of post-translational histone modifications in the marine diatom *Phaeodactylum tricornutum*. Genome Biol. 16: 102. Doi: 10.1186/s13059–015–0671–8.

Verhoeven, K.J.F., Jansen, J.J., van Dijk, P.J. and Biere, A. 2010. Stress-induced DNA methylation changes and their heritability in asexual dandelions. New Phytol. 185: 1108–1118.

Vidalis, A., Živković, D., Wardenaar, R., Roquis, D., Tellier, A. and Johannes, F. 2016. Methylome evolution in plants. Genome Biol. 17: 264. Doi: 10.1186/s13059-016-1127-5.

Villavicencio, A., Rueda, M.S., Turin, C.G. and Ochoa, T.J. 2016. Factors affecting lactoferrin concentration in human milk: How much do we know? Biochemistry and Cell Biology 95: 12–21. http://Doi.org/10.1139/bcb-2016-0060.

von Bertalanffy, L. 1932. Theoretische Biologie I. Borntraeger, Berlin.

Voss, T.C. and Hager, L. 2014. Dynamic regulation of transcriptional states by chromatin and transcription factors. Nat. Rev. Genet. 15: 69–81. Doi: 10.1038/nrg3623.

Wacey, D. 2009. Early Life on Earth. A Practical Guide. Springer.

Wächtershäuser, G. 1990. The case for the chemoautotrophic origin of life in an iron-sulfur world. Ori. Life Evol. Biosphere 20: 173–176.

Wächtershäuser, G. 1994. Life in a ligand sphere. Proc. Nat. Acad. Sci. USA 91: 4283–4287.

Waddington, C.H. 1939. An Introduction to Modern Genetics. Allen and Unwin, London.

Waddington, C.H. 1940. Organisers and Genes. Cambridge University Press, Cambridge.

Waddington, C.H. 1942. Canalization of development and the inheritance of acquired characters. Nature 150: 563–565.

Waddington, C.H. 1953. Genetic assimilation of an acquired character. Evolution 7: 118–126.

Waddington, C.H. 1975. The Evolution of an Evolutionist. Edinburgh University Press, Edinburgh.

Wagner, G.P., Mezey, J. and Calabretta, R. 2005. Natural selection and the origin of modules. pp. 33–49. *In*: Callebaut, W. and Rasskin-Gutman, D. [eds.]. Modularity. Understanding the Development and Evolution of Natural Complex Systems. The Vienna Series in Theoretical Biology. MIT Press, Cambridge, MA.

Wagner, G.P. 2007. The development genetics of homology. Nat. Rev. Genet. 8: 473–479. Doi: 10.1038/nrg2099.

Walkley, C.R. and Li, J.B. 2017. Rewriting the transcriptome: adenosine-to-inosine RNA editing by ADARs. Genome Biol. 18: 205. Doi: 10.1186/s13059-017-1347-3.

Wan, C., Borgeson, B., Phanse, S., Tu, F., Drew, K., Clark, G. et al. 2015. Panorama of ancient metazoan macromolecular complexes. Nature 525: 339–344. Doi: 10.1038/nature14877.

Washburn, J.O., Gross, M.E., Mercer, D.R. and Anderson, J.R. 1988. Predator-induced trophic shift of a free-living ciliate: parasitism of mosquito larvae by their prey. Science 240: 1193–1195. Doi: 10.1126/science.3131877.

Waterland, R.A. and Jirtle, R.L. 2003. Transposable elements: targets for early nutritional effects on epigenetic gene regulation. Mol. Cell. Biol. 23: 5293–30.

Watson, J.D. and Crick, F.H.C. 1953a. Molecular structure of nucleic acids. A structure for deoxyribose nucleic acid. Nature 171: 737–738.

Watson, J.D. and Crick, F.H.C. 1953b. Genetical implications of the structure of deoxyribonucleic acid. Nature 171: 964–967.

Weaver, I.C.G., Cervoni, N., Champagne, F.A., D'Allessio, A.C., Sharma, S., Seckl, J.R. et al. 2004. Epigenetic programming by maternal behavior. Nat. Neurosci. 7: 847–854.

Weaver, I.C.G., Champagne, F.A., Brown, S.E., Dymov, S., Sharma, S., Meaney, M.J. et al. 2005. Reversal of maternal programming of stress responses in adult offspring through methyl supplementation: altering epigenetic marking later in life. J. Neurosci. 25: 11045–11054. Doi: 10.1523/JNEUROSCI.3652-05.2005.

Webb, J.S., Givskov, M. and Kjelleberg, S. 2003. Bacterial biofilms. Prokaryotic adventures in multicellularity. Curr. Opin. Microbiol. 6: 578–585.

Webster, G. and Goodwin, B. 1996. Form and Transformation. Generative and Realational Principles in Biology. Cambridge University Press, Cambridge, UK.

Weeks, A.R., Turelli, M., Harcombe, W.R., Reynolds, K.T. and Hoffmann, A.A. 2007. From parasite to mutualist: rapid evolution of *Wolbachia* in natural populations of *Drosophila*. PLoS Biol. 5: e114. Doi: 10.1371/journal.pbio.0050114.

Weismann, A. 1893. The Germ-plasm: A Theory of Heredity. Charles Scribner's Sons, New York.

Weiss, P. 1926. Morphodynamik. Schaxels Abhandl. Theor. Biol. 23: 1–43.

Weng, M. and Walker, W.A. 2013. The role of gut microbiota in programming the immune phenotype. J. Dev. Orig. Health Dis. 4: 203–214. Doi: 10.1017/S2040174412000712.

Went, F.W. 1971. Parallel evolution. Taxon 20: 197–226. Doi: 10.2307/1218877.

Wesemann, D.R., Portuguese, A.J., Meyers, R.M., Gallagher, M.P., Cluff-Jones, K., Magee, J.M. et al. 2013. Microbial colonization influences early B-lineage development in the gut lamina propia. 501: 112–5. Doi: 10.1038/nature12496.

West, S.A., Griffin, A.S., Gardner, A. and Diggle, S.P. 2006. Social evolution theory for microorganisms. Nat. Rev. Microbiol. 4: 597–607. Doi: 10.1038/nrmicro1461.

West, S.A. 2007. The social lives of microbes. Annu. Rev. Ecol. Evol. Syst. 38: 53–77. Doi: 10.1146/annurev.eclsys.38.091206.093740.

West, S.A., Diggle, S.P., Buckling, A., Gardner, A. and Griffin, A.S. 2007. The social lives of microbes. Annu. Rev. Ecol. Evol. Syst. 38: 53–77.

West-Eberhard, M.J. 2003. Developmental Plasticity and Evolution. Oxford: Oxford University Press.

Whitehead, A.N. 1929. Process and reality. Free Press New York.

Whitehead, A.N. 1997[1925]. Science and the Modern World. Free Press (Simon & Schuster). p. 52.

Whiteley, M., Diggle, S.P. and Greenberg, E.P. 2017. Progress in and promise of bacterial quorum sensing research. Nature 551: 313–320. Doi: 10.1038/nature24624.

Wier, A.M., Nyholm, S.V., Mandel, M.J., Massengo-Tiassé, R.P., Schaefer, A.L., Koroleva, I. et al. 2010. Transcriptional patterns in both host and bacterium underlie a daily rhythm of anatomical and metabolic change in a beneficial symbiosis. Proc. Natl. Acad. Sci. USA 107: 2259–2264. Doi: 10.1073/Proc. Nat. Acad. Sci. USA .0909712107.

Williams, T.A., Foster, P.G., Cox, C.J. and Embley, T.M. 2013. An archaeal origin of eukaryotes supports only two primary domains of life. Nature 504: 231–236. Doi: 10.1038/nature12779.

Willyard, C. 2015. The boom in mini stomachs, brains, breasts, kidneys and more. Nature 523: 520–522. Doi: 10.1038/523520a.

Willyard, C. 2018. Could baby's first bacteria také root before birth? Nature 553: 264–266. Doi: 10.1038/d41586-018-00664-8.

Wilson, E.O. 1975. Sociobiology: The New Synthesis. Harvard University Press.

Wimsatt, W.C. and Schank, J.C. 2004. Generative entrenchment, modularity, and evolvability. When genic selection meets the whole organism. pp. 359–394. *In*: Schlosser, G. and Wagner, G.P. [eds.]. Modularity in Development and Evolution. The University of Chicago Press, Chicago.

Wochner, A. 2011. Ribozyme-catalyzed transcription of an active ribozyme. Science 332: 209–212. Doi: 10.1126/science.1200752.

Woese, C. 1998. The universal ancestor. Proc. Nat. Acad. Sci. USA 95: 6854–6859.

Woese, C. 2002. On the evolution of cells. Proc. Nat. Acad. Sci. USA 99: 8742–8747. Doi: 10.1073/Proc. Nat. Acad. Sci. USA .132266999.

Woese, C. 2004. A new biology for a new century. Microbiol. Mol. Biol. Rev. 68: 173–86. Doi: 10.1128/MMBR.68.2.173-186.2004.

Wolfner, M.F. and Miller, D.E. 2016. Alfred Sturtevant walks into a *Bar*: Gene dosage, gene position, and unequal crossing over in *Drosophila*. Genetics 204: 833–835.

Wolpert, L. 1969. Positional information and the spatial pattern of cellular differentiation. J. Theor. Biol. 25: 1–47.

Wolpert, L. and Lewis, L.J. 1975. Towards a theory of development. Federation Proceedings 34: 14–20.

Wommack, K.E. and Colwell, R.R. 2000. Virioplankton: viruses in aquatic ecosystems. Microbiol. Mol. Biol. Rev. 64: 69–114.

Wright, S. 1932. The roles of mutation, inbreeding, crossbreeding and selection in evolution. pp. 356–366. *In*: Proc. 6th Int. Congr. Genetics. Brooklyn Botanic Garden, New York.

Wu, J., Huang, B., Chen, H., Yin, Q., Liu, Y., Xiang, Y. et al. 2016. The landscape of accessible chromatin in mammalian preimplantation embryos. Nature 534: 652–657. Doi: 10.1038/nature18606.

Yachi, S. and Loreau, M. 1999. Biodiversity and ecosystem productivity in a fluctuating environment: The insurance hypothesis. Proc. Nat. Acad. Sci. USA 96: 1463–1468. Doi: 10.1126/science.aan1121.

Yager, E.J., Szaba, F.M., Kummer, L.W., Lanzer, K.G., Burkum, C.E., Smiley, S.T. et al. 2009. γ-Herpesvirus-induced protection against bacterial infection is transient. Viral Immunol. 22: 67–71. Doi: 10.1089/vim.2008.0086.

Yang, H., Berry, S., Olsson, T.S.G., Hartley, M., Howard, M. and Dean, C. 2017. Distinct phases of Polycomb silencing to hold epigenetic memory of cold in Arabidopsis. Science 357: 1142–1145. Doi: 10.1126/science.aan1121.

Yano, J.M., Yu, K., Donaldson, G.P., Shastri, G.G., Ann, P., Ma, L. et al. 2015. Indigenous bacteria from the gut microbiota regulate host serotonin biosynthesis. Cell 161: 264–276.

Yassour, M., Vatanen, T., Siljander, H., Hämäläinen, A.M., Härkönen, T., Ryhänen, S.J. et al. 2016. Natural history of the infant gut microbiome and impact of antibiotic treatment on bacterial strain diversity and stability. Sci. Transl. Med. 8: 343ra81. Doi: 10.1126/scitranslmed.aad091710.

Yin, Y., Morgunova, E.1., Jolma, A., Kaasinen, E., Sahu, B., Khund-Sayeed, S. et al. 2017. Impact of cytosine methylation on DNA binding specificities of human transcription factors. Science 356: eaaj2239. Doi: 10.1126/science.aaj2239.

Yuan, X., Chen, Z., Xiao, S., Zhou, C. and Hua, H. 2011. An early Ediacaran assemblage of macroscopic and morphologically differentiated eukaryotes. Nature 470: 390–393. Doi: 10.1038/nature09810.

Yus, E., Maier, T., Michalodimitrakis, K., van Noort, V., Yamada, T., Chen, W.H. et al. 2009. Impact of genome reduction on bacterial metabolism and its regulation. Science 326: 1263–1268. Doi: 10.1126/science.1177263.

Zaremba-Niedwiedzka, K., Caceres, E.F., Saw, J.H., Bäckström, D., Juzokaite, L., Vancaester, E. et al. 2017. Asgard archaea illuminate the origin of eukaryotic cellular complexity. Nature 541: 354–358. Doi: 10.1038/nature21031.

Zehr, J.P. 2015. How single cells work together. Science 349: 1163–1164. Doi: 10.1126/science.aac9752.

Zenk, F., Loeser, E., Schiavo, R., Kilpert, F., Bogdanović, O. and Iovino, N. 2017. Germ line-inherited H3K27me3 restricts enhancer function during maternal-to-zygotic transition. Science 357: 212–216. Doi: 10.1126/science.aam5339.

Zhalnina, K., Louie, K.B., Hao, Z., Mansoori, N., da Rocha, U.N., Shi, S. et al. 2018. Dynamic root exudate chemistry and microbial substrate preferences drive patterns in rhizosphere microbial community assembly. Nat. Microbiol. 3: 470–480. Doi: 10.1038/s41564-018-0129-3.

Zhang, B., Zheng, H., Huang, B., Li, W., Xiang, Y., Peng, X. et al. 2016. Allelic reprogramming of the histone modification H3K4me3 in early mammalian development. Nature 537: 53–557. Doi: 10.1038/nature19361.

Zhang, Y., Wong, C.H., Birnbaum, R.Y., Li, G., Favaro, R., Ngan, C.Y. et al. 2013. Chromatin connectivity maps reveal dynamic promoter-enhancer long-range associations. Nature 504: 306–310. Doi: 10.1038/nature12716.

Zhu, S., Zhu, M., Knoll, A.H., Yin, Z., Zhao, F., Sun, S. et al. 2016. Decimetre-scale multicellular eukaryotes from the 1.56-billion-year-old Gaoyuzhuang Formation in North China. Nat. Commun. 17: 11500. Doi: 10.1038/ncomms11500.

Zhu, W., Gaetani, G.A., Fusseis, F., Montési, L.G. and De Carlo, F. 2011. Microtomography of partially molten rocks: three-dimensional melt distribution in mantle peridotite. Science 332: 88–91. Doi: 10.1126/science.1202221.

Zhu, P., Guo, H., Ren, Y., Hou, Y., Dong, J., Li, R. et al. 2018. Single-cell DNA methylome sequencing of human preimplantation embryos. Nat. Genet. 50: 12–19. Doi: 10.1038/s41588-017-0007-6.

Zilber-Rosenberg, I. and Rosenberg, E. 2008. Role of microorganisms in the evolution of animal and plants: the hologenome theory of evolution. FEMS Microbiol. Rev. 32: 723–35. Doi: 10.1111/j.1574-6976.2008.00123.x.

Zipfel, C. and Oldroyd, G.E. 2017. Plant signalling in symbiosis and immunity. Nature 543: 328–336. Doi: 10.1038/nature22009.

Zipperer, A., Konnerth, M.C., Laux, C., Berscheid, A., Janek, D., Weidenmaier C. et al. 2016. Human commensals producing a novel antibiotic impair pathogen colonization. Nature 535: 511–516. Doi: 10.1038/nature18634.

Zola, E. 2009[1893]. Dr. Pascal. The Project Gutenberg EBook.

Zola, E. 2012[1971]. The Fortune of the Rougons. Oxford World's Classics.

Index

E

F

G

H

I

L

M